OPTICAL, ACOUSTIC, MAGNETIC, and MECHANICAL SENSOR TECHNOLOGIES

Devices, Circuits, and Systems

Series Editor

Krzysztof Iniewski

CMOS Emerging Technologies Inc., Vancouver, British Columbia, Canada

OPTICAL, ACOUSTIC, MAGNETIC, and MECHANICAL SENSOR TECHNOLOGIES

Edited by
Krzysztof Iniewski

CRC Press
Taylor & Francis Group
Boca Raton London New York

CRC Press is an imprint of the
Taylor & Francis Group, an **informa** business

CRC Press
Taylor & Francis Group
6000 Broken Sound Parkway NW, Suite 300
Boca Raton, FL 33487-2742

Version Date: 20111222

International Standard Book Number: 978-1-4398-6975-8 (Hardback)

Library of Congress Cataloging-in-Publication Data

Optical, acoustic, magnetic, and mechanical sensor technologies / editor, Krzysztof Iniewski.
    p. cm. -- (Devices, circuits, and systems)
    "A CRC title."
    Includes bibliographical references and index.
    ISBN 978-1-4398-6975-8 (hardcover : alk. paper)
    1. Detectors. I. Iniewski, Krzysztof.

TA165.O64 2012
681'.2--dc23
                                                               2011046927

Visit the Taylor & Francis Web site at
http://www.taylorandfrancis.com

and the CRC Press Web site at
http://www.crcpress.com

Contents

PART 1   *Optical and Acoustic Sensors*

PART 2   Magnetic and Mechanical Sensors

Preface

Sensor technologies are a rapidly growing topic in science and product design, embracing developments in electronics, photonics, mechanics, chemistry, and biology. Their presence is widespread in everyday life; they sense sound, movement, optical, or magnetic signals. The demand for portable and lightweight sensors is relentless, filling various needs in consumer electronics, biomedical engineering, or military applications.

The book is divided into two parts. The first part deals with optical and acoustic sensors. Rogério Nogueira starts with optical fiber sensors while Christian-Alexander Bunge and Hans Poisel discuss sensors based on polymer optical fibers. Jeff Chamberlain and Daniel M. Ratner, from the University of Washington, discuss the potential of integrated optical biosensors and silicon photonics. This is followed by chapters by Joey Talghader and Merlin L. Mah on luminescent thermometry and by Andreas Stadler on solar cell analyses. The first part ends with Ellen Holthoff and Paul Pellegrino from the United States Army Research Laboratory describing sensing applications using photoacoustic spectroscopy while Bridget Benson and Ryan Kastner cover the design of underwater acoustic modems.

The second part of the book deals with magnetic and mechanical sensors. Hendrik Husstedt starts with the topic of magnetic field scanning. Researchers from Université Catholique de Louvain describe artificial microsystems for sensing airflow, temperature, and humidity that they accomplished by combining MEMS and CMOS technologies. Jürgen Hildenbrand, Andreas Greiner, and Jan Korvink present MEMS-based micro hot-plate devices while Marcin Marzencki and Skandar Basrour discusses vibration energy harvesting with piezoelectric MEMS. The second part concludes with Anurag Kasyap and Alexander Edrington describing self-powered wireless sensing.

With such a wide variety of topics covered, I am hoping that the reader will find something stimulating to read and discover that the field of sensor technologies is both exciting and useful in science and everyday life. Books like this one would not have been possible without many creative individuals meeting together in one place to exchange thoughts and ideas in a relaxed atmosphere. I would like to invite you to attend CMOS Emerging Technologies events that are held annually in beautiful British Columbia, Canada, where many topics covered in this book are discussed. See http://www.cmoset.com for presentation slides from the previous meeting and announcements about future ones. If you have any suggestions or comments about the book, please email me at kris.iniewski@gmail.com.

Krzysztof (Kris) Iniewski
Vancouver, British Columbia, Canada

Preface

Editor

Krzysztof (Kris) Iniewski manages R&D at Redlen Technologies Inc., a start-up company in Vancouver, Canada. Redlen's revolutionary production process for advanced semiconductor materials enables a new generation of more accurate, all-digital, radiation-based imaging solutions. Kris is also a president of CMOS Emerging Technologies (www.cmoset.com), an organization of high-tech events covering communications, microsystems, optoelectronics, and sensors.

Dr. Iniewski has held numerous faculty and management positions at the University of Toronto, the University of Alberta, SFU, and PMC-Sierra Inc. He has published over 100 research papers in international journals and conferences. He holds 18 international patents granted in the United States, Canada, France, Germany, and Japan. He is a frequent invited speaker and has consulted for several organizations internationally. He has written and edited several books for IEEE Press, Wiley, CRC Press, McGraw-Hill, Artech House, and Springer. His personal goal is to contribute to healthy living and sustainability through innovative engineering solutions. In his leisure time, Kris can be found hiking, sailing, skiing, or biking in beautiful British Columbia, Canada. He can be reached at kris.iniewski@gmail.com.

Contributors

Nélia Alberto
Department of Physics
University of Aveiro
Campus Universitário de Santiago
Aveiro, Portugal

Nicolas André
Institute of Information and
    Communication Technologies
Université catholique de Louvain
Louvain-la-Neuve, Belgium

Udo Ausserlechner
Sense and Control
Infineon Technologies Austria AG
Villach, Austria

Skandar Basrour
TIMA Laboratory
CNRS-UJF-INPG
Grenoble, France

Bridget Benson
Department of Electrical Engineering
California Polytechnic State
    University
San Luis Obispo, California

Lúcia Bilro
Department of Physics
University of Aveiro
Campus Universitário de Santiago
Aveiro, Portugal

Christian-Alexander Bunge
Hochschule für Telekommunikation
    Leipzig
Deutsche Telekom AG
Leipzig, Germany

Jeffrey W. Chamberlain
University of Washington
Seattle, Washington

Alexander Edrington
Parasitx LLC
Norfolk, Virginia

Denis Flandre
Institute of Information and
    Communication Technologies
Université catholique de Louvain
Louvain-la-Neuve, Belgium

Laurent A. Francis
Institute of Information and
    Communication Technologies
Université catholique de Louvain
Louvain-la-Neuve, Belgium

Andreas Greiner
University of Freiburg
Freiburg im Breisgau, Germany

Jürgen Hildenbrand
University of Freiburg
Freiburg im Breisgau, Germany

Ellen L. Holthoff
Sensors and Electron Devices
    Directorate
United States Army Research
    Laboratory
Adelphi, Maryland

Hendrik Husstedt
Department of Applied Mechatronics
Alps-Adriatic University
Klagenfurt, Austria

Manfred Kaltenbacher
Department of Applied Mechatronics
Alps-Adriatic University
Klagenfurt, Austria

Ryan Kastner
Department of Computer Science and
  Engineering
University of California, San Diego
San Diego, California

Anurag Kasyap
GE Global Research
General Electric Company
Niskayuna, New York

Jan G. Korvink
University of Freiburg
Freiburg im Breisgau, Germany

Hugo Lima
Instituto de Telecomunicações
Universidade de Aveiro
Aveiro, Portugal

Merlin L. Mah
Department of Electrical and Computer
  Engineering
University of Minnesota
Minneapolis, Minnesota

Marcin Marzencki
Simon Fraser University
Burnaby, British Columbia, Canada

Rogério Nogueira
Instituto de Telecomunicações
Universidade de Aveiro
Aveiro, Portugal

Paul M. Pellegrino
Sensors and Electron Devices Directorate
United States Army Research Laboratory
Adelphi, Maryland

João Lemos Pinto
Department of Physics
University of Aveiro
Campus Universitário de Santiago
Aveiro, Portugal

Hans Poisel
POF-Application Center
Ohm-Hochschule Nürnberg
Nuernberg, Germany

Jean-Pierre Raskin
Institute of Information and
  Communication Technologies
Université catholique de Louvain
Louvain-la-Neuve, Belgium

Daniel M. Ratner
University of Washington
Seattle, Washington

Bertrand Rue
Institute of Information and
  Communication Technologies
Université catholique de Louvain
Louvain-la-Neuve, Belgium

Andreas Stadler
Paris Lodron University of Salzburg
Salzburg, Germany

Joseph J. Talghader
Department of Electrical and Computer
  Engineering
University of Minnesota
Minneapolis, Minnesota

Part 1

Optical and Acoustic Sensors

1 Optical Fiber Sensors

Rogério Nogueira, Lúcia Bilro, Nélia Alberto,
Hugo Lima, and João Lemos Pinto

CONTENTS

INTRODUCTION

The laser invention in the 1960s and the advances toward low-loss optical fiber in the 1970s stimulated further scientific advances, both in telecommunications and in optical fiber sensors. With the first sensing applications of optical fibers the interest of the scientific community quickly grew for this new technology, and the number of research groups in optical fiber sensing rapidly increased.

The research on optical fiber sensors produced and continues to give life to a variety of measurement techniques for different applications, competing with traditional sensing methods, mainly in niche areas, from the airspace to the medical industry. The success of this technology relies on the intrinsic flexibility, low weight,

3

immunity to electromagnetic interference, passive operation, and high dynamic range, associated with remote monitoring and multiplexing capabilities, which allows optical fiber sensors to succeed in difficult measurement situations where conventional sensors fail.

The technology is now in a mature state, with different applications already using commercial optical fiber sensors as a standard. This includes not only massive deployment for real-time structural health monitoring in airspace, civil and oil industry but also more specific applications such as environment monitoring, biochemical analyses, or gas leak monitoring in hazardous environments.

Optical fiber sensors operate by modifying one or more properties of the light passing through the sensor, when the parameter to be measured changes. An interrogation scheme is then used to evaluate the changes in the optical signal by converting them to a signal that can be interpreted. In this way, depending on the light property that is modified, optical fiber sensors can be divided into three main categories: intensity-, phase-, and wavelength-based sensors.

INTENSITY-BASED SENSORS

Of the range of optical fiber sensors reported in the literature, intensity-based sensors represent one of the earliest and perhaps the simplest type of optical fiber sensors. In applications where the precise signal intensity measurement is not critical or required, it has been shown that intensity-based systems are a valid solution for biomedical, structural health, and environmental applications. Generally, the system is based on a light source, an optical fiber, and a photodetector (or optical spectrum analyzer—OSA). Miniature solid-state light sources and photodetectors are available commercially, allowing the construction of rugged and portable hardware acquisition systems.

Intensity-based sensors offer the advantages of ease of fabrication, low price–performance ratio, and the simplicity of signal processing. These make them highly attractive, particularly in applications where the cost of implementation frequently excludes the use of other significantly more expensive optical fiber systems. Although high resolution and refined measurement capability are achievable using grating sensors and interferometric fiber sensors, it is not always necessary and, as such, less costly intensity-based sensing methods may offer an option in industry.

A wide number of intensity-based sensors are being presented and developed using different schemes; they can be grouped into two major classes: intrinsic- and extrinsic-type sensors. In the extrinsic type, the optical fiber is used as a means of transporting light to an external sensing system. In the intrinsic scheme, the light does not have to leave the optical fiber to perform the sensing function. In this class of sensors, the fiber itself plays an active role in the sensing function and this may involve the modification of the optical fiber structure.

TRANSMISSION AND REFLECTION SCHEMES

An additional classification scheme usually used is related to the way the optical signal is collected. If the receiver and emitter are at opposite ends of the fiber or fibers, the sensor is of a transmission kind, otherwise it is of reflection.

A straightforward example of the first situation is the intensity modulation based on the dependence of the power transmitted from one fiber to another on their separation. This basic sensing principle was used for structural health monitoring purposes by Kuang et al. [1]. The authors presented a comprehensive study where the performance of this sensor was evaluated in quasistatic tensile tests. The optical fiber sensor was surface-attached to an aluminum alloy specimen and revealed a high degree of strain linearity. Free vibration tests based on a cantilever beam configuration were also conducted to assess the dynamic response of the sensor. An impulse-type loading test was also performed to evaluate the ability to detect the various modes of vibration.

With respect to reflection methods, there are some variations, but most of them use reflecting surfaces to couple the light in the fiber, as presented by Binu et al. [2] for their fiber optic glucose sensor based on the changes in the refractive index (RI) with glucose concentration. Other sensors are based on Fresnel reflection mechanisms [3,4] or special geometries of the fiber tip [5]. One interesting application is described by Baldini et al. [6] with their optical fiber sensor for dew detection inside organ pipes. The working principle is based on the change in the reflectivity observed on the surface of the fiber tip when a water layer is formed on its distal end. Intensity changes around 35% were measured.

Macrobending or Microbending Sensors

Several mechanisms in an optical fiber weaken the propagated signal, such as absorption and diffusion by impurities, Rayleigh scattering, ultraviolet and infrared absorption, and microbending and macrobending. Despite many efforts to minimize the power losses in an optical fiber, its dependence on environmental physical parameters is strongly exploited by optical sensing technology.

Modulation due to an environmental effect can be transduced in the form of microbend or macrobend loss. Generally, a microbend is defined as a sharper bend in the optical fiber whose radius of curvature is smaller than the fiber radius and a macrobend is one with a bending radius much larger than the fiber radius. If the bending radius is reduced below a critical value, the loss in transmitted signal increases very rapidly, allowing the construction of a relatively sensitive macrobending fiber optic sensor. Large bending loss occurs at and below a critical bending radius, r_c, given by

$$r_c = \frac{3n_{core}^2 \lambda}{4\pi(n_{core}^2 - n_{clad}^2)^{3/2}} \tag{1.1}$$

where n_{core} and n_{clad} are the refractive indices of the core and cladding, respectively, and λ is the operating wavelength [7]. Macrobending sensors are relatively few and measure parameters such as deformation [7], pressure [8], and temperature [9].

Rajan et al. [9] presented an all-fiber temperature sensor based on a macrobending single-mode fiber loop exploiting the thermo-optic coefficient of the cladding and core. Since the cladding and core are made of silica material and have a positive thermo-optic coefficient, the thermally induced change in RI of the core and cladding is linear, resulting in a linear variation of bend loss with temperature. The temperature sensitivity of the sensor can be varied by changing the bending radius or the operating wavelength. An absorption layer is applied over the cladding to absorb the radiation modes and to reduce the reflections from the air-cladding boundary. In this way, the fiber structure is approximately equivalent to a core-infinite cladding structure. The temperature information is extracted using a simple ratiometric power measurement system.

Among the innumerous transmission and reflection systems reported, there are several transduction mechanisms, namely spectrally based sensors and evanescent wave sensors.

Spectrally Based Sensors

For many applications, spectroscopic detection has been a reliable method for the design of fiber optic sensors and is popularly used for chemical, biological, and biochemical sensing [10]. This method examines the optical signal obtained and the related absorption, fluorescence measurements, or RI to the concentration of the target analyte. When a properly designed sensor reacts to changes in a physical quantity like RI or fluorescence intensity, a simple change of light intensity can possibly be correlated to the concentration of a measurand, which can be a biological or chemical species [11].

Generally, as shown in Figure 1.1, the design of the sensors can simply comprise optical fibers with a sample cell, for direct spectroscopic measurements, or be configured as fiber optrodes, where a chemical selective layer comprising chemical reagents in suitable immobilizing matrices is deposited onto the optical fiber.

In its simplest form, the technique involves confining a sample between two fibers and the quantification of the light transmitted through the sample. The attenuation in the optical path is related to the absorption or scattering properties of the medium. This detection procedure was applied for the purpose of environmental monitoring. A low-cost water turbidity sensor was presented by Bilro et al. [12,13], where the concentration of the total suspended solids in a liquid was determined by the attenuation of the light beam caused by the suspended particles (clay, ashes, and flour). A similar system configuration was used by Yokota et al. [14], but with a multi-wavelength approach for the analysis of soil nutrients. The wavelength of the light-emitting diodes (LEDs) was chosen to fit the absorption band of chemical reagents whose color develops by reaction with soil nutrients. The sensor is applied to detect six soil nutrients including ammonia nitrogen, nitrate nitrogen, and available phosphorus.

The fiber itself can play an active role acting as a sensing probe. The activation can be accomplished replacing the original cladding material, on a small section or end of the fiber, with a chemical agent or an environmentally sensitive material, in order to cause attenuation of the propagated light when the material is exposed to different chemicals or environments. A wide number of sensors reported in the literature make use of this technique. Goicoechea et al. [15], using the reflection method, developed an optical fiber pH sensor based on the indicator Neutral Red. Different strategies for the fabrication of the nanostructured pH-sensitive overlays

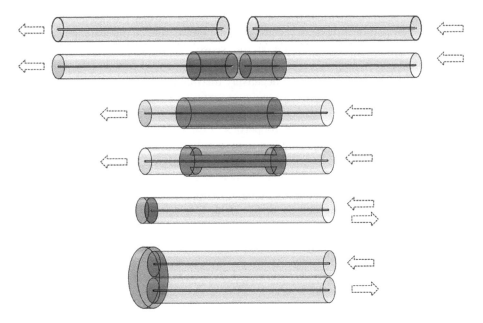

FIGURE 1.1 Schematic diagrams of different sensing methods for spectrally based sensors.

were presented. Some of them revealed high nonlinearity, because the matrices where the dye was immobilized were also sensitive to pH variations. However, the authors found an optimal solution for the fabrication of optical fiber pH sensors. A multilayer structure composed of poly(acrylic acid) and Neutral Red showed a faster response time (below 10 s), high repeatability, low hysteresis, and a dynamic range of 2.5 dB from pH 3–9. Regarding humidity sensing, most spectroscopic-based configurations are based on moisture-sensitive reagents (such as cobalt chloride, cobalt oxide, rhodamine B) attached to the tip of the sensing fiber, usually with the aid of a polymeric material to form the supporting matrix [11]. Silicon dioxide nanoparticles were used as a possible humidity reagent [16]. A nanostructured coating with silicon dioxide nanoparticles is superhydrophilic and, thus, the films are highly sensitive to humidity. The optical fiber humidity sensor demonstrated a good linearity in the range from 40% to 98% of relative humidity and a very fast response. The system is intended to be applied to human breath analysis and monitoring.

A small variation of the method described in the previous paragraph appears with the use of fluorescent materials. Parameters that are measured in such sensors include intensity, decay time, anisotropy, quenching efficiency, and luminescence energy transfer [17]. A usual measurand in fluorescent methods is oxygen. Although many different oxygen-sensitive dyes can be used in optical oxygen sensors, organic dyes, Ru complexes and Pt complexes are among those most commonly used ones. These dye complexes are easily excited using compact and low-cost LED light sources. Furthermore, the phosphorescence wavelengths of Pt [18] and Ru [19] are well separated from the excitation LED wavelength, and hence the influence of the excitation light source can be easily eliminated. One source of error in these systems is the temperature, because the diffusion of oxygen depends on this parameter. To compensate for temperature-induced variations in the luminescence intensity, it is necessary to determine the temperature at the sensor when measuring the oxygen concentration and to apply an appropriate calibration factor. Lo et al. [18] proposed a variable attenuator design using a negatively thermal expansion material as a temperature compensation method, for gaseous oxygen measurements. Ganesh and Radhakrishnan [20] coated the sensitive area with black silicone, avoiding interferences due to changes in the optical properties of the surroundings of the sensor tip (e.g., RI, turbidity, reflectivity), ambient light, and background fluorescence in sediments and biofilms. Their system was for water-dissolved oxygen determination. The oxygen determination by measuring the fluorescence lifetime of metal organic ruthenium complexes can be an indirect method of determining other chemical species. A glucose sensor was designed by Scully et al. [21] using the oxygen-consuming enzymatic conversion of glucose to gluconic acid. The sensitive element consisted of an optical substrate in the form of an optical fiber coated with a matrix containing glucose oxidase as a sensitive and reactive compound as well as the ruthenium complex [21]. With respect to physical sensors, Aiestaran et al. [22] proposed a fluorescent optical fiber linear position sensor. A fluorescent fiber is side illuminated and the transmitted power collected at each end of the fiber is related to the position of the incident light [22].

Evanescent Wave-Based Sensors

When preparing a fiber for sensing the light in the fiber core, it must be able to be modulated by the measurand surrounding the fiber. To achieve this aim, several methods to enhance the interaction between the light guided in the fiber core and the measurand are used. The commonly used methods are side polishing, etching to thin or to strip off the cladding, and tapering or shaping to a D-shape [23].

These are often needed to promote light scattering and evanescent field coupling. Aside from the large portion of light guided in the core, there is a small component of light, known as the evanescent field that decays exponentially away from the core surface and propagates in the cladding. The regular fibers do not lead to the interaction of the evanescent field with the environment outside the cladding. The evanescent field strength is a function of discontinuities of the interface, refractive indices, launching angle of the light beam, and dimensions of the fiber core.

Partial or Total Removal of the Cladding

The three major methods of removing the fiber cladding are chemical etching, thermal tapering, and side polishing. The chemical etching method, both technically and economically, is advantageous over the other two methods. The etching method allows for uniform production of a very long (up to meters) section of fiber. A recent method also comprises the use of commercial carbon dioxide laser systems for rapid construction of discontinuities in silica and in polymer fibers [23].

Wang and Herath [24] reported the fabrication and characterization of fiber loop ring-down evanescent field sensors using partially uncladded single-mode fibers as a sensing element. The adopted technique allowed the online control of fiber etching, in terms of the fiber diameter, to be up to submicrometers. A scattering sensor based on RI difference was demonstrated for the detection of water at 1515.25 nm [24].

Generally, evanescent wave fiber optic sensors use a single source and detector showing changes in output optical power independent of the interacting species, that is, they lack selectivity. Varghese et al. [25] presented a study on the design, development, and characterization of a fiber optic evanescent wave-based sensor with selectivity suitable for the measurement of silica in water. The design used two sources: one corresponding to the absorption peak of the analyte (815 nm) and the second to a nonabsorbing peak (640 nm). The introduction of the analytical wavelength helps to provide selectivity and the use of the nonabsorbing wavelength supports repeatability. The sensing head is a multimode polymer optical fiber (POF) whose jacket was removed, and the exposed region was fine polished.

TAPERS

With tapered fibers, not only does the evanescent field extend beyond the cladding but its magnitude is also enhanced in that tapered region. When a liquid medium is placed at the tapered region, the evanescent field interacts with the medium affecting the transmission. The evanescent field strength is also determined by the diameter and the taper geometry. This is due to the mode coupling that occurs at the tapered region from the changes in the cladding and the RI of the analyte. Mode coupling causes the magnitude of the evanescent field and transmission of the fiber to change and as a result changes in the RI at the tapered region are reflected in the sensor output. This fact is popular among the scientific community performing RI sensing.

With regard to tapers as biosensors, Souza et al. [26] performed a chemical treatment of a taper in order to allow the attachment of the covalent protein (isolated from the cell surface of *Staphylococcus aureus*) [24]. This protein allowed the connection of an antibody to increase cell capture. The objective was the development of a POF biosensor to detect bacteria in water and other fluids. The bacteria are detected through the changes in intensity of the evanescent field resulting from the antibody attachment.

It is also common to combine the use of tapers and surface plasmon resonance (SPR) methods. As an example, Leung et al. [27] presented a biosensor that consisted of a taper coated with gold and housed in a flow cell. For the detection of single-stranded DNA, authors showed that it was feasible to directly detect the hybridization of this DNA to its complementary strand immobilized on the sensor surface. Detection was performed under flow conditions because flow reduces nonspecific binding to sensor surface, eliminates optical transmission changes due to mechanical movements, and allows for instantaneous switching of samples when needed. The sensor also showed selectivity against a single-nucleotide mismatch. After the optimization of detection limit, authors pointed out that further development of the system would be the detection of DNA from live cells. The same authors presented a very similar study but for the continuous detection of various concentrations of bovine serum albumin and the detection of the target serum in the presence of a contaminating protein, ovalbumin [28].

SIDE POLISHING WITH CORE EXPOSURE

As mentioned before, side polishing with core exposure (Figure 1.2) enhances the sensitivity of an optical fiber sensor to a certain physical parameter. This method is valuable for physical, chemical, and biological sensing.

With respect to physical sensors, the most frequent is the curvature sensors. Bilro et al. [29] presented a curvature sensor for the development of a wearable and wireless system to quantitatively evaluate the human gait. The principle behind the sensor was the intrinsic relation between the attenuation of the transmitted signal power and the bending angle of the fiber.

With regard to sensors based on RI variations at the polished interface, there are some interesting applied works, namely for the development of a multipoint liquid level measurement sensor [30], spectroelectrochemical characterization [31], and resin cure monitoring [13].

The combination of a side-polished section in an optical fiber and spectrally based techniques was the basis for the development of an electroactive fiber optic chip. To create the sensor, a side-polished fiber optic was coated with a thin film of indium-tin oxide as the working electrode and used to probe electrochemically driven changes in absorbance for surface-confined redox species [31]. The sensor was used to probe the redox properties of an electrodeposited thin film. A sensitivity enhancement of near 40 times higher than a transmission measurement was demonstrated by the authors.

Another interesting variation of the method is the incorporation of the SPR technique. An immunosensor using side-polished optical fibers based on SPR was proposed for the detection of *Legionella pneumophila* with an 850-nm LED and halogens light source sensing system [32]. The sensing fiber was side polished down to closing half the core and coated with a 37-nm gold thin film by DC sputter. The sensitivity of the SPR fiber for the *L. pneumophila* could be confirmed with the detection limit of 10^1 CFU/mL and the detection range of 10^1–10^3 CFU/mL. The same results shown on spectrum and power meter were demonstrated for both light sources. Another study presented a fiber optic ammonia sensor using a zinc oxide nanostructure grown on the side-polished section [33]. A zinc oxide planar waveguide was grown by pulsed laser deposition on the flat polished fiber surface, and, thus, a distributed evanescent wave coupling between the fiber and the planar waveguide was implemented. The sensor element operation principle is based on a distributed coupling between the fiber mode and the corresponding mode of the metal oxide planar waveguide. The gas sensing performance of the sensor element was

FIGURE 1.2   Side-polished optical fiber.

tested under ammonia gas exposure at room temperature. The sensitivity reported yield a low concentration limit of approximately 50 ppm for ammonia in air.

SELF-REFERENCE TECHNIQUES

The major disadvantage of intensity-based optical fiber sensors is related to the stability of light sources. However, to overcome this problem several referencing techniques can be used, solving this intrinsic limitation of these systems. The most common method is the ratiometric method that uses a reference path not affected by environmental conditions. In its simplest form, one can use an optical element that splits the optical signal in different paths that are equally affected by power fluctuations. Another self-referencing method is given by Lo et al. [18] that used two different fluorescent indicators immobilized in the same matrix: one was designed for sensing and the other for reference. From the same viewpoint, another practice consists of the transmission of multi wavelengths, where only one signal is attenuated, or excites the fluorescent dye, according to the measurand quantity [25]. The fiber loop ring-down scheme [24] measures the light intensity decay rate, not the absolute intensity change. Therefore, the measurement of a quantity (in this case the presence of water) is insensitive to fluctuations of the incident light intensity. With this method, the sensitivity of a sensor can be enhanced by up to several orders of magnitude.

PHASE-BASED SENSORS

Phase-modulated sensors, also known as interferometric sensors, offer sensitivities as high as 10^{-13} m by comparing the phase difference of coherent light traveling along two different paths. The light is provided by a coherent laser source and is generally injected into two single-mode fibers and recombined later. If one fiber is perturbed relatively to the other, a phase shift occurs and can be detected with high precision using an interferometer.

PHASE DETECTION

When a light wave of a given wavelength, λ, propagates inside an optical fiber of length L, the phase angle, Φ at the end of the fiber is given by

$$\Phi = \frac{2\pi L}{\lambda} = \frac{2\pi n_{\text{core}} L}{\lambda_0} \tag{1.2}$$

where n_{core} is the RI of the fiber core and λ_0 is the wavelength of the light in vacuum. If an external perturbation causes a change in the RI or in the length of the fiber, a phase change occurs, and can be defined by

$$\Delta\Phi = \frac{2\pi}{\lambda}(n_{\text{core}} \Delta L + L \Delta n_{\text{core}}) \tag{1.3}$$

Assuming that the RI remains constant, a length variation will induce a phase change of

$$\Delta\Phi = \frac{2\pi}{\lambda} n_{core} \Delta L \tag{1.4}$$

When a light wave is injected into two equal single-mode fibers, the power is split but the phase remains the same. If the two optical fibers experience the same conditions, the light waves will recombine at the same phase angle and constructive interference will occur, giving the maximum intensity output. On the other hand, if the fibers experience different thermal or mechanical strains, they will recombine with a phase difference proportional to the different lengths the light waves traveled. Destructive interference will occur in this situation, causing the output intensity to decrease.

Some interferometer configurations that can be used to detect phase shifts are: Mach–Zehnder, Michelson, Fabry–Perot, and Sagnac interferometers. These will now be discussed in detail.

MACH–ZEHNDER

The Mach–Zehnder interferometer, represented in Figure 1.3a, consists of two 3-dB fiber-to-fiber couplers, separated by two optical fibers, where one is the reference fiber and the other is the sensing fiber. In the first coupler, 50% of the power is injected into each fiber, and the fibers recombine at the second coupler. If both fiber lengths are the same, or differ by an integral number of wavelengths, the light waves will recombine in the exact phase and the output intensity will be maximum. However, if the fiber lengths differ by half a wavelength, the recombined beams will be in the opposite phase and the output intensity will be minimum.

MICHELSON

The Michelson interferometer configuration may be considered as an inverted-over-itself version of the Mach–Zehnder interferometer. This configuration is schematically represented in Figure 1.3b and uses a single optical fiber coupler and fibers with mirrored ends that back-reflect the laser beams, which recombine at the coupler and are directed to the detector. Once again, one fiber is kept unperturbed, the reference fiber, while the sensing fiber is subjected to perturbations. This interferometer has the advantage of requiring a single 3-dB coupler but has the drawback of having the coupler feed light into both the detector and the laser, which causes noise and limits the performance of the system.

FABRY–PEROT

The Fabry–Perot interferometer differs from the previously presented interferometers by not requiring a reference fiber. In this configuration, shown in Figure 1.3c, the fiber presents two partially reflective mirrors. The partially transmitting mirrors cause the light to travel multiple passes inside the cavity transmitted at the second mirror and

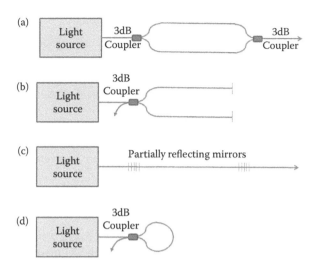

FIGURE 1.3   Illustration of the four main interferometer configurations: (a) Mach–Zehnder, (b) Michelson, (c) Fabry–Perot, and (d) Sagnac.

reaching the detector, which magnifies the phase difference, doubling the sensitivity to phase differences when compared with other interferometer configurations.

SAGNAC

The Sagnac interferometer uses counterpropagating light beams in a ring path. This is achieved by connecting the laser source into a 3-dB optical fiber coupler injecting light into both ends of the same optical fiber in a coiled configuration, as shown in Figure 1.3d. This causes light to travel along the fiber in both directions, and both directions are the sensing fibers.

If the fiber coil is rotated in an axis perpendicular to the coil plane, the light propagation time in one direction will be shortened while the light in the other path will require more time because it will need to travel a longer distance, resulting in a relative phase shift (Sagnac effect). On the other hand, if the coil is kept stationary, light travels the same distance in both directions and no phase shift occurs.

This configuration allows one to measure rotation with high precision, such as the rotation of the earth around its axis. The sensitivity for rotation measurement depends on the area covered by the coil multiplied by the number of turns.

POLARIZATION CONTROL

A potential difficulty associated with optical fiber interferometers is the control of light polarization. Despite the use of single-mode fibers, these fibers are in fact dual-mode fibers as the fundamental mode can assume two degenerate polarization modes: horizontal and vertical. To accomplish high-precision measurements, it is necessary to preserve the polarization state of light. This is achieved using polarization-maintaining fibers and often polarization controllers.

WAVELENGTH-BASED SENSORS

Most wavelength-modulated sensors are based on fiber Bragg gratings (FBGs). An FBG can be described as a periodic modulation of the RI of the fiber core. It is generally obtained when a photosensitive optical fiber is exposed to an ultraviolet radiation in a periodic pattern. For sensor applications, the typical modulation of the RI is $\Delta n \approx 10^{-4}$ with a few millimeters length.

When an FBG is illuminated by a broadband light source, the spectral component that satisfies the Bragg condition is reflected by the grating. In the transmission spectrum, this component is missing. The Bragg condition is given by the following expression [34]:

$$\lambda_B = 2n_{eff,core}\Lambda \qquad (1.5)$$

where λ_B is the wavelength of the back-reflected light (Bragg wavelength), $n_{eff,core}$ is the effective RI of the core, and Λ is the periodicity of the RI modulation (≈ 0.5 μm). A schematic representation of an FBG is shown in Figure 1.4.

Equation 1.5 denotes that the signal reflected by the grating is dependent on the physical parameters of the FBG. When the grating is subjected to mechanical deformation or temperature variation, a shift in wavelength of the Bragg signal is obtained:

$$\Delta\lambda_B = 2\left(\Lambda\frac{\partial n_{eff,core}}{\partial \varepsilon} + n_{eff,core}\frac{\partial \Lambda}{\partial \varepsilon}\right)\Delta\varepsilon + 2\left(\Lambda\frac{\partial n_{eff,core}}{\partial T} + n_{eff,core}\frac{\partial \Lambda}{\partial T}\right)\Delta T \quad (1.6)$$

The first term in Equation 1.6 represents the mechanical perturbation on the Bragg wavelength given by the alteration of the grating pitch and the changes in the RI caused by the strain-optic effect. At constant temperature, the wavelength shift caused by a mechanical perturbation can be expressed as

$$\Delta\lambda_B = \lambda_B(1 - p_e)\varepsilon_z \qquad (1.7)$$

where p_e is an effective strain-optic constant.

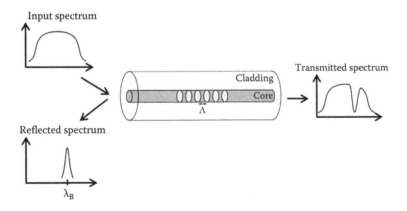

FIGURE 1.4   Schematic representation of a fiber Bragg grating.

The second term in Equation 1.6 is related to the thermal perturbation on the Bragg wavelength. The resulting wavelength shift is due to the changes in the grating pitch, due to thermal expansion, and by changes in the RI. Assuming the strain is constant, Equation 1.6 can be written as

$$\Delta\lambda_B = \lambda_B(\alpha_\Lambda + \alpha_n)\Delta T \tag{1.8}$$

where α_Λ ($\approx 0.55 \times 10^{-6}$ K^{-1} for silica) and α_n ($\approx 8.6 \times 10^{-6}$ K^{-1} for silica) are the thermal expansion coefficient and the thermo-optic coefficient, respectively. For a Bragg grating written in a germanium-doped optical fiber, strain and thermal sensitivities of 1.2 pm/$\mu\varepsilon$ and 13.7 pm/°C are expected, respectively [35]. Although the main application of FBGs as sensors are in temperature and strain monitoring, it is possible to monitor other parameters, such as pressure, acceleration, or the presence of certain chemicals, by adapting the FBG to a structure that can translate these parameters into changes of temperature and/or strain. The RI can also be monitored if the fiber cladding diameter along the grating region is reduced through an etching hydrofluoric acid treatment, increasing the sensitivity of the environmental medium to the RI [36], as opposed to FBGs in standard fibers, where the interaction of the evanescent field with the external medium is almost negligible. This procedure presents as a drawback the fragility of the resulting sensor. Other designs of optical fiber sensors, as for instance tilted fiber Bragg gratings (TFBGs) and long-period gratings (LPGs), can also be used in RI sensing. TFBGs are FBGs where the RI modulation is purposely tilted to the fiber axis, in order to enhance the coupling of the light from the forward-propagating core mode to the backward-propagating cladding mode [37]. This is obtained by tilting the fiber or the phase mask during the grating inscription process. The reflected Bragg wavelength λ_{TFBG} and the cladding mode resonances λ_{clad}^i are determined by the phase matching condition, through the following equations:

$$\lambda_{TFBG} = \frac{2n_{eff,core}\Lambda}{\cos\theta} \tag{1.9}$$

$$\lambda_{clad}^i = \frac{\left(n_{eff,core}^i + n_{eff,clad}^i\right)\Lambda}{\cos\theta} \tag{1.10}$$

where $n_{eff,core}$, $n_{eff,core}^i$, and $n_{eff,clad}^i$ are the effective indices of the core mode at λ_{TFBG} and the core mode and the ith cladding mode at λ_{clad}^i, respectively. The grating period along the axis of the fiber, Λ_{TFBG}, is given by $\Lambda_{TFBG} = \Lambda\cos\theta$, where θ is the tilt angle. The cladding modes attenuate rapidly, being observed only in the transmission spectrum as numerous resonances. In Figure 1.5, a schematic representation of a tilted grating is shown.

In an LPG the RI is periodically modulated to produce a grating; the periodicity of this modulation is typically in the range 100 μm to 1 mm, instead of ≈ 0.5 μm as in FBGs. Normally, the grating length is between 2 and 4 cm. The small grating wave vector, $2\pi/\Lambda_{LPG}$, where Λ_{LPG} is the periodicity of the RI modulation, promotes

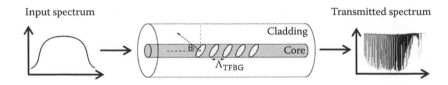

FIGURE 1.5 Schematic representation of a tilted fiber Bragg grating.

the coupling of light from the guided fundamental core mode (the LP_{01} mode) to different forward-propagating cladding modes (LP_{0i} mode with $i = 2, 3, 4, \ldots$). The cladding modes are quickly attenuated as they propagate along the fiber axis, due to scattering losses at the cladding–air interface [38]. As a result, the transmission spectrum of an LPG has several loss bands, at different wavelengths, given by

$$\lambda_{LPG,i} = \left(n_{eff,core} - n_{eff,clad}^{i} \right) \Lambda_{LPG} \tag{1.11}$$

where λ_i is the coupling wavelength, $n_{eff,core}$ is the effective RI of the core mode, and $n_{eff,clad}^{i}$ is the RI of the ith cladding mode. In Figure 1.6, a schematic representation of an LPG is presented.

Since the first LPG was produced in 1996 by Vengsarkar et al. [38], by exposing a hydrogen-loaded germanosilicate fiber to a krypton fluoride laser (248 nm), using an amplitude mask made of chrome-plated silica, the number of methods to obtain the modulation of the RI has been growing. The most typical techniques to produce LPGs include the use of ultraviolet lasers, carbon dioxide lasers, electrical discharges, and irradiation by femtosecond pulses in the infrared. However, it is also possible to obtain LPGs by diffusion of dopants into the fiber core, ion implantation, and deformation of the fiber [39]. A combination of two methods, namely femtosecond laser and carbon dioxide laser, to produce LPGs has also been demonstrated [40]. Such a technique exhibits the advantages of high fabrication flexibility and good thermal stability, without hydrogen loading of the fiber.

In both LPGs and TFBGs, the attenuation bands obtained in the transmission spectrum have been widely explored in the development of sensors, since they are sensitive to a range of parameters. The response of the resonances to the measurand is dependent on the order of cladding modes. This feature offers the possibility of developing multiparameter sensing systems using a single sensor element. For instance, using an 8° TFBG and monitoring the wavelengths of the core mode and a specific cladding mode, Miao et al. [41] measured, simultaneously, the RI and temperature. The sensing principle is based on the fact that the core mode is only

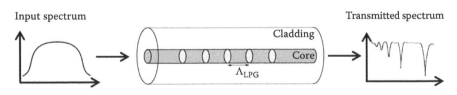

FIGURE 1.6 Schematic representation of a long period grating.

sensitive to temperature variations and the cladding modes are affected by both parameters [41]. By also measuring the change in the wavelength of the core mode, three parameters (temperature, strain, and RI) have also been demonstrated [42].

Apart from the schemes already exposed, in sensing applications it is also common to use chirped FBGs and gratings written in tapers. A chirped FBG is a grating that has a nonuniform period along its length. The most common chirped FBGs have a linear or quadratic evolution of the period of the grating along the length. This category is particularly attractive in situations where a large number of individual sensors are required to cover an extended area, acting as multipoint sensing. A taper consists of a gradual decrease of the fiber diameter in the grating region, which can be obtained through chemical or thermal methods.

All gratings that were introduced before were written in silica fibers. However, a major drawback of using silica is that only 3% of elongation is achieved. This fact motivated the scientific community to think of new fibers for FBG inscription, namely in POF. The first Bragg grating in a step-index fiber was reported in 1999 [43] and the first in a microstructured POF only in 2005 [44]. The properties of POFs are quite different from those of silica and offer some significant potential advantages. Young's modulus of poly(methyl methacrylate) is approximately 25 times less than that of silica and POFs can undergo much higher strains. The most common method for FBG inscription in POFs is the phase mask method with a 325-nm helium–cadmium laser [45].

In the last years, the optical fiber sensors have been tailored to improve their resistance, working range, and sensitivity to specific parameters of interest, leading to a considerable increase in the number of sensors available. Common techniques used to carry out these aims include the deposition of thin films on the fiber, use of fibers with different profiles/compositions, and use of more than one grating, according to the desired sensing application. This is the case of the sensor proposed by Lu et al. [46] for simultaneous monitoring of soluble analytes and temperature. The sensor head consisted of two FBGs coated with different polymers. Onto one of them a polyimide film was deposited, allowing sensing of both temperature and concentrations of soluble analytes. The other one was coated with acrylate, being only sensitive to the environmental temperature. As a result of the deposition of a thin diamond-like carbon film on an LPG, Smietana et al. [47] obtained an RI sensitivity 15 times higher than that achieved in the case of uncoated gratings. Taking advantage of Young's modulus of POFs, that is lower than that of silica fibers, and their inherent fracture resistance and flexibility, Chen et al. [45] developed a bend sensor based on an FBG written in an eccentric cored polymer fiber.

MULTIPARAMETER SENSORS

Multiparameter measurement using fiber sensors is a challenging topic, while it enables the minimization of size, cost, and complexity of sensing systems. FBG sensors are intrinsically sensitive to strain and temperature; however, they present cross sensitivity between these two parameters, because changes in both the parameters are encoded in the peak wavelength shift. The majority of FBG applications handle this cross sensitivity using a pair of FBGs, where one grating is isolated from strain, measuring only temperature variations, while the other FBG measures both strain

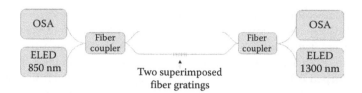

FIGURE 1.7 Experimental setup for strain–temperature discrimination using a dual-wavelength fiber Bragg grating.

and temperature. However, the first sensing head, proposed by Xu et al. in 1994 [48], used a different principle of operation. The reported scheme (Figure 1.7) was based on two superimposed FBGs written at two very distant wavelengths, 850 and 1300 nm, and explored the wavelength dependence of the photoelastic and thermo-optic coefficients to obtain four different strain and temperature coefficients.

The concept of using different strain and temperature sensitivities was then widely explored. In 1996, James et al. [49] spliced two FBGs written in fibers with different diameters achieving distinct strain coefficients. The distinct temperature sensitivities of FBG sensors written in the splice region of fibers with different dopants was later presented [50]. Also, a sensing head was accomplished based on fibers doped with different concentrations [51] or formed by splicing different FBG types [52].

Lu et al. [53] proposed a sensor based on two FBGs with distinct polymeric coatings. From the different optical responses of the gratings due to the coating, the sensitivity of the sensor to individual parameters can be exactly determined [53]. Mondal et al. [54] employed an embedded dual FBG mounted on opposite sides of an arch-shaped steel strip. The compressive and tensile strain effects were explored for thermal and strain discrimination [54].

In 2010, Lima et al. [55] designed a sensing head and presented the necessary interrogation parameters to perform strain and temperature discrimination. They used a single FBG written in an optical fiber taper with a linear diameter variation, as represented in Figure 1.8 [55]. When subjected to tension and due to the different cross sections of the fiber along its length, different values of strain arise, causing the broadening of the FBG signal and allowing the use of the information contained in both peak wavelength and spectral width.

Using a single TFBG, Chehura et al. [56] discriminated these two parameters based on the fact that core mode resonance and cladding mode resonance exhibit different thermal sensitivities but approximately equal strain sensitivities. The application of high-birefringence fibers and more complex configurations based on Fabry–Perot cavities with FBG mirrors [57], sampled FBGs [58], superstructured FBGs [59], or the combination of single-mode/multi mode fibers [60], holey fibers [61], and photonic crystal fibers [62], as well as many other sensing configurations involving fiber-grating devices were also proposed. At present, the temperature and strain discrimination continues to attract attention of the optical sensing community.

Beyond temperature and strain discrimination, research on FBG sensors evolved toward the measurement of other parameters. FBG sensors capable of monitoring displacement, pressure, curvature, load, or RI were also proposed and many of them

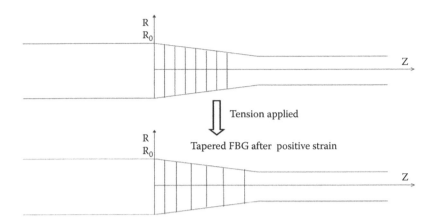

FIGURE 1.8 Illustration of a tapered fiber Bragg grating (FBG) (above) and tapered FBG after positive strain (below).

were also able to measure simultaneously other parameters, such as, generally, temperature.

Several solutions for displacement/temperature [63], pressure/temperature [64], and transverse load/temperature [65] discrimination were proposed based on the grating technology with specific properties. The combination of an FBG and an LPG allows discrimination of strain and curvature [66] and RI and temperature [67]. Beside temperature, other pairs of parameters can also be measured, such as stress/cracks [68] and RI/strain [69].

Rao et al. [70] proposed the first sensor to discriminate three physical parameters. This device was designed for structural health monitoring and measured strain, temperature, and vibration using a combination of an FBG and a Fabry–Perot interferometer. Later, Abe et al. [71] reported a sensing head based on superimposed Bragg gratings in high-birefringence optical fiber capable of discriminating axial strain, temperature, and transverse load. More recently, Men et al. [72] measured saccharinity, salinity, and temperature. A four-parameter fiber sensor for discrimination of RI, surface tension, contact angle, and viscosity was proposed by Zhou et al. [73].

FBG INSCRIPTION METHODS

The most common FBG inscription methods include the phase mask, interferometric, and point-by-point techniques. In the first case, a phase mask is used; this is an optical diffractive device that is designed to diffract light under normal incidence. The superposition of the diffracted light occurs in front of the phase mask, where the photosensitive optical fiber needs to be placed, creating an interference pattern on the core of the optical fiber, which will be responsible for RI modulation (Figure 1.9a). Using this grating inscription technique, it is possible to achieve some tuning capability by applying strain to the fiber during the inscription process; however, the tuning ability of this method is limited by the elastic and mechanical limits of the optical fiber. Moreover, to inscribe gratings with different wavelengths, a different phase mask for each wavelength is necessary. This tuning drawback is compensated for by the reproducibility in the grating inscription. The phase mask method can also be used in the production of more advanced profiles such as chirped gratings.

In the interferometric method, generically, the ultraviolet beam is split into two beams and recombined in the core of the fiber after being reflected by two ultraviolet mirrors, creating an interference pattern. Adjusting the angle of incidence, by changing the angle of the mirrors, is possible to control the wavelength of the inscribed grating, from wavelengths close to the ultraviolet source to virtually infinity. The separation of the beams is usually made with an amplitude beam splitter (Figure 1.9b) or a phase mask (Figure 1.9c).

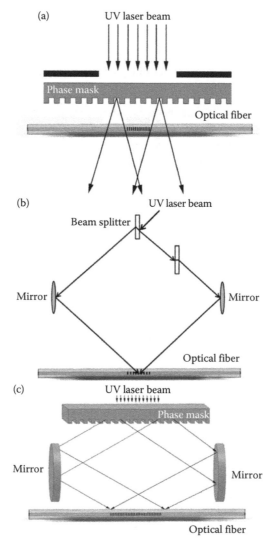

FIGURE 1.9 Schematic representation of gratings inscription techniques: (a) phase mask, (b) interferometric with amplitude beam splitter, and (c) interferometric with phase mask.

In the point-by-point method, each grating plane is produced individually by a focused single pulse from an ultraviolet laser. After the first grating plane is inscribed, the fiber is then translated and a new point is written. The distance between two planes corresponds to the grating pitch. The process is repeated to form the grating structure in the fiber core. More in-depth information regarding the FBG inscription methods can be found in Reference [74].

INTERROGATION OF FBG SENSORS

The principle of operation of a wavelength-based sensor system consists of monitoring the resonance wavelength shift caused by a perturbation. An interrogation system is required to measure this change. The ideal interrogation method should have a high resolution over a large measurement range, typically from subpicometer to a few picometers, be cost effective to compete with conventional optical or electrical sensors, and must also be compatible with multiplexing techniques.

A common setup to interrogate FBGs, especially in a laboratory environment, is composed by a broadband light source, such as a superluminescent LED, an OSA, and an optical circulator or optical coupler, as shown in the scheme in Figure 1.10. However, conventional spectrometers/OSAs have typical resolutions of 0.1 nm, and so they are normally used for inspection of the optical properties of the FBGs rather than for high-precision wavelength shift detection.

Agile tunable lasers and simple photodiode detectors allow one to obtain the transmission and reflection spectrum of an FBG, offering an improvement of several orders of magnitude in both output power and signal linewidth when compared with broadband sources and spectrometers. These advantages are recognized for some time, but the high cost of these systems has discouraged their use.

One alternative approach to interrogate an FBG signal is based on the use of a tunable passband filter. The tunable filter is used to scan the wavelength range of interest, where the FBG signal is located and the output is the convolution of both the spectrum of the FBG and the tunable filter. When the spectra of both the tunable filter and the FBG match, the maximum output occurs. This method has a relatively

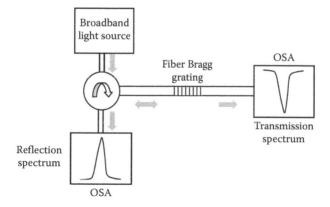

FIGURE 1.10 Schematic representation of a fiber Bragg grating interrogation setup.

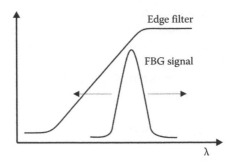

FIGURE 1.11 Principle of operation of the edge filter interrogation method.

high resolution over a large working range, allowing one to interrogate, simultaneously, several FBGs. The resolution is mainly dependent on the linewidth of the tunable filter and on the signal-to-noise ratio of the FBG signal. The most common tunable filters used in this configuration are Fabry–Perot filters, acousto-optic filters, and FBG-based filters.

The methods described above can be used both for analysis of the optical properties of FBGs and for wavelength shift detection, since they allow one to obtain the complete spectrum of one or more FBGs from which the peak wavelengths are easily obtained. The edge filter technique only permits the identification of the peak wavelength of one FBG, but is a cost-efficient method when high sampling rates are demanded. This is based on the use of an edge filter with a linear dependence between wavelength shifts and the output intensity variations of the filter. Using this setup, as represented in Figure 1.11, a wavelength change can be converted to an intensity change, easily detected using a photodetector. However, as the measurement accuracy can be affected by both optical power fluctuations of the source and alignment stability, it is a common practice to compare the transmittance through the edge filters with a reference path. The measurement range of this setup is also inversely proportional to the detection resolution.

REFERENCES

1. Kuang, K., Quek, S., and Maalej, M., Assessment of an extrinsic polymer-based optical fibre sensor for structural health monitoring, *Measurement Science and Technology*, 15, 2133–2141, 2004.
2. Binu, S., Pillai, V., Pradeepkumar, V., Padhy, B., Joseph, C., and Chandrasekaran, N., Fibre optic glucose sensor, *Materials Science and Engineering: C*, 29(1), 183–186, 2009.
3. Chen, J. and Huang, X., Fresnel-reflection-based fiber sensor for on-line measurement of ambient temperature, *Optics Communications*, 283(9), 1674–1677, 2010.
4. Zhou, A., Liu, Z., and Yuan, L., Fiber-optic dipping liquid analyzer: theoretical and experimental study of light transmission, *Applied Optics*, 48(36), 6928–6933, 2009.
5. Nath, P., Singh, H., Datta, P., and Sarma, K., All-fiber optic sensor for measurement of liquid refractive index, *Sensors and Actuators A: Physical*, 148(1), 16–18, 2008.
6. Baldini, F., Falciai, R., Mencaglia, A., Senesi, F., Camuffo, D., Valle, A., and Bergsten, C., Miniaturised optical fibre sensor for dew detection inside organ pipes, *Journal of Sensors*, 2008, 321065(5pp), 2008.

7. Vijayan, A., Gawli, S., Kulkarni, A., Karekar, R., and Aiyer, R., An optical fiber weighing sensor based on bending, *Measurement Science and Technology*, 19, 105302(8pp), 2008.
8. Regez, B., Sayeh, M., Mahajan, A., and Figueroa, F., A novel fiber optics based method to measure very low strains in large scale infrastructures, *Measurement*, 42(2), 183–188, 2009.
9. Rajan, G., Semenova, Y., and Farrell, G., All-fibre temperature sensor based on macrobend singlemode fibre loop, *Electronics Letters*, 44(19), 1123–1124, 2008.
10. McDonagh, C., Burke, C., and MacCraith, B., Optical chemical sensors, *Chemical Reviews*, 108(2), 400–422, 2008.
11. Yeo, T., Sun, T., and Grattan, K., Fibre-optic sensor technologies for humidity and moisture measurement, *Sensors and Actuators A: Physical*, 144(2), 280–295, 2008.
12. Bilro, L., Prats, S., Pinto, J., Keizer, J., and Nogueira, R., Design and performance assessment of a plastic optical fibre-based sensor for measuring water turbidity, *Measurement Science and Technology*, 21(10), 107001(4pp), 2010.
13. Bilro, L., Alberto, N., Pinto, J., and Nogueira, R., Simple and low-cost cure monitoring system based on side-polished plastic optical fibre, *Measurement Science and Technology*, 21(11), 117001, 2010.
14. Yokota, M., Okada, T., and Yamaguchi, I., An optical sensor for analysis of soil nutrients by using LED light sources, *Measurement Science and Technology*, 18(7), 2197–2201, 2007.
15. Goicoechea, J., Zamarreno, C., Matias, I., and Arregui, F., Optical fibre pH sensor based on layer-by-layer electrostatic self-assembled Neutral Red, *Sensors and Actuators B*, 132(1), 305–311, 2008.
16. Corres, J., Matias, I., Hernaez, M., Bravo, J., and Arregui, F., Optical fiber humidity sensors using nanostructured coatings of SiO_2 nanoparticles, *IEEE Sensors Journal*, 8(3), 281–285, 2008.
17. Borisov, S. and Wolfbeis, O., Optical biosensors, *Chemical Reviews*, 108, 423–461, 2008.
18. Lo, Y.-L., Chu, C.-H., Yur, J., and Chang, Y., Temperature compensation of fluorescence intensity-based fiber-optic oxygen sensors using modified Stern-Volmer model, *Sensors and Actuators B: Chemical*, 131(2), 479–488, 2008.
19. Chu, F., Yang, J., Cai, H., Qu, R., and Fang, Z., Characterization of a dissolved oxygen sensor made of plastic optical fibre coated with ruthenium-incorporated sol gel, *Applied Optics*, 48, 338–342, 2009.
20. Ganesh, A. and Radhakrishnan, T., Fiber-optic sensors for the estimation of oxygen gradients within biofilms on metals, *Optics and Lasers in Engineering*, 46, 321–327, 2008.
21. Scully, P., Betancor, L., Bolyo, J., Dzyadevych, S., Guisan, J., Fernández-Lafuente, R., Jaffrezic-Renault, N. et al., Optical fibre biosensors using enzymatic transducers to monitor glucose, *Measurement Science and Technology*, 18(10), 3177–3186, 2007.
22. Aiestaran, P., Dominguez, V., Arrue, J., and Zubia, J., A fluorescent linear optical fiber position sensor, *Optical Materials*, 31(7), 1101–1104, 2009.
23. Irawan, R., Chuan, T., Meng, T., and Ming, T., Rapid constructions of microstructures for optical fiber sensors using a commercial CO_2 laser system, *Biomedical Engineering Journal*, 2, 28–35, 2008.
24. Wang, C. and Herath, C., Fabrication and characterization of fiber loop ringdown evanescent field sensors, *Measurement Science and Technology*, 21(8), 085205(10pp), 2010.
25. Varghese, P.B., John, S., and Madhusoodanan, K.N. Fiber optic sensor for the measurement of concentration of silica in water with dual wavelength probing, *Reviews of Scientific Instruments*, 81, 035111(5pp), 2010.

26. Souza, N., Beres, C., Yugue, E., Carvalho, C., Neto, J., Silva, M., Werneck, M., and Miguel, M., Development of a biosensor based in polymeric optical fiber to detect cells in water and fluids, *Proceedings of the 18th International Conference on Plastic Optical Fibre*, Sydney, 4pp, 2009.

27. Leung, A., Shankar, P., and Mutharasan, R., Label-free detection of DNA hybridization using gold-coated tapered fiber optic biosensors (TFOBS) in a flow cell at 1310 nm and 1550 nm, *Sensors and Actuators B: Chemical*, 131(2), 640–645, 2008.

28. Leung, A., Shankar, P., and Mutharasan, R., Model protein detection using antibody-immobilized tapered fiber optic biosensors (TFOBS) in a flow cell at 1310 nm and 1550 nm, *Sensors and Actuators B: Chemical*, 129(2), 716–725, 2008.

29. Bilro, L., Oliveira, J., Pinto, J., and Nogueira, R., Gait monitoring with a wearable plastic optical sensor, *Proceedings of IEEE Sensors Conference*, 787–790, 2008.

30. Lomer, M., Arrue, J., Jauregui, C., Aiestaran, P., Zubia, J., and Lopez-Higuera, J., Lateral polishing of bends in plastic optical fibres applied to a multipoint liquid-level measurement sensor, *Sensors and Actuatuators A: Physical*, 137, 68–73, 2007.

31. Beam, B., Armstrong, N., and Mendes, S., An electroactive fiber optic chip for spectroelectrochemical characterization of ultra-thin redox-active films, *Analyst*, 134, 454–459, 2009.

32. Lin, H.-Y., Tsao, Y.-C., Tsai, W.-H., Yang, Y.-W., Yan, T.-R., and Sheu, B.-C., Development and application of side-polished fiber immunosensor based on surface plasmon resonance for the detection of *Legionella pneumophila* with halogens light and 850 nm-LED, *Sensors and Actuators A: Physical*, 138(2), 299–305, 2007.

33. Dikovska, A., Atanasova, G., Nedyalkov, N., Stefanov, P., Atanasov, P., Karakoleva, E., and Andreev, A., Optical sensing of ammonia using ZnO nanostructure grown on a side-polished optical-fiber, *Sensors and Actuators B: Chemical*, 146(1), 331–336, 2010.

34. Kersey, A., Davis, M., Patrick, H., LeBranc, M., Koo, K., Askins, C., Putnam, M., and Friebele, E., Fiber grating sensors, *Journal of Lightwave Technology*, 15(8), 1442–1463, 1997.

35. Othonos, A. and Kalli, K., *Fiber Bragg Gratings—Fundamentals and Applications in Telecommunications and Sensing*, Artech House, Norwood, MA, 1999.

36. Ladicicco, A., Cusano, A., Campopiano, S., Cutolo, A., and Giordano, M., Thinned fiber Bragg gratings as refractive index sensors, *IEEE Sensors Journal*, 5(6), 1288–1295, 2005.

37. Erdogan, T. and Sipe, J., Tilted fiber phase gratings, *Journal of the Optical Society of America*, 13(2), 296–313, 1996.

38. Vengsarkar, A., Lemaire, P., Judkins, J., Bhatia, V., Erdogan, T., and Sipe, J., Long-period fiber gratings as band-rejection filters, *Journal of Lightwave Technology*, 14(1), 58–65, 1996.

39. James, S. and Tatam, R., Optical fibre long-period grating sensors: characteristics and applications, *Measurement and Science Technology*, 14, R49–R61, 2003.

40. Fang, X., He, X., Liao, C., Yang, M., Wang, D., and Wang, Y., A new method for sampled fiber Bragg grating fabrication by use of both femtosecond laser and CO_2 laser, *Optics Express*, 18(3), 2646–2654, 2010.

41. Miao, Y., Liu, B., Tian, S., and Zhao, Q., Simultaneous measurement of surrounding temperature and refractive index by tilted fiber Bragg grating, *Proceedings of Passive Components and Fiber-Based Devices V, SPIE*, 7134, 71343W, 2008.

42. Alberto, N., Marques, C.A.F., Pinto, J.L., and Nogueira, R.N., Three-parameter optical fiber sensor based on a tilted fiber Bragg grating, *Applied Optics*, 49(31), 6085–6091, 2010.

43. Xiong, Z., Peng, G., Wu, B., and Chu, P.L., Highly tunable Bragg gratings in single-mode polymer optical fibers, *IEEE Photonics Technology Letters*, 11(3), 352–354, 1999.

44. Dobb, H., Webb, D.J., Kalli, K., Argyros, A., Large, M.C., and van Eijkelenborg, M., Continuous wave ultraviolet lightinduced fiber Bragg gratings in few and single-mode microstructured polymer optical fibers. *Optics Letters* 30(24); 3296–3298, 2005.
45. Chen, X., Zhang, C., Webb, D., Peng, G.-D., and Kalli, K., Bragg grating in a polymer optical fibre for strain, bend and temperature sensing, *Measurement and Science Technology*, 21, 094005(5pp), 2010.
46. Lu, P., Men, L., and Chen, Q., Polymer-coated fiber Bragg grating sensors for simultaneous monitoring of soluble analytes and temperature, *IEEE Sensors Journal*, 9(4), 340–345, 2009.
47. Smietana, M., Pawlowski, M., Bock, W., Pickrell, G., and Szmidt, J., Refractive index sensing of fiber optic long-period grating structures coated with a plasma deposited diamond-like carbon thin film, *Measurement and Science Technology*, 19, 085301(7pp), 2008.
48. Xu, M., Archambault, J., Reekie, L., and Dakin, J., Discrimination between strain and temperature effects using dual-wavelength fibre grating sensors, *Electronics Letters*, 30(13), 1085–1087, 1994.
49. James, S., Dockney, M., and Tatam, R., Simultaneous independent temperature and strain measurement using in-fibre Bragg grating sensors, *Electronics Letters*, 32(12), 1133–1134, 1996.
50. Cavaleiro, P., Araújo, F., Ferreira, L., Santos, J., and Farahi, F., Simultaneous measurement of strain and temperature using Bragg gratings written in germanosilicate and boron-codoped germanosilicate fibers, *IEEE Photonics Technology Letters*, 11(12), 1635–1637, 1999.
51. Frazão, O. and Santos, J., Simultaneous measurement of strain and temperature using a Bragg grating structure written in germanosilicate fibre, *Journal of Optics A: Pure and Applied Optics*, 6(6), 553–556, 2004.
52. Shu, X., Liu, Y., Zhao, D., Gwandu, B., Floreani, F., Zhang, L., and Bennion, I., Dependence of temperature and strain coefficients on fiber grating type and its application to simultaneous temperature and strain measurement, *Optics Letters*, 27(9), 701–703, 2002.
53. Lu, P., Men, L., and Chen, Q., Resolving cross sensitivity of fiber Bragg gratings with different polymeric coatings, *Applied Physics Letters*, 92, 171112(3pp), 2008.
54. Mondal, S., Mishra, V., Tiwara, U., Poddar, G., Singh, N., Jain, S., Sarkar, S., and Kapur, P., Embedded dual fiber Bragg grating sensor for simultaneous measurement of temperature and load (strain) with enhanced sensitivity, *Microwave and Optical Technology Letters*, 51(7), 1621–1624, 2009.
55. Lima, H., Antunes, P., Nogueira, R., and Pinto, J., Simultaneous measurement of strain and temperature with a single fibre Bragg grating written in a tapered optical fibre, *IEEE Sensors Journal*, 10(2), 269–273, 2010.
56. Chehura, E., James, S., and Tatam, R., Temperature and strain discrimination using a single fibre Bragg grating, *Optics Communications*, 275(2), 344–347, 2007.
57. Du, W., Tao, X., and Tam, H., Fiber Bragg grating cavity sensor for simultaneous measurement of strain and temperature, *IEEE Photonics Technology Letters*, 11(1), 105–107, 1999.
58. Frazão, O., Romero, R., Rego, G., Marques, P., Salgado, H., and Santos, J., Sampled fibre Bragg grating sensors for simultaneous strain and temperature measurement, *Electronics Letters*, 38(14), 693–695, 2002.
59. Guan, B., Tam, H., Tao, X., and Dong, X., Simultaneous strain and temperature measurement using a superstructure fiber Bragg grating, *IEEE Photonics Technology Letters*, 12(6), 675–677, 2000.

60. Zhou, D., Wei, L., Liu, W., Liu, Y., and Lit, J., Simultaneous measurement for strain and temperature using fiber Bragg gratings and multimode fibers, *Applied Optics*, 47(10), 1668–1672, 2008.
61. Han, Y., Song, S., Kim, G., Lee, K., Lee, S., Lee, J., Jeong, C., Oh, C., and Kang, H., Simultaneous independent measurement of strain and temperature based on long-period fiber gratings inscribed in holey fibers depending on air-hole size, *Optics Letters*, 32(15), 2245–2247, 2007.
62. Ju, J. and Jin, W., Photonic crystal fiber sensors for strain and temperature measurement, *Journal of Sensors*, 2009, 476267(10pp), 2009.
63. Dong, X., Liu, Y., and Liu, Z., Simultaneous displacement and temperature measurement with cantilever-based fiber Bragg grating sensor, *Optics Communications*, 192(3–6), 213–217, 2001.
64. Chen, G., Liu, L., Jia, H., Yu, J., Xu, L., and Wang, W., Simultaneous pressure and temperature measurement using Hi-Bi fiber Bragg gratings, *Optics Communications*, 228(1–3), 99–105, 2003.
65. Abe, I., Frazão, O., Kalinowski, H., Schiller, M., Nogueira, R., and Pinto, J., Characterization of Bragg gratings in normal and reduced diameter HiBi fibers, *International Microwave and Optoelectronics Conference*, 2, 887–891, 2003.
66. Gwandu, B., Shu, X., Liu, Y., Zhang, W., Zhang, L., and Bennion, I., Simultaneous measurement of strain and curvature using superstructure fibre Bragg gratings, *Sensors and Actuators A: Physical*, 96(2–3), 133–139, 2002.
67. Shu, X., Gwandu, B., Liu, Y., Zhang, L., and Bennion, I., Sampled fiber Bragg grating for simultaneous refractive-index and temperature measurement, *Optics Letters*, 26(11), 774–776, 2001.
68. Watekar, P., Ju, S., and Han, W., A multi-parameter sensor system using concentric core optical fiber, *Optical and Quantum Electronics*, 40(7), 485–494, 2008.
69. Alberto, N., Marques, C., Pinto, J., and Nogueira, R., Simultaneous strain and refractive index sensor based on a TFBG, *Proceedings of SPIE*, 7653, 765324(1–4), 2010.
70. Rao, Y., Henderson, P., Jackson, D., Zhang, L., and Bennion, I., Simultaneous strain, temperature and vibration measurement using a multiplexed in-fibre-Bragg grating/fibre Fabry-Perot sensor system, *Electronics Letters*, 33(24), 2063–2064, 1997.
71. Abe, I., Kalinowski, H., Frazão, O., Santos, J., Nogueira, R., and Pinto, J., Superimposed Bragg gratings in high birefringence fibre optics: Three-parameter simultaneous measurements, *Measurement Science and Technology*, 15(8), 1453–1457, 2004.
72. Men, L., Lu, P., and Chen, Q., Intelligent multiparameter sensing with fiber Bragg gratings, *Applied Physics Letters*, 93, 071110(3pp), 2008.
73. Zhou, A., Yang, J., Liu, B., and Yuan, L., A fiber-optic liquid sensor for simultaneously measuring refractive index, surface tension, contact angle and viscosity, *Proceedings of the SPIE*, 7503, 75033B1–75033B4, 2009.
74. Othonos, A., Fiber Bragg gratings, *Review of Scientific Instruments*, 68(12), 4309–4341, 1997.

2 Sensors Based on Polymer Optical Fibers

Microstructured and Solid Fibers

Christian-Alexander Bunge and Hans Poisel

CONTENTS

OVERVIEW

Modern technical systems need more and more sensors that are versatile, simple to use, and immune to electromagnetic interferences. Optical sensors fulfill all these requirements and are being more widely installed. While many sensors rely on glass optical fibers, which are mostly single mode and thus relatively delicate in terms of handling and operation, multimode fibers, especially robust, large-core polymer optical fibers (POFs), are a cost-effective alternative. For quite a while, POF sensors have tried to enter the market, but only now they are considered seriously. Apparently, the first applications of mass-produced POF sensors will be in automotive and industrial areas such as impact sensors or seat occupancy sensors.

In this chapter, we will present an overview of sensors based on POFs. First, we will categorize the most common sensor types according to their fiber type and sensing effect. Here, we will differentiate between microstructured and solid POF sensors. Following this line of thought, we will then present several applications for microstructured POFs, their advantages, and application scenarios. We will then provide an overview of the principles of solid POF sensors and exemplary applications.

General Requirements and Sensing Effects

Several optical effects can be used for sensing. Often, influences of the environment— usually tried to be avoided—are exploited for the generation of a sensing signal. We will try to structure this magnitude of effects in order to give an organized overview of the fundamental sensing principles.

Optical sensors may be classified in different ways:

- Intrinsic or extrinsic sensors
- Multimode or single-mode sensors
- Parameters influenced (e.g., transmitted or reflected amplitude, phase, polarization, frequency, mode distribution, spectral distribution, etc.)

Figure 2.1 provides a hierarchical structure of fiber optical sensors in general. They can be classified into glass and polymer fibers. Very many glass optical sensors are in use today. Most of them are single-mode sensors making use of interferometric or polarization effects. These effects are well known and a lot of literature deals with these sensors. We will therefore concentrate on polymer fiber sensors for the remainder of the chapter. As in their glass counterparts, polymer fibers are available as solid fibers as well as microstructured fibers with tiny holes in the cross section running along the whole length of the fiber. These holes alter the optical properties of the material they are embedded in and usually cause a decrease in the material's effective refractive index so that the material acts like being doped with air (see, e.g., Reference [1]). Another approach is to form a crystal-like structure around the fiber core, in which energy bands develop as in semiconductors. These bands stand for allowed energy states for the photons so that a fiber can be constructed with energy states that are allowed in the core but not in the cladding. Then, photons cannot help remain in the core region. This effect is called band-gap guidance and can even be used for fibers with air cores that are otherwise impossible to produce (see, e.g., Reference [2]). The main advantage of this type of fiber is that the holes can be changed in terms of geometry (e.g., by pressure) or refractive index (e.g., by blowing in gases or liquids), or that they provide a means to bring the material to be studied closer to the light in order to interact. Solid POFs are usually large in diameter and numerical aperture and thus massively multimode. There are also a couple of single-mode POF applications,

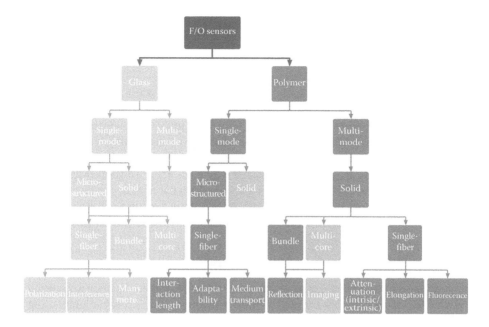

FIGURE 2.1 Classification of the most common sensor concepts.

but the main advantages of POF sensors are their robustness, ease of handling, and the easy reception of light (e.g., for reflectance measurements). One can also differentiate between single fibers, multicore fibers, and bundles. Most of the sensor effects themselves are based on single fibers, but these sensors can be easily extended or multiplexed by fiber bundles or in a more compact form as multicore structures. In this way, one can design distributed sensors that can sense at many different locations and bring their sensing signal to one centralized detector. Some effects, however, make direct use of fiber bundles or multicore fibers; for example, one can use a bundle of fibers in order to receive the reflected light of a surface or use an additional fiber for calibration purposes. Solid POF sensors usually fall into several distinct categories. One can use their attenuation by a simple power measurement. Other sensors rely on refraction, which finally also leads to an increased attenuation. Since standard POF are multimode, one can also measure the modal power distribution and detect for instance bends that change the modal structure of the fiber. Although multimode, one can also use phase measurements for the detection of elongation or other mechanical changes in the fiber. In this case, the optical phase cannot be measured, but a lower-frequency sine signal can be used and its phase can be compared with a reference signal.

In general, almost every physical parameter can be measured by an optical system, making use of the attractive properties inherent in these systems, such as small dimensions, electromagnetic immunity, or high bandwidth. POF sensors offer cost-effective alternatives to otherwise ultra-precise, but expensive, glass optical sensors. In some cases, where a large diameter is needed, they provide the only viable solution for a fiber sensor.

SENSORS BASED ON MICROSTRUCTURED POFs

While standard polymer fibers provide many potential sensor applications that combine low cost and ruggedness, some well-known applications that require single-mode operations cannot be applied. While solid single-mode POFs will be available for a couple of years, they do not share the main advantages of the POFs. Especially the widely used fiber gratings (both Bragg and long period) provide simple high-resolution sensor applications. In addition, microstructured fibers open completely new possibilities for the detection of liquids or gases, which can be filled into the holes or have a much stronger interaction due to the largely increased surface area. While most of these sensor principles can be realized with glass fibers as well, several applications forbid glass altogether (all applications for which the fiber has to be put inside the body or several automotive applications due to vibrations). In the following section, we will provide a short report on applications based on microstructured POFs.

MECHANICAL SENSING WITH FIBER USING GRATINGS

Polymers have some advantageous mechanical properties that make microstructured POFs better for mechanical sensing in some situations. Preliminary studies using microstructured POF sensors based on long-period gratings (LPGs) [3] show that the use of polymer fiber increases the range of repeatable strain measurements by several times and the yield limit by an order of magnitude, compared to a silica-based sensor. Figure 2.2 shows an example for an LPG-based sensor. The visco-elastic properties of the polymer means there are time-dependent effects relating to strain rate and magnitude. These effects are small when the sensor is intermittently strained up to 2% and are relatively small at strains of up to 4–5%. Further testing is ongoing to characterize these effects at very high strains. The effect of stress relaxation has a

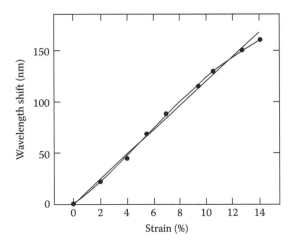

FIGURE 2.2 Strain results using a long-period grating in a microstructured polymer optical fiber, in which the strain is removed rapidly after application. There is a small deviation from linearity at higher strains.

small effect on the change in the wavelength of the loss features used in the measurement of strain.

The use of high strains complicates the response of the sensor due to the viscoelastic properties that can experience a partly nonelastic deformation requiring careful calibration. This technology is the only one that allows the use of fiber strain sensors operating at strains of up to 30–45%. These very-high-strain sensors are currently being developed.

SENSING OF FLUIDS

The fact that microstructured POFs have cladding holes, and may even have a hollow core (photonic band-gap fibers), opens up significant new opportunities for fluid sensing. Fluid sensing in conventional optical fibers is difficult because of their low refractive indices, which means guidance by total internal reflection (TIR) is impractical. The fluids have to be brought near the optical core of the fiber, which also provides some challenges [4].

Microstructured optical fibers offer two elegant solutions to this problem. In Figure 2.3, one can see a liquid-filled core and a photonic band-gap fiber, both of which have been experimentally demonstrated. A selectively filled core fiber can be used in conjunction with a high air fraction in the cladding region, or a fully filled band gap can be used [5]. In both cases in Figure 2.3, the microstructure confines both the light and the fluid, and critically allows almost a complete overlap of the optical field with the fluid, a dramatic improvement on evanescent field sensors that are the conventional fiber alternative. An alternative approach uses a microstructured core [6] though this relies on the evanescent field being maximized and produces a mode profile that makes efficient launching a challenge. In both cases, the main benefit of the microstructure is the reduced interaction length that is needed because of the strong evanescent fields at the sensing region, which is accomplished

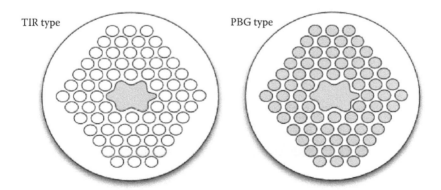

FIGURE 2.3 Liquid sensing is possible in liquid-filled microstructured polymer optical fiber: Open circles are air filled, closed circles are liquid filled. On the left, a selectively filled core with total internal reflection with an air-filled cladding is used, while on the right a photonic band gap is used, in which both core and cladding holes are liquid filled.

by the special field structure induced by the holes and the possibility of bringing the fluid close to the waveguide.

SPECTROSCOPY USING EVANESCENT-FIELD INTERACTION

Another use of the evanescent field is important for many surface-sensitive effects such as, for example, gas sensing or the detection of antibodies. These are an area where microstructured fibers are likely to be extremely important because of their high surface area/volume ratio. The selective detection of antibodies in microstructured POFs has been studied in References [7,8], and metal coating of the holes allows them to be used as a platform for Surface Enhanced Raman Spectroscopy (SERS), a technique that potentially combines very high sensitivity with detailed molecular spectra. SERS has already been demonstrated in microstructured POFs using silver colloidal solution, easily detecting concentrations of 200 nM of rhodamine 6 G [9]. Aqueous sensing for medical or environmental applications makes this area a very active one, with continuous online monitoring being one of the targets.

In order to bring the liquid to investigate near the sensitive area, a "side-hole" fiber [10] has been developed. It simplifies the filling of the fiber and separates it from the coupling of light in and out of the ends (see Figure 2.4). While the side-hole fiber allows the rapid ingress of fluids close to the sensing region, it also allows efficient surface treatment on the core. Most surface treatments on the interior surfaces of the holes in microstructured fibers require the fluids to pass over the full length of the fiber. This approach is problematic for some applications because it intrinsically treats a very long section of the fiber and also does not allow the full range of processing techniques to be applied. In particular, Surface Plasmon Resonance (SPR) requires the application of a very-high-quality metal layer. Using the side-hole approach, one can apply this layer externally using sputtering, an approach that is known for high-quality films.

With this geometry the first experimental demonstration of an SPR sensor in a polymer microstructured fiber [11] was realized recently. The width and position

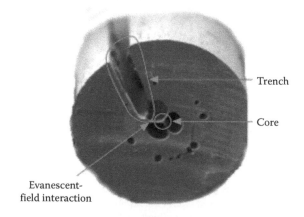

FIGURE 2.4 Fiber cross section with the microstructured cross section.

of these shifts are consistent with theory, given the effect of measured surface roughness [12].

SOLID POF SENSORS

Despite the fact that microstructured POFs allow for interesting sensing concepts, most POF sensors rely on solid fibers with a large diameter and a high numerical aperture. They often use so-called standard step-index POFs according to IEC 60793-2 A4a.2 due to their robustness, easy handling and preparation, their possibility to receive light from a large range of angles, and a big cross section. Since these are highly multimode fibers, they cannot make use of, for example, interference or other very sensitive, but single-mode, effects; they usually rely on easier-to-implement methods such as, for example, attenuation measurements, where the sensing signal is transformed into an additional attenuation, reflections that can be easily captured because of POF's good light-capture properties and so on. In the following sections, several sensor concepts will be presented that use different methods to generate the sensor signal.

ATTENUATION

Attenuation can be measured relatively easily by power meters. Thus, it is an ideal candidate for POF sensor concepts. Here, we will present two sensor applications that use the induced additional attenuation for sensing purposes: One relies on microbending losses of the fiber once it is subject to side pressure and deformation, the other uses the refractive index difference between core and cladding, which can lead to losses if the index difference becomes too small.

Attenuation due to Microbending

Two attenuation-based sensors have been developed for the car industry. According to the European Union directive 2003/102/EC, pedestrian protection has to be provided for every new car since 2005. This can be fulfilled either by

- Passive means, that is, structural measures such as "soft" front ends and sufficient deformation room between the hood and the engine; or by
- Active means, that is, sensors that identify a pedestrian's impact and then trigger protective means such as lifting the hood by means of actuators.

Currently, there are two solutions in use by European car manufacturers: one is supplied by Leoni[*] and the other one by Magna.[†]

The Leoni system is based on attenuation of the evanescent electromagnetic field, which is restricted to a small range outside the core of a Leoni-proprietary POF core. When the vicinity of the core is changed, for example, due to an absorbing material

---

[*] http://www.leoni.com/fileadmin/leoni.com/downloads/pdf/produkte/en_pinchguard_flyer.pdf
[†] http://www.magnasteyr.com/xchg/en/43-66/electronic_systems/Products+%26+Services/Driver+Assistance+%26+Safety+Systems

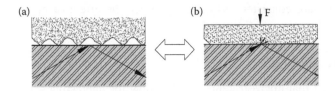

FIGURE 2.5 (a) Light rays totally reflected due to air outside the core. (b) Total internal reflection no longer possible due to the presence of absorbing material.

such as a human finger coming close, the evanescent field will be absorbed leading to an absorption of the signal in the fiber itself without the need of physical contact [13]. Practically, this can be realized through foam surrounding the bare core, ensuring that in most parts of the surface air is outside leading to TIR. When this foam is compressed, the conditions for TIR are no longer fulfilled and absorption occurs (Figure 2.5).

In a different approach from Magna, the sensor is located in the front bumper of a car. A POF is fixed between two structures that enhance the bending of the fiber thus leading to increased attenuation (Figure 2.6). The temporal behavior of the signal is characteristic of each collision partner and thus allows identification if a pedestrian has been hit.

Attenuation due to Change of Refractive Index

Attenuation-based sensors can also be used for liquid detection. A flexible quasidistributed liquid-level sensor based on the changes in the light transmittance in a POF cable has been proposed in Reference [14]. The measurement points are constituted by small areas created by side polishing on a curved fiber and the removal of a portion of the core. These points are distributed on each full turn of a coil of fiber built on a cylindrical tube vertically positioned in a tank (Figure 2.7).

The changes between the refractive indices of air and liquid generate a signal power proportional to the position and level of the liquid. The sensor system has been successfully demonstrated in the laboratory, and experimental results of two prototypes with 15 and 18 measurement points and with bend radii of 5 and 8 mm, respectively, have been achieved; the results of the 5-mm type are shown in

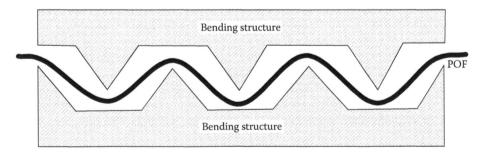

FIGURE 2.6 Polymer optical fiber pedestrian impact sensor principle currently used in several European cars.

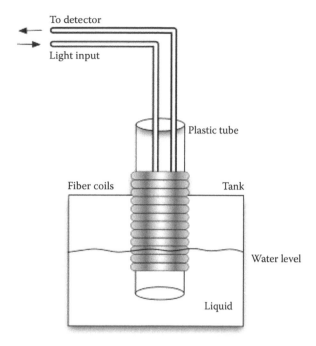

FIGURE 2.7   Schematic of the quasidistributed level sensor (one fiber/detector).

Figure 2.8. The number of measurement points can be easily increased with the only limitation due to the dynamic range of the attenuation-measurement equipment.

REFLECTION

Another sensing approach uses reflections either outside or within the fiber. The POF's large-core diameter and high numerical aperture make it an ideal fiber to capture reflected light and to guide it to a central location where the evaluation of the sensor signal takes place. Reflections within the fiber, for example, due to side pressure, localized elongation with consequent change of geometry, and so on, can be detected by optical time-domain reflectometry (OTDR), a measurement method widely used in fiber optics. With this approach even distributed sensing is possible along the total length of the fiber.

Capture of Reflected Light

The changing intensity of reflected light from a mirror at a certain distance can be used for a displacement sensor. The power transmitted from the output end face of one fiber to the input end face of another depends on their separation distance and closely follows an inverse square law. Utilizing this fact, for example, Augousti [15] and Ioannides et al. [16,17] developed a linear displacement sensor usable between 15 mm and 80 m. In this case, two receiving, axially separated fibers are put in front of the surface whose displacement is to be measured, with a third fiber acting as the

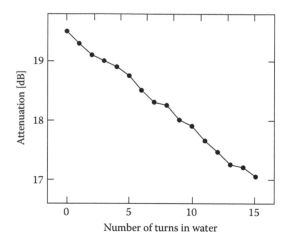

FIGURE 2.8   Experimental loss obtained as a function of turns immersed in water.

emitter (Figure 2.9). The accuracy, resolution, and stability of the sensor are better than 1%.

This sensor has also been shown to be suitable for the measurement of the frequency of large-amplitude vibrating surfaces [18]. The main features of the POF that make this sensor possible are its high numerical aperture and its large core size. The measurement of displacement can be made to relate to changes in the liquid level [19,20], and has been used to relate to water uptake by a plant with a resolution of $2.5 \cdot 10^{-9}$ L. The same sensor has also been used to monitor the reflectivity of a mirror surface as a function of condensation, as part of a condensation dew-point hygrometer [21]. There are also other cases where the POF displacement sensor can be used to indirectly probe into conditions of the environment: for example, detect the liquid nitrogen level present under quench conditions in a super-conducting fault–current–limiter

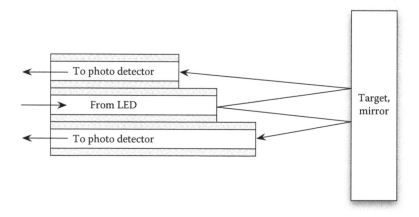

FIGURE 2.9   Schematics of the displacement sensor.

device using POF and not affected in performance up to 420 kV, as well as the sensor being inexpensive [22].

Mechanical Properties Using OTDR

Mechanical stress can induce changes of the refractive index or the fiber geometry, which can lead to local reflections within the fiber. The evaluation of these reflections can be used for a strain sensor. Because of their high elasticity and high breakdown strain, POFs are well suited for integration into technical textiles. Smart textiles with integrated POF sensors can be realized, which are able to sense various mechanical, physical, and chemical quantities and can react and adapt themselves to environmental conditions.

Medical textiles with fiber optic sensors (Figure 2.10) that can measure vital physiological parameters, such as respiratory movement, cardiac activity, pulse oxymetry, and temperature of the body, are needed for wearable health monitoring, for patients requiring continuous medical assistance and treatment.

The monitoring of anesthetized patients under Medical Resonant Imaging (MRI) requires pure fiber optic solutions due to the immunity of fiber optics against electromagnetic radiation. POF sensors are advantageous because of their biocompatibility in case of fiber breakage. The breathing movements of patients during an MRI examination can be monitored by POF sensors embedded into a textile yarn and placed on an efficient area of the thorax or the abdomen [23,24]. Using an OTDR technique, the cyclic strain induced in the POF due to the respiratory movement can be measured distributed in the range between 0% and 3% (Figure 2.11). The distributed measurement provided by the OTDR technique allows one to focus only on a special part of the fiber and so to differentiate between abdominal and thoracic respiration, and to neglect contributions from nonsensing parts.

POFs can also be woven into geotextiles in order to add a sensing functionality to them. Geotextiles are commonly used for reinforcement of geotechnical structures such as dikes, dams, railway embankments, landfills, or slopes. The incorporation of optical sensor fibers into geotextiles leads to additional functionalities of the textiles, for example, monitoring of mechanical deformation, strain, temperature, humidity, pore pressure, detection of chemicals, measurement of the structural integrity, and the health of the geotechnical structure [25]. Especially solutions for the distributed measurement of mechanical deformations over extended areas of some 100 m up to

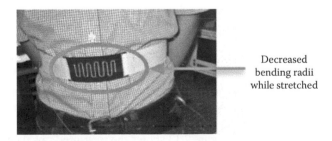

Decreased
bending radii
while stretched

FIGURE 2.10 Monitoring of respiratory abdominal movement of an adult by polymer optical fiber optical time-domain reflectometry.

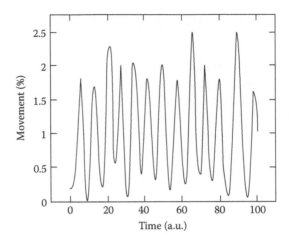

FIGURE 2.11 Respiratory abdominal movement recorded by polymer optical fiber optical time-domain reflectometry.

several kilometers are needed. Textile-integrated, distributed fiber optic sensors can provide for any position of extended geotechnical structures' information about critical soil displacement or slope slides via distributed strain measurement along the fiber with a high spatial resolution of less than 1 m. So the early detection of failure in geotechnical structures with a high risk potential can be ensured.

The integration of POFs as sensors into geotextiles is very attractive because of their high elasticity and capability of measuring high strain values of more than 40% [26]. Especially the monitoring of relatively small areas with expected high mechanical deformations such as endangered slopes takes advantage of the outstanding material properties of POFs.

ELONGATION

As a low-cost alternative to fiber Bragg-grating sensors targeting the lower sensitivity range, POF elongation sensors have been proposed, for example, by Doering [27] and Poisel et al. [28]. A recently recovered detection system known from laser distance meters turned out to be very sensitive while staying simple. The approach is based on measuring the phase shift of a sinusoidal light signal guided in a POF under different tensions resulting in different transit times. One of the setups under investigation is shown in Figure 2.12. The reference fiber is used for compensating temperature influence. Its output signal serves as a reference in order to obtain a phase difference by comparison.

This sensor was used to monitor the rotor blades of a windmill. Under hefty wind conditions the blades can bend beyond an allowed degree and finally break. Thus, they should be monitored and once wind conditions become too severe the blades will turn out of wind direction. Figure 2.13 shows the measured raw data taken during a normal day. The 39-m blade was fixed on the ground and oscillations could be excited through an external mechanism. One can see that the intensity of the sensor

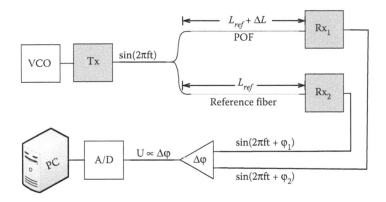

FIGURE 2.12 Schematic of the phase-measurement setup.

signal drifts during the day because of irradiation through the sun. Using a digital high-pass filter, this drift can be eliminated. The bottom of Figure 2.13 shows the processed data. One can observe the elongation of the fiber when the blade is excited with different amplitudes. Fast Fourier transform processing of the signals yields a very precise value of the oscillation frequency, thus offering the possibility of observing frequency shifts, for example, due to layers of ice or snow which might cause dangerous situations in the winter. Oscillation frequencies of more than 100 Hz can be measured easily.

We also investigated the resolution of the strain sensor for different elongations. For this measurement, a reference frequency of 2 GHz was used. Figure 2.14 shows the results for a 1-m POF that was stretched in 50-μm steps. One can clearly observe the remarkable accuracy and repeatability that could be achieved.

These sensors are thought to be implemented in structures such as the rotor blades of a wind power generator, aircraft wings, or, in general, in structures that have to be monitored for their integrity (structural health monitoring). Transmitters and receivers have been directly adopted from POF transceivers developed for Gbit/s applications.

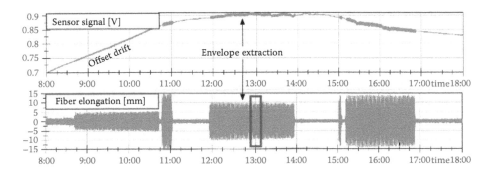

FIGURE 2.13 Sensor raw data (above) and processed data (below) over a typical day.

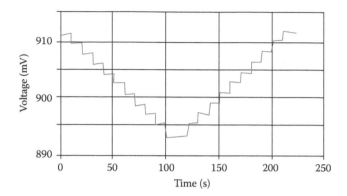

FIGURE 2.14 Resolution of the strain sensor when a 1-m polymer optical fiber sample is stretched in 50-μm steps ($f = 2$ GHz).

TEMPERATURE SENSORS BASED ON FLUORESCENCE

Temperature is a very important parameter for the electric power industry because insulators, copper conductors, core iron of transformers, insulator oil, and all equipments are very sensitive to temperature, which has to be strictly controlled at all times. Nevertheless, when dealing with high voltage, sometimes one cannot use conventional electric sensors because they work in near-ground level and therefore need to be kept at a creep distance from energized parts. Thus, some important parts, such as copper conductors, simply cannot be monitored in a conventional way. For this reason, optical fiber sensors can offer many advantages over conventional sensors, such as high immunity to electromagnetic interference, electrical isolation, no need of electric power to work, and therefore can be placed at high electric potentials. Such fiber-optical temperature sensors can be realized using ruby fluorescence when pumped by a pulsed ultra-bright green light-emitting diode (LED). In a field test, we used a 10-m single-fiber probe and could achieve temperature measurements in the range of 25–75°C with ±1°C accuracy with a time response of a few seconds. The prototype with four probes was installed on a substation harmonic filter at 75 kV for temperature monitoring. The fluorescence-based optical sensors use a green LED to pump the ruby, which returns red light. Due to the distance between these two wavelengths, the sensor is potentially more sensitive and error-immune than other fiber-optic temperature sensors that only rely on absorption difference when the temperature varies [29]. Previously, experiments were performed with commercial polystyrene fluorescent fibers [30] and ruby [31] as temperature sensors. Ruby has already been used for fluorescence thermometry [32]. It is low cost, easily available, POF-compatible, and can be driven with low-cost light sources (blue or green ultra-bright LEDs). Si-based photo detection and simple electronics can be used because of the strong intensity and long lifetime of the fluorescence signal. The fluorescence peaking at a wavelength of 694 nm features a long decay time of 2–4 ms. Figure 2.15 depicts the spectra of the components used. The system was installed in March 2007, with a fully automatic transfer of the sensor data via the Internet.

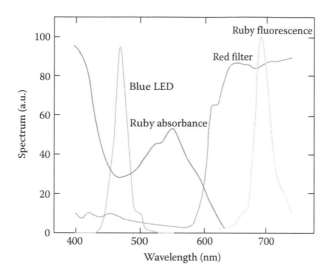

FIGURE 2.15 Spectra of all components involved in the system.

SENSOR MULTIPLEXING BY FIBER BUNDLES

Fiber optical sensors not only provide interesting applications and sensor principles, they can also be multiplexed easily and the sensor signals can be combined into compact fiber bundles or even more compact multicore fibers. Under certain conditions it is even possible to multiplex the sensor signals onto different frequencies or wavelengths so that only one single fiber may be used. We realized several multiple-fiber sensors with POF bundles. For instance, in the liquid-level sensor system we used a commercial USB camera at a central location, limited in temporal and spatial resolution and bad coupling efficiency that was still sufficiently good to receive the sensor signals accurately.

In a similar way, we developed a so-called multifiber receiver (MFR) [33]. A commercial complementary metal–oxide–semiconductor sensor array was used with some modifications: In order to obtain efficient butt coupling of the fiber bundle to the sensitive chip we had to remove the protective window. Thus, the bundle could be put as close as possible to the array's surface (without breaking the bond wires). A sketch of a sensor application is shown in Figure 2.16.

Each sensing point consists of a pair of POFs (c.f. inset in Figure 2.16), one transporting the light of a single LED to the region of interest, the other one catching the scattered light and transporting it to the MFR. The intensity of the back-scattered light depends on the compression of the foam and serves as the sensor signal. For a first demonstration, we coupled 12×12 POFs of $250\,\mu m$ diameter each from a Kinotex™ sensor mat [34,35] to our MFR thus resulting in 12×12 pressure-sensitive points distributed over an area of 50 cm by 50 cm. Using this mat, the shape of objects putting pressure on it can be recognized: By increasing the number of sensing points combined with a proper pressure calibration one could further improve the resolution (Figure 2.17). Even dynamic processes can be detected through the fast

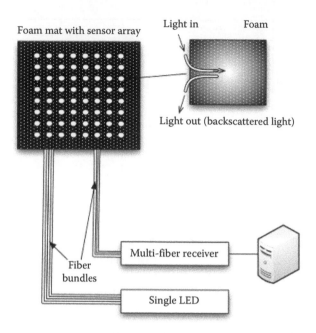

FIGURE 2.16   Application of the multifiber receiver in a Kinotex sensor mat.

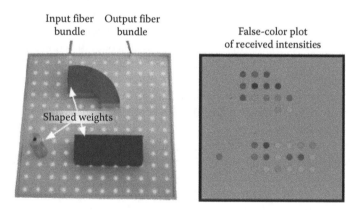

FIGURE 2.17   Shape sensing with a Kinotex sensor mat and a multifiber receiver (Adapted from H. Poisel, *18th POF Conference 2009*, Sydney, Australia, September 2009.)

response of the complementary metal–oxide–semiconductor electronics, allowing at least 500 frames per second in our demo system.

CONCLUSION

Fiber sensors offer a great variety of applications. While glass-based fiber sensors are often single mode and sometimes critical in terms of handling and robustness, POF sensors can easily be used in harsh environments. In addition, they provide cost-efficient alternatives to already-existing sensor applications that often rely on simpler-to-implement methods. We have given a rough overview of available POF sensor concepts with microstructured and solid fibers. While microstructured fiber sensors take advantage of a stronger interaction with the media to sense and allow additional functionalities because of their capillaries, solid POF sensors offer more robust and often simpler sensors for low-cost applications based on effects like attenuation or reflection, which can be easily measured.

REFERENCES

1. M.A. van Eijkelenborg, M.C.J. Large, A. Argyros, J. Zagari, S. Manos, N.A. Issa, I. Bassett et al., Microstructured polymer optical fibre, *OSA Optics Express*, 9(7), 319–327, 2001.
2. T.P. White, R.C. McPhedran, L.C. Botten, G.H. Smith, and C.M. de Sterke, Calculations of air-guided modes in photonic crystal fibers using the multipole method, *OSA Optics Express*, 9(11), 721–732, 2001.
3. M.C.J. Large, J.H. Moran, and L. Ye, The role of viscoelastic properties in strain testing using microstructured polymer optical fibres (mPOF), *Measurement Science and Technology*, 20(3), 034014, 2009.
4. M.C.J. Large and C.-A. Bunge, Microstructured polymer optical fibres compared to conventional POF: Novel properties and applications, *IEEE Sensors Journal, Special Issue on Photonic Crystal-Based Sensors*, 10(7), 1213–1217, 2010.
5. F.M. Cox, A. Argyros, and M.C.J. Large, Liquid-filled hollow core microstructured polymer optical fiber, *Optics Express*, 14(9), 4135–4140, 2006.
6. C.M.B. Cordeiro, M.A.R. Franco, G. Chesini, E.C.S. Barretto, R. Lwin, C.H. Brito Cruz, and M.C.J. Large, Microstructured-core optical fibre for evanescent sensing applications, *Optics Express*, 14, 13056–13066, 2006.
7. J.B. Jensen, J.P. Hoiby, G. Emiliyanov, O. Bang, L. Pedersen, and A. Bjarklev, Selective detection of antibodies in microstructured polymer optical fibers, *Optics Express*, 13(15), 5883–5889, 2005.
8. G. Emiliyanov, J.B. Jensen, O. Bang, P.E. Hoiby, L.H. Pedersen, E.M. Kjær, and L. Lindvold, Localized biosensing with Topas microstructured polymer optical fiber, *Optics Letters*, 32(5), 460–462; erratum: p. 1059, 2007.
9. F.M. Cox, A Argyros, M.C.J. Large, and S. Kalluri, Surface enhanced Raman scattering in a hollow core microstructured optical fiber, *Optics Express*, 15(21), 13675–13681, 2007.
10. F.M. Cox, R. Lwin, M.C.J. Large, and C.M.B. Cordeiro, Opening up optical fibres, *Optics Express*, 15(19), 11843–11848, 2007.
11. A. Wang, A. Docherty, B.T. Kuhlmey, F. Cox, and M.C.J. Large, Surface plasmon resonance in slotted microstructured polymer optical fibres, *International Conference on Materials for Advanced Technologies, Symposium P: Optical Fiber Devices and their Applications*, Singapore, pp. 41–43, June 2003.
12. M. Kanso, S. Cuenot, and G. Louarn, Roughness effect on the SPR measurements for an optical fibre configuration: Experimental and numerical approaches, *Journal of Optics A: Pure and Applied Optics*, 9, 586–592, 2007.
13. G. Kodl, Large area optical pressure sensor based on evanescent field, *Proceedings of the 12th POF Conference 2003*, Seattle, p. 64, 2003.
14. M. Lomer, J. Arrue, C. Jauregui, P. Aiestaran, J. Zubia, and J.M. López-Higuera, Lateral polishing of bends in plastic optical fibers applied to a multipoint liquid-level measurement sensor, *Sensors and Actuators A: Physical*, 137(1), 68–73, 2007.

15. T. Augousti, *Sensors VI: Technology Systems and Applications*, IOP Publishing Ltd, Bristol, UK, pp. 291–293, 1993.
16. N. Ioannides, D. Kalymnios, and I. Rogers, A POF-based displacement sensor for use over long ranges, *Proceedings of the 4th POF 1995*, Boston, USA, pp. 157–161, 1995.
17. N. Ioannides, D. Kalymnios, and I.W. Rogers, An optimised plastic optical fiber (POF) displacement sensor, *Proceedings of the 5th POF 1996*, Paris, France, pp. 251–255, 1996.
18. N. Ioannides and D. Kalymnios, A plastic fiber (POF) vibration sensor, *Proceedings of the Applied Optics Divisional Conference of the IOP*, Brighton, UK, pp. 163–168, ISBN 0 7503 0456 1, 1998.
19. J.D. Weiss, The pressure approach to fiber liquid level sensors, *Proceedings of the 4th POF 1995*, Boston, USA, pp. 167–170, 1995.
20. S. Vargas, C. Vázquez, A.B. Gonzalo, and J.M.S. Pena, A plastic fiber-optic liquid level sensor, *Proceedings of the Second European Workshop on Optical Fiber Sensors*, SPIE volume 5502, pp.148–151, 2004.
21. S. Hadjiloucas, S. Karatzas, D.A. Keating, and M.J. Usher, Optical sensors for monitoring water uptake in plants, *Journal of Lightwave Technology*, 13(7), 1421–1428, 1995.
22. J. Niewisch, POF sensors for high temperature superconducting fault current limiters, *Proceedings of the 6th POF 1997*, Kauai/Hawaii, USA, pp. 130–131, 1997.
23. A. Grillet, D. Kinet, J. Witt, M. Schukar, K. Krebber, F. Pirotte, A. Depre, N.V, Barco, and N.V.Kortrijk, Optical fiber sensors embedded into medical textiles for healthcare monitoring, *IEEE Sensors Journal*, 8(7), 1215–1222, 2008.
24. J. Witt, C.-A. Bunge, M. Schukar, and K. Krebber, Real-time strain sensing based on POF OTDR, *Proceedings of the 15th International POF Conference*, Paper SEN-II-4, pp. 210–213, Torino, Italy, September 2007.
25. N. Nöther, A. Wosniok, K. Krebber, and E. Thiele, Dike monitoring using fiber sensor-based geosynthetics, *Proceedings of the ECCOMAS Conference on Smart Structures and Materials*, Gdansk/Poland, July 2007.
26. P. Lenke, K. Krebber, M. Muthig, F. Weigand, and E. Thiele, Distributed strain measurement using polymer optical fiber integrated in technical textile to detect displacement of soil, *Proceedings of the ECCOMAS Conference on Smart Structures and Materials*, Gdansk/Poland, July 2007.
27. H. Doering, High resolution length sensing using PMMA optical fibers and DDS technology, *Proceedings of the 15th POF Conference 2006*, Seoul, Korea, September 11–14, 2006.
28. H. Poisel, POF strain sensor using phase measurement techniques, *Proceedings of the 16th POF Conference 2007*, Turin, Italy, September 10–13, 2007.
29. K. Asada and H. Yuuki, Fiber optic temperature sensor, *3rd POF Conference 1994*, Yokohama, Japan, pp. 49–51, 1994.
30. R.M. Ribeiro, L.A. Marques-Filho, and M.M. Werneck, Fluorescent plastic optical fibers for temperature monitoring, *12th POF Conference 2003*, Seattle, USA, pp. 282–285, 2003.
31. R.M. Ribeiro, L.A. Marques-Filho, and M.M. Werneck, Simple and low cost temperature sensor using the ruby fluorescence and plastic optical fibers, *14th POF Conference 2005*, Hong Kong, pp. 291–294, 2005.
32. K.T.V. Grattan and Z.Y. Zhang, *Optic Fluorescence Thermometry*, Chapman & Hall, London, 1995.
33. T. Hofmann, Multi-Faser-Receiver, Master's thesis, GSO University of Applied Sciences Nuernberg, 2009.
34. E.M. Reimer and L.H. Baldwin, Cavity sensor technology for low cost automotive safety & control devices, Paper presented at Air Bag Technology 1999, Cobo Convention Center, Detroit. Available at: http://www.canpolar.com/principles.shtm
35. H. Poisel, POF sensors—an update, *18th POF Conference 2009*, Sydney, Australia, September 2009.

3 Label-Free Biosensors for Biomedical Applications
The Potential of Integrated Optical Biosensors and Silicon Photonics

Jeffrey W. Chamberlain and Daniel M. Ratner

CONTENTS

INTRODUCTION

Sensors, in their most general form, measure signals from the environment for interpretation and analysis—with the objective of providing actionable information to the user. Such is the case for biosensors, which have found widespread applications in medical research, healthcare, environmental monitoring, chem-bio defense, and food safety. In biomedicine, biosensors are used for discovering new drugs, elucidating biological pathways, and studying biomolecular interactions. Within the healthcare setting, biosensors can be implemented as biometric assays and can serve as diagnostic tools, potentially predicting disease or suggesting disease susceptibility via genetic screening. Biosensors are also used for environmental monitoring, such as detecting the presence of specific allergens or environmental contaminants, and for defense against threat agents such as anthrax, ricin, or other toxins. While this chapter focuses on biosensor design considerations in medicine (research, diagnostics, etc.), nearly all the devices discussed have far-reaching applications beyond the exclusive domain of biomedical research. For instance, a biosensor for detecting malaria could also be implemented to detect pathogenic organisms in food, or as a research tool to screen for inhibitors to prevent disease.

In this chapter, we address the requirements and desired features of biosensors for biomedical applications, emphasizing the goal of realizing fully integrated and distributable lab-on-a-chip (LOC) devices. We highlight silicon photonics as an advantageous technology for such LOC applications, particularly for label-free optical biosensing. In addition, we address other promising label-free electrochemical and mechanical biosensors, with a brief survey of current research in the field. Finally, we provide a perspective on the remaining challenges that need to be addressed for biosensors to inform researchers and clinicians, with the ultimate objective of improved healthcare outcomes.

DESIRED BIOSENSOR CHARACTERISTICS

The ideal biosensor, within the realm of biomedicine, should: rapidly detect any analyte of choice from a low-volume unprocessed sample; be disposable or contain reusable low-cost components; require minimal training; integrate the source, transducer, and detector into a single portable device; allow long-term storage in ambient conditions; and produce a clear and quantitative readout of the amount of target analyte in the sample. While such an ideal biosensor has not been realized, it is useful to keep the ultimate goal in mind when discussing the desired characteristics of biosensors. It is also important to note that all these qualities are not required for a technology to have value, but they are the characteristics needed to achieve true point-of-care (POC) biosensing. Incremental steps toward such a system could still have important applications. For example, simplifying the operator requirements and allowing use with unprocessed samples would significantly improve the utility of diagnostics in a clinical setting.

Sensitivity and Selectivity

A biosensor must, first and foremost, be both sensitive and selective for the target agent or agents it is designed to detect. With the hope of using unprocessed samples such as saliva, blood, urine, or other bodily fluids, the analyte to be detected will usually be at low relative concentrations within the complex biological milieu of cells, proteins, lipids, and salts found within a typical biological sample. A comprehensive study of potential cancer biomarkers reported that 88% of cancer biomarkers found in plasma (out of 211 surveyed) are below 10 μg/mL, and 49% are below 10 ng/mL [1]. As a comparison, common plasma proteins such as albumin and fibrinogen exist at milligram per milliliter levels, and the salinity is roughly 150 mM. Even when sample processing is an option in a properly equipped laboratory, it can be laborious and time consuming, and the motivation for using unprocessed or minimally processed samples reemerges. Rapid testing and analysis are of utmost importance in a clinical setting, where diagnostics can be used to guide prophylactic or therapeutic intervention, significantly influence patient outcome, and dramatically reduce the time and cost associated with patient care.

Label-Free

Along these lines, it is desirable to detect target analyte without the need for a label. In biosensors and diagnostic assays, the label is usually a chromophore, fluorophore, or enzyme that is either directly attached to one of the interacting molecules or attached to a secondary reporter molecule in order to amplify the binding event signal (Figure 3.1a). While assays that use labels continue to find widespread use in research and medicine—take, for instance, the ubiquitous enzyme-linked immunosorbant assay (ELISA)—there are many reasons as to why it is desirable to eliminate the indicator (Figure 3.1b). Labeling not only increases the time and cost of an assay, but it also inherently alters binding interactions [2] and obscures quantification [3]. Labels are also, in general, unstable molecules that require careful storage; this makes assay standardization difficult and limits their use in a POC setting. Finally, when the target analyte is unknown, such as high-throughput screening of molecular libraries, labeling is simply not an option. Label-free biosensors can decrease cost and assay time; provide quantitative information of

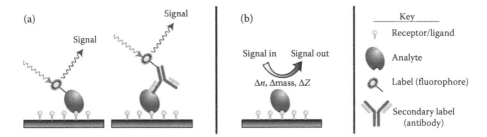

FIGURE 3.1 Label-based and label-free biosensing differ in terms of the means by which the analyte is detected. (a) Example of label-based sensing via a fluoscently labeled analyte or secondary probe (antibody). (b) Example of label-free biosensing, where an inherent property of the analyte, such as refractive index (n), mass, or impedance (Z), alters the input signal such that detection can occur.

unaltered binding interactions in real time; and potentially enable portable bio-sensing of unprocessed samples. The lack of a label, however, increases the requirements for sensitivity and necessitates on-chip controls to ensure that signals being analyzed are due to the analyte and not due to nonspecific interactions with the sensor.

Multiplexing

The need for on-device controls is complementary with the desire for biosensors to be multiplexible, where multiplexing describes a biosensor's ability to run multiple experiments simultaneously (Figure 3.2). For screening applications and disease diagnostics, the ability to test for multiple binding interactions is imperative. Disease states are most often described by multiple biomarkers, and the ability to provide a conclusive diagnosis is reliant on taking a systems approach for

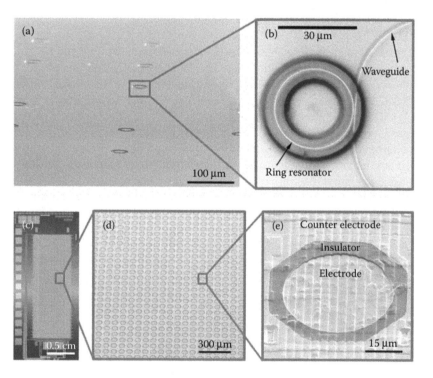

FIGURE 3.2 Two examples of multiplexed biosensor platforms. (a) Each donut-shaped depression seen in the scanning electron micrograph represents an individual microring resonator (manufactured by Genalyte, Inc., San Diego, California) (b) exposed through lithographic etching of a polymer coating. Microrings are interrogated with a bus waveguide that comes within 100 nm of the resonator. (c–e) Example of a commercial microelectrode array manufactured by CustomArray, Inc. (Mukilteo, Washington). The array contains 12,544 individually addressable microelectrodes. Each electrode is 44 μm across and is separated from the reference electrode by an insulating layer (e). Photo credits: (a) Nanophotonics Laboratory, University of Washington; (b) Tate Owen; (c) CustomArray, Inc.; (d,e) Authors.

measuring the concentrations of more than one biomolecule [4–7]. A good example of the dangers of relying on a single biomarker for disease can be found in prostate-specific antigen (PSA), a Food and Drug Administration (FDA)-approved cancer biomarker, which was once widely thought to be a direct indicator for prostate cancer. As a result, researchers developed a number of diagnostic assays and recommendations for diagnosing cancer based on the detected concentration of PSA. However, it became apparent that PSA levels varied widely between individuals and it served as a poor indicator for the presence or prognosis of cancer [4,8]. Investigators have found similar poor diagnostic values for the remaining eight FDA-approved cancer biomarkers, highlighting a shortcoming of single biomarker-based diagnostics [1]. In contrast, a systems approach operates on the hypothesis that disease introduces genetic or environmental perturbations that alter biological networks, resulting in widespread changes being reflected in the levels of multiple proteins and other biomarkers present in the body. Considering the complexity of a physiologic system, it is clear why multiple biomarkers may be needed to diagnose disease.

In addition to accurately diagnosing single diseases, multiplexible biosensors could be used to screen for multiple diseases or pathogens at once, which would be especially useful for POC applications where resources are limited and visits to the clinic are rare. In the case of pathogen detection, multiple tests are often needed to positively identify a pathogenic organism. Currently, diagnosticians use various culture and biochemical tests, requiring days to reach a positive conclusion. The most accurate tests rely on a polymerase chain reaction, but this process is prohibitively expensive and unavailable in areas of the world where such tests are most needed. As an alternative to pathogen detection based on nucleic acid assays, many pathogens can be uniquely characterized by their ligand-binding affinities and antigenic profiles [9–14]; a quantitative biosensor containing a panel of known pathogen-binding ligands and antibodies could detect the presence of these pathogens from patient samples.

In biosensor design, multiplexibility requires multiple sensors or multiple sensing regions that can be differentially functionalized and interpreted. Microarrays using DNA [15,16], proteins [17,18], and carbohydrates [19–22] provide an excellent example for the potential of high-throughput screening, yet they almost always rely on a fluorescent readout. Nonfluorescent microarrays are increasingly being used with surface plasmon resonance imaging (SPRi) [23,24]. In addition to being high-throughput and label-free, SPRi allows real-time analysis of binding interactions.

GENERAL LABEL-FREE BIOSENSOR SETUP AND OPERATION

A biosensor is any device that converts a biological event—typically, but not limited to, binding between complementary biomolecules—to an output signal that can be analyzed to describe the sensed event. When no label is used, detecting the binding event relies on a transducer, the mechanism of which can be electrochemical, mechanical, or optical. The transducer must be functionalized with a bioactive surface that captures the target biomolecules or changes in response to the target, such that these changes on the surface of the transducer elicit a detectable signal (Figure 3.1b). The bioactive surface is generally composed of one or multiple bio-recognition molecules such as oligonucleotides, peptides, antibodies, aptamers, phages, and carbohydrates. These molecules can be attached to the transducers using nonspecific adsorption or by specific covalent attachment via reactive groups that are either native to the biomolecules or introduced through molecular biology or synthetic techniques. Regardless of the mechanism of attachment, the bioactive surface must be resilient to degradation and prevent nonspecific binding to other biomolecules or contaminants. In addition to the transducer element of a biosensor, its operation relies on a variety of support instrumentation and additional components. First, a biosensor requires some way to deliver the sample to the sensing region. With the exception of gas-based sensors, this usually requires a fluidic handling system (pumps, tubing, channels, etc.) that can effectively deliver aqueous samples while simultaneously minimizing reagent consumption. Next, the signal generated by the transducer must be analyzed and correlated with the sample, whether quantifying analyte concentration, binding kinetics, or simply the presence of the target. These additional aspects of the biosensor, beyond initial signal transduction, require instrumentation that: actuates (e.g., signal generator, light source), detects (e.g., oscilloscope, photodetector), and processes (e.g., microprocessor) the signal. In most cases, all these components require power. Thus, considerations for constructing and implementing integrated POC biosensors extend beyond the sensor itself.

TOWARD FULLY INTEGRATED BIOSENSORS

Significant effort has been dedicated to miniaturization and integration of biosensing components to construct LOC devices. The motivation is obvious: fully integrated chip-based biosensors would allow biosensor applications to expand beyond research laboratories into clinics, households, and the POC. In addition to the aforementioned desired characteristics of a biosensor, the devices need to be cheap, robust, reliable, easy to use, and consume low amounts of power. While many technologies and disciplines are converging to achieve this goal, two stand out as especially enabling for the future of biosensors. The first is the fabrication of miniaturized and integrated electronic devices, an effort spearheaded by the microelectronics industry, whose techniques have had far-reaching applications in microtechnologies and nanotechnologies. The other is microfluidics (a beneficiary of microelectronic fabrication processes), which enables the handling and manipulation of small volumes of fluidic samples and is particularly amenable to device integration. In order to realize an integrated LOC biosensor, researchers will, without a doubt, need to leverage both of these technologies. It follows that these devices would greatly benefit from having planar chip-based components in order to integrate microfluidics while capitalizing on semiconductor fabrication technologies [25].

Silicon photonics has become a focal point of many parallel efforts to achieve fully integrated biosensors complete with on-chip light sources, detectors, and data processors [26–29]; the high-throughput, cost-effective, and scalable manufacturing techniques developed by the microelectronics industry and the sensitivity, efficiency, and pervasiveness of optical biosensing are united by their ability to be implemented

on silicon. Jokerst et al. [25] point out the components necessary for a fully integrated planar photonic biosensor. Importantly, all the components—a light source
(e.g., thin film III–V edge emitting laser) [30], a sensor (e.g., microring resonators)
[31], and a photodetector (e.g., InGaAs metal–semiconductor–metal photodetector)
[32]—have been fabricated in planar formats and tested independently by different
groups, so all that remains is piecing everything together. This is no trivial task,
especially considering the other hurdles such as microfluidic integration, sensor
functionalization, and device characterization, yet the technologies exist and efforts
are underway to make fully integrated and distributable biosensors. Steps toward
this goal have been made. Microresonators, optical biosensing devices, have been
integrated with chip-based photodetectors that were able to monitor the resonant
condition of the microresonators [33,34]. Chip-based light sources (e.g., thin-film
edge emitting lasers [35] and dye lasers [36]) have also been integrated with an interferometric coupler [35], as well as a waveguide and a photodetector in series [30].
A variety of traditional optical components such as microlenses [37], mirrors [38],
filters [39,40], laser diodes, and photodiodes [39], as well as sensing components
such as interferometers and microresonators, have been integrated into microfluidic
devices [25–27,41]. Microfluidic devices have also incorporated valves [42,43],
pumps [44–46], and sample processing capabilities such as mixers [47,48], target
concentrators [49,50], and target separators [51–53].

SILICON PHOTONICS FOR DEVICE INTEGRATION

A Note on Classification

As with many broad classification systems, defining the scope of silicon photonics is
a difficult task, and the silicon photonic community remains divided on means of
classification. A decent approach is classification based on the materials used in the
system; some in the field argue that silicon photonics is strictly limited to devices
composed of only silicon waveguides, while others include materials such as silicon
oxides, silicon nitride (Si_3N_4), and silicon in oxynitride (SiON). We take the latter
approach of using a more broad definition of silicon photonics, and place emphasis on
planar devices that leverage complementary metal–oxide–semiconductor (CMOS)-
based microfabrication techniques and thus have potential for component integration.
Some of the devices used as examples do not use silicon-based waveguides, while
others do not use waveguides at all. Nevertheless, it is useful to introduce some of the
basics of silicon photonics at this point in the discussion.

Why Silicon?

Traditional optical components are bulky and expensive; they require exotic materials such as indium phosphide (InP), gallium arsenide (GaAs), and lithium niobate
($LiNbO_3$). In addition, compound optical devices are usually assembled by hand
[54], such that the difficulty of assembly increases exponentially with the number of
components. This makes the large-scale production of complex free-space optical
systems onerous. Silicon, on the other hand, has a history of automated processing
and the microelectronics industry has developed extensive manufacturing infrastructure. This industry has invested hundreds of billions of dollars to the processing

and implementation of silicon in microelectronic devices, driven by silicon's advantageous electronic properties and its low cost. More recently, much attention has been directed toward determining ways of implementing silicon as an optical material to address the cost and manufacturing limitations of current optical devices and to advance the microelectronics industry [55,56]. Importantly, silicon has properties that also make it attractive for photonics, most notably its transparency to wavelengths of light greater than 1100 nm and its high refractive index (RI = 3.5). These properties, coupled with CMOS-processing techniques, allow optical devices to be defined in silicon substrates such that fabricating thousands of silicon photonic components has become a trivial task. Device alignment, typically the most critical and time-consuming step of assembling traditional compound optical systems, becomes a fully passive process because the photolithographic masks define the device layout.

An important feature of silicon-based devices is the facile modification by oxidation, doping, and metallization. The incorporation of oxygen into silicon (forming SiO_2) is the most common modification, and it is ubiquitously used as the insulator in silicon-on-insulator (SOI) microelectronic devices. As applied to silicon photonic structures, the lower RI of SiO_2 (1.46) compared to silicon allows it to serve as a cladding material and confine the light modes within features, known as waveguides, defined in the silicon. Doping is necessary to impart electrical functionality, such as diodes, into silicon. A simple example that demonstrates this feature is a p–i–n diode, where group III and group V ions are implanted in discrete regions on the silicon to introduce p-type and n-type behavior, respectively [54]. Finally, the metallization of silicon (i.e., deposition of metallic structures on silicon) allows for the inclusion of device interconnects and contact pads for external manipulation and interrogation. These common processes utilized by the microelectronics industry can also be incorporated into the fabrication of silicon photonic devices to act as, for example, optical switches and modulators [57].

Waveguides can shuttle and manipulate light on the devices in a way that is analogous to how metallic wiring guides electrical signals, and since using light in place of electrons significantly increases potential data transfer rates and decreases loss, silicon photonics is of particular interest to the communications industry and assures continued investment in this area. Fiber optic cables are already used to rapidly transmit data over long distances with little loss, and the transition to replace on-chip electronics with optical components has begun [58]. Further, because of silicon's excellent electrical properties, the extensive tools available for modifying silicon (e.g., silicon doping and metallization) [54], and the demonstrations of on-chip optical components discussed above, it is clear that fully integrated optoelectronic chips are achievable in the near future.

Waveguides: Fabrication and Basic Principles

Waveguides are defined on silicon substrates using standard lithographic processes that are widely implemented in CMOS fabrication techniques. A variety of lithographic techniques exist for creating silicon structures, including wet and dry chemical etching approaches and e-beam lithography. Dry etching is accepted as the more precise and reproducible technique for large-scale fabrication, achieving critical dimensions of 10 nm. This is smaller than what is required for many basic waveguide structures (with the exception of slot waveguides), but the capability remains, and very high densities of waveguides can be attained. E-beam lithography, while not amenable to high-throughput fabrication, allows rapid prototyping of device structures which enables research in silicon photonics. Both planar and two-dimensional waveguides have been used in biosensors, but, in most practical applications that extend beyond biosensing, light must be confined in two dimensions so as to manipulate and direct light in a controlled manner while minimizing loss [54]. One of the most basic two-dimensional waveguide structures, a ridge waveguide, is shown in Figure 3.3.

Light is coupled into and confined within the waveguide core on account of its higher RI relative to the cladding layers that surround it. The lower cladding layer is almost always silicon dioxide, while the upper cladding can be air (RI = 1), water (RI = 1.33), more silicon dioxide, or just about any material with an RI below that of the waveguide core. Rectangular rib waveguides can support multiple light propagation modes, the number of which is determined by the dimensions of the waveguide. In general, the larger the waveguide cross-section, the more modes it will support. However, given the correct waveguide geometry, only the fundamental mode will propagate because higher-order modes will leak out over very short distances [59]. A portion of the light, the amount of which depends on the waveguide structure and the cladding material, propagates outside of the core. The light outside of the waveguide is known as the evanescent field; the intensity of this field decays exponentially away from the core and it is influenced by the RI of the cladding (Figure 3.3b). Thus, the propagation of light through the waveguide is sensitive to changes outside of the core within the evanescent field, a property that is exploited for applications in biosensing. The distance that the evanescent field extends into the dielectric can be tuned by altering the dimensions of the waveguide, the wavelength of light, and the dielectric itself, but it is typically on the order of tens to a few hundred nanometers [60,61]. Like surface plasmon resonance (SPR) biosensors, this gives waveguide-based biosensors the ability to sense localized biomolecular interactions at the waveguide surface while being largely insensitive to changes in the bulk fluid.

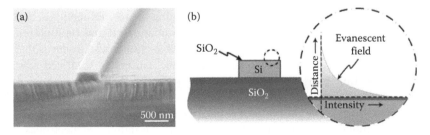

FIGURE 3.3 Basic ridge waveguide. (a) A scanning electron micrograph showing the cross section of a silicon ridge waveguide on top of a silicon oxide substrate. (Photo courtesy of Nanophotonics Laboratory, University of Washington.) (b) A schematic illustrating the evanescent field associated with the light mode traveling through a ridge waveguide.

LABEL-FREE BIOSENSORS

As previously discussed, biosensors that do not require labeling of the target molecules significantly increase the potential applications and the amount of information that can be obtained. This facilitates quantitative real-time binding analysis of native molecules for extracting kinetic binding parameters and makes it possible to use unprocessed samples, thus enabling high information content POC and distributed devices. Instead of using a label, these biosensors rely on inherent properties of the target such as impedance, mass, or RI to measure binding. Although the focus of this chapter is silicon-based optical biosensors, we would be remiss not to acknowledge the role of electrochemical and mechanical biosensors in label-free sensing. These devices have shown potential for realizing fully integrated devices and stand to benefit from some of the same fabrication and integration capabilities provided by silicon.

ELECTROCHEMICAL BIOSENSORS

Generally speaking, electrochemical biosensors operate by detecting the change in the resistance or capacitance on an electrical sensing component in response to the formation of binding complexes or to environmental perturbations. The most prevalent and well-known examples of electrochemical biosensors are those used in the majority of glucose monitors [62]. However, all of these devices, along with most electrochemical biosensors for other applications, possess limited sensitivity and require the use of an electroactive indicator to generate a detectable signal [62,63]. Nonetheless, electrochemical biosensors remain attractive because they can be mass produced at low costs, they have low power requirements, and they can be scaled down to allow miniaturization and multiplexing [64]. Given these benefits, there are significant ongoing efforts to improve label-free electrochemical biosensors, through the construction of more sensitive nanoscale devices based on nanowires, nanotubes, and nanofibers [65–67].

Electrical Impedance Spectroscopy with Microelectrodes

Electrical impedance spectroscopy (EIS), in the case of affinity biosensors, measures the change in the impedance of an electrical circuit due to the binding of analyte to a functionalized electrode. EIS is favored over voltammetry and amperometry because the measurement technique is less damaging to the biofunctional capture layer [64]. The electrodes used for EIS can be miniaturized and multiplexed, as indeed they have been [68,69]. A significant advantage of microelectrodes is that, by applying a current or voltage at the electrodes, one can create localized reaction environments—this opens up the possibility of on-chip synthesis and functionalization in a highly multiplexed fashion. A microelectrode microarray developed by CombiMatrix Corp., now CustomArray, Inc., enabled electrochemically controlled synthesis of unique peptides and sequences of DNA on individual electrodes [70]. The initial devices containing 1024 electrodes were then scaled up to contain 12,544 electrodes (Figure 3.2c–e), a device that the company has used to develop a completely automated, high-throughput assay to screen for and identify subtypes of influenza A by way of genotyping [69]. In addition to directly synthesizing DNA probes on the electrodes, they could be used to polymerize pyrrole, which could then be used to immobilize capture agents such as DNA and antibodies, the latter of which was used to develop an assay for staphylococcal enterotoxin B [71,72]. Despite these examples, with EIS sensing regimes, device performance is tied to the properties of the electrodes, and it remains unclear how binding can be quantified and how miniaturization may alter the equivalent circuits used to describe binding [64]. Further, due to the lower sensitivity of label-free EIS, there exists a limit to how small the electrodes can be before the surface area available for biomolecule functionalization is too small to generate a detectable signal upon target binding.

Nanofield Effect Transistors

Nanoelectrochemical sensors primarily act as field effect transistors (FETs), a sensing technique that is free from the size limitations discussed for the electrodes used for EIS. In a standard transistor setup, a semiconducting material attaches a source and drain electrode; a third electrode, known as a gate electrode, separated from the semiconductor by a thin dielectric, controls the conductance of the semiconductor by applying positive and negative voltages. In this manner, the gate electrode acts as a switch for the current flowing from the source to the drain. In an FET-based biosensor, biomolecules take the place of the gate electrode, whereby the conductance of the semiconductor is altered by the binding of biomolecules (Figure 3.4). Thus, changes in current between the source and the drain are correlated with binding events. Researchers have investigated nanowires [73–77], nanofibers [78], and carbon nanotubes [74,79,80] as semiconducting materials to use in nanoelectrochemical biosensors. A primary motivation is the potential sensitivity of these materials to biomolecular binding events—these one-dimensional materials have similar sizes to biomolecules, such that extremely small amounts of bound analyte will significantly affect the electrical properties of the transistor. Investigators have demonstrated detection of proteins [74,75,79], single viruses [73], and DNA with concentrations reported down into the picomolar range [80] and even down to 10 fM [76,77]. The limited size of these sensors, coupled with the pre-established microelectronics infrastructure, also make them good candidates for multiplexing, and devices containing up to 200 sensors have been reported [73,81]. Some hurdles remain before FET-based nanoelectrochemical biosensors can be used as reliable research instruments, let alone as biosensors in a healthcare setting. Despite the apparent potential to achieve highly multiplexed devices, there are no good

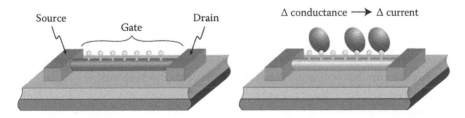

FIGURE 3.4 A nano field effect transistor biosensor is shown schematically. The intrinsic electrical properties of biomolecules that bind to the functionalized semiconducting material (e.g., nanowire, carbon nanotube) changes its conductance and therefore influences the amount of current that flows from the source to the drain.

solutions for high-throughput fabrication. Carbon nanotubes must be synthesized off-chip and then placed in position; nanowires, to their credit, can be patterned lithographically [77,82] or grown on-chip [75,76], but they still suffer from fabrication inconsistencies. Given their size and high sensitivities, heterogeneity between sensors can result in altered performance and poor reproducibility. FET sensors are also ion-sensitive, and ions in solution will act as a gating mechanism and dramatically reduce sensitivity. This, along with the structural fragility of the FETs, limits the potential samples and experimental setups that can be used. Lastly, the mechanisms by which biomolecules influence electrical properties are not well understood, and different biomolecules alter the sensor response in ways that do not correlate predictably with the size of the biomolecule or its concentration in solution.

MECHANICAL BIOSENSORS

Mechanical biosensors directly detect the change in mass on the sensor surface due to the binding of biomolecules, viruses, or cells. Mechanical sensors represent the most sensitive of sensing techniques, with noise floors and mass resolutions reported as low as 20 and 7 zg, respectively [83]. Given their potential sensitivity, mechanical biosensor research has largely been directed toward reaching very low detection limits for applications such as rare analyte sensing and weighing individual viruses and cells. Surface acoustic wave sensors, including the quartz crystal microbalance (QCM) [84–86], utilize the sensitivity of piezoelectric crystal resonance to perturbations in the surrounding environment. In QCM, the quartz surface is usually coated with an anchoring layer to which biological receptors are immobilized. Electrodes attached to the quartz apply an alternating voltage that elicits a resonant mechanical oscillation. Tracking changes in the oscillatory frequency in response to binding at the sensor surface produces the signal, with reported limits of detection (LOD) down to 10 pg/mm^2 [61]. QCM biosensors have been used to detect binding interactions of proteins [87], oligonucleotides [88,89], carbohydrates [90–92], lipids [93,94], viruses [95,96], and cells [97,98]. A distinct advantage of QCM over both electrical and optical biosensors is the wide range of materials that can be deposited on top of the quartz. Since the sensing mechanism does not rely on the transmission of an optical signal or the propagation of light, QCM supports the study of interfacial interactions using a wide variety of materials [84,99]. However, QCM is not without its limitations. While QCM is amenable to performing binding experiments in a liquid environment, sensitivity is reduced and it can be difficult to separate the effects of mass, density, and viscosity in the QCM signal [100]. It is also difficult to fabricate dense arrays of acoustic wave devices, although it has been demonstrated [101].

Microelectromechanical and Nanoelectromechanical Systems

Researchers have investigated microbiosensors and nanobiosensors using mechanical transduction mechanisms in an effort to increase sensitivity and allow multiplexing. These devices use standard photolithography techniques and are usually made out of silicon or silicon nitride, allowing high densities of devices and the possibility of integration with electronic components and flow cells. The majority of these devices are based on analyte binding to functionalized cantilevers, which either changes the deflection of the cantilever (static devices) or its resonant frequency of oscillation (dynamic devices). Static devices (Figure 3.5a) have the advantage of being able to operate in both gas and liquid environments, but they have decreased sensitivity because it requires the attachment of a near-monolayer of analyte to deflect the cantilever [67]. Nevertheless, static cantilever devices have been shown to detect single base-pair mismatches of 12-mer DNA strands and picomolar limits of detection of oligonucleotides [102] and nanomolar concentrations for proteins [102,103]. An impressive study also showed detection limits of PSA down to 0.2 ng/mL (6 nM) in a background of both 1 mg/mL bovine and human serum albumin, which matches detection limits of ELISA for PSA and is physiologically relevant [104]. Dynamic devices (Figure 3.5b) have significantly higher potential sensitivities and researchers have shown the ability to detect viruses [105] down to the single virus [106–108], single cells [109,110], DNA with a single 1587-mer strand [111], and PSA down to 10 pg/mL [112]. The Bashir group has also used their devices for weighing single viruses [107] and cells [110]. As with QCM, an important drawback of most dynamic mechanical biosensors is that their sensitivity is limited by the dampening effects of liquid, so detection must be done in vacuum or air. However, several recent papers have implemented nanofluidic channels fully confined within the cantilevers (Figure 3.5c) [113,114]. In this configuration the cantilevers are maintained in a vacuum and the channels within the cantilevers allow biological interactions to occur in solution—not only are the interactions measured in a physiologic (aqueous) environment, but they are detected in real time. Studies using these devices have measured the changing masses of individual cells during growth [115,116],

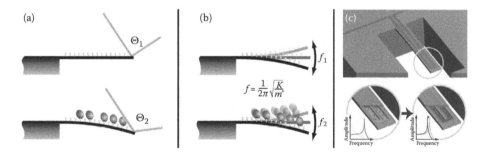

FIGURE 3.5 The two different modes of operation for microcantilever-based mechanical biosensors are demonstrated schematically. In both cases, the cantilevers are functionalized with the appropriate receptor or capture molecule. (a) Static microcantilever biosensors correlate the deflection of the cantilever arm upon the binding of biomolecules. (b) Dynamic microcantilever biosensors sense binding of biomolecules through changes in the oscillatory frequency. The mass of bound material can be calculated because the deflection or the change in the frequency (f) of oscillation can be related to the spring constant (K) and the effective mass (m^*) of the cantilever. (c) Since viscous damping of liquids leads to a dramatic decrease in sensitivity for dynamic microcantilever biosensors, Burg et al. [113] have fabricated devices that contain a nanofluidic channel inside of the cantilever arm. (With permission from Macmillan Publishers Ltd., *Nature*, Burg, T.P. et al., Weighing of biomolecules, single cells and single nanoparticles in fluid, 446(7139): 1066–1069, copyright 2007.)

detected cancer biomarkers in serum down to 10 ng/mL [117], and immunoglobulin G (IgG) protein below nanomolar concentrations [113].

OPTICAL BIOSENSORS

Optical biosensors are the most widely used label-free biosensing platforms to study biomolecular interactions because of their relative ease of use, high sensitivity, and the high information content of the data they generate. In 2008 alone, there were over 1400 articles published on optical biosensors [118]. Compared to many electrochemical and mechanical platforms, optical biosensors are more flexible and easy to use from an operational standpoint. Additionally, the sensitivity of optical biosensors is not drastically reduced by physiological salinity or viscosity in the analyte buffer, making them amenable to a wide range of samples. Label-free detection methods using optical biosensors include RI detection, optical absorbance detection, and Raman spectroscopic detection [119], with the most common form being RI detection. RI detection is based on the sensitivity of light to changes in RI; biomolecules have a higher RI than buffer solutions (e.g., 1.45 for proteins versus 1.33 for water), allowing their detection by monitoring the properties of the interacting light. A number of RI-based optical biosensors exist, including SPR, optical fibers, planar waveguides, interferometers, photonic crystals, and resonant cavities.

SPR and SPRi

First reported in 1983 [120], biosensors based on SPR are among the most widely used optical biosensors. As of 2008, a total of 24 manufacturers offered commercial platforms, including instruments made by GE, Bio-Rad, Biosensing Instruments, and Reichert [118]. SPR detection relies on the sensitivity of evanescent fields to changes in the local RI of the dielectric. In most SPR instruments, the evanescent field is associated with surface plasmon modes that are created from the coupling of light with a metallic film, usually gold, via total internal reflection within a prism (Figure 3.6). The conditions of total internal reflection (the wavelength of light and the incident angle of light that couples with the metallic film) vary with the RI of the dielectric above the metal film. A flow cell delivers biomolecules to the surface of the metallic film, where binding of analyte to immobilized receptors causes changes in the local RI. The instrument tracks these changes in real time and reports them as a shift in resonant wavelength (angular SPR) or as changes in the intensity of reflected light (SPRi). Traditional angular SPR has superior limits of detection compared to SPRi—the RI detection limit for SPR typically ranges from 10^{-6} to 10^{-8} refractive index units (RIU), whereas SPRi is usually in the range of 10^{-5} to 10^{-6} RIU [119,121]—but SPR is only able to monitor binding in a single region at once for each light source. SPRi, on the other hand, uses a CCD array to detect the intensity of reflected light from the entire chip surface, allowing arrayed and simultaneous sensing of multiple binding interactions. The number of interactions that can be monitored simultaneously is limited only by the spatial resolution of the instrument (~4 μm) [24] and the arraying density of functionalization, the latter of which has been widely addressed by the microarray community. Realizing 100 interaction regions on a single SPRi chip is common, with reports of systems that allow up to 10,000 spots [122].

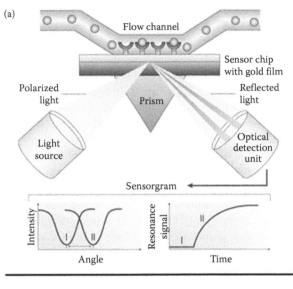

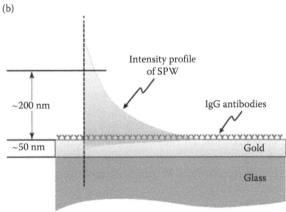

FIGURE 3.6 (a) In a surface plasmon resonance biosensor, the binding of biomolecules to immobilized receptors changes the coupling properties (in this case, the coupling angle) of light reflected off of a metallic film. (With permission from Macmillan Publishers Ltd., *Nature Reviews Drug Discovery*, Cooper, M.A., Optical biosensors in drug discovery, 1(7): 515–528, copyright 2002.) These coupling angle changes, documented over a given time period, generate a sensorgram that describes the binding events. (b) The coupling properties of light are sensitive to changes in refractive index within the surface plasmon wave that extends ~200 nm (intensity = 1/e) from the metal–dielectric interface. The schematic includes immunoglobulin G antibodies drawn to scale for perspective.

Given its high-throughput potential, SPRi appears to have significant potential as a multiplexed POC device. Most studies using SPRi have used it for multiplexed interaction screening and characterization rather than for detection [123], favoring angular SPR for the higher sensitivity measurements. Several commercial instruments exist, including the FlexChip from BIA-CORE [124,125], GWC's SPRimager®II

[126,127], and Texas Instruments' Spreeta system [128,129]. In addition to the Spreeta instrument, other groups have developed portable SPRi systems that make important in-roads to realizing POC applications [130–132]. Unlike the aforementioned multiplexed devices, SPRi benefits from the fact that different sensing areas can be defined by the user based on the locations of the functionalized regions, and the metal surface (usually gold) of SPRi sensor chips is relatively robust. This greatly reduces the alignment difficulties that are encountered when attempting to functionalize specific devices on multiplexed electrochemical and mechanical devices. As a biosensing platform, SPR benefits from the extensive literature on the biofunctionalization of gold surfaces—one of the best understood surfaces for functionalization and the standard for biological surface analysis [133]. Among the important properties are gold's biocompatibility and its ability to bind strongly to thiol groups (at near-covalent strength), which allows for facile tethering of biomolecules and non-fouling self-assembled monolayers [134,135]. However, despite these advantages, SPRi has yet to realize widespread use in the clinic or in POC settings.

As mentioned, the popularity of SPR is partly a result of the detailed binding information that the technique can produce. Properly designed experiments can yield both qualitative and quantitative information, such as the selectivity, the strength, the kinetic binding parameters, and the thermodynamic parameters of a binding interaction, as well as identify the active concentration of the target molecule. However, emphasis must be placed on the careful design, execution, and analysis of optical biosensing experiments if meaningful information is to be extracted. An unfortunate reality of the widespread use of SPR, and optical biosensors more generally, is that many investigators make incorrect conclusions from their data as a result of poorly run experiments or faulty analyses. Rich and Myszka [118,136,137] reviewed the optical biosensor literature every year from 1998 to 2008 and found that a large majority contains major flaws in some aspect. These reviews are excellent sources for understanding the proper utilization of optical biosensing technologies and they also communicate the wide range of applications of these instruments.

Grating-Based Sensors

Optical fibers and planar waveguides can both be incorporated into surface plasmon wave (SPW) biosensors, and they operate similarly to SPR. In these cases, an optical fiber or a waveguide acts in place of the prism to couple light with a metallic layer, which generates the SPW and the corresponding evanescent wave that is used for sensing RI changes in the dielectric. Alternatively, in non-SPW biosensing conformations, optical fibers and planar waveguides often rely on coupling light with a grating structure. A grating consists of a periodic physical perturbation on the surface of the sensor; light couples into the grating at a specific angle and wavelength that are determined by the effective RI (n_{eff}) of the fiber or waveguide and the grating period. Binding of biomolecules changes n_{eff}, enabling real-time detection. In the case of fibers, the gratings are etched into the optical core or into the cladding immediately surrounding the core, such that a biofunctionalized grating provides the sensing region (Figure 3.7a). While these devices are more widely used for sensing load, strain, temperature, and vibration [138], examples of their application for biosensing include: the detection of DNA 20 base pairs in length down to 0.7 µg/mL using a

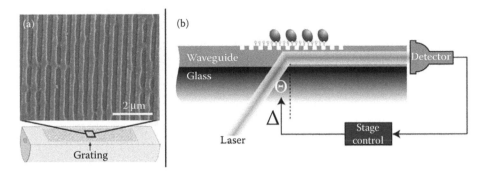

FIGURE 3.7 Grating-based biosensors have been shown in a variety of conformations, two of which are depicted here. (a) A Bragg grating is etched into the cladding of a D-shaped fiber. (Smith, K.H. et al., Surface-relief fiber Bragg gratings for sensing applications. *Applied Optics*, 2006, 45(8): 1669–1675. With permission of Optical Society of America.) (b) An optical waveguide light-mode spectroscopy biosensor changes the angle of the stage in response to biomolecule binding to maintain coupling. These changes are recorded over time to generate the sensorgram.

device with an RI detection limit of 7×10^{-6} RIU [139], real-time monitoring of antibody binding with a dynamic range from 2 to 100 μg/mL and antigen detection from crude *Escherichia coli* lysate [140], and the detection of hemoglobin in sugar solutions with an inferred sensitivity to a change in hemoglobin concentration of 0.005% [141]. While fiber gratings are inexpensive and straightforward to manufacture, they suffer from relatively poor sensitivities [119].

Grating-coupled planar waveguides are also easy and cheap to fabricate, as they consist of a thin-film waveguide deposited on a glass support into which a grating can be etched using photolithography or imprinting [142]. Optical waveguide light-mode spectroscopy (OWLS) is one well-known implementation of this sensing modality; these devices measure the change in the coupling angle due to changes in the RI on the grating (Figure 3.7b). They have been used for biosensing applications including antibody capture of the herbicide trifluralin down to 100 ng/mL [143] and the detection of mycotoxins down to 0.5 ng/mL [144]. OWLS has been more widely applied to studying biomolecular adsorption kinetics and conformation on a variety of material surfaces [145–147]. OWLS does not permit multiplexing capabilities, but a very similar technique that uses planar waveguide gratings known as wavelength-interrogated optical sensors addresses this issue, and a device with 24 different sensing sites has been used to simultaneously monitor four different classes of veterinary antibiotics in milk with a detection limit ranging from 0.5 to 34 ng/mL, depending on the class of antibiotic [148].

Interferometric Sensors

Mach–Zehnder Interferometers

In a Mach–Zehnder interferometer (MZI), a single frequency, coherent polarized light source is split into two paths. The sample is placed in one of these paths where light interactions with the sample cause a shift in the phase of the light, and the other path acts as the reference. The light is then recombined and the phase shift caused by

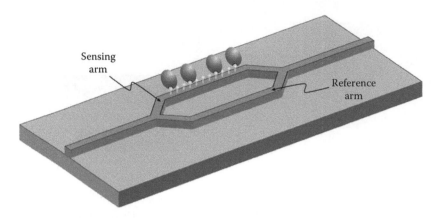

FIGURE 3.8 Schematic representation of a chip-based Mach–Zehnder interferometer biosensor. Changes in the refractive index surrounding the sensing arm induce a phase change, resulting in interference upon recombination with the reference arm.

the sample in the sensing arm leads to interference that can be detected by a change in the light's intensity. Although traditionally done in free space, MZIs can be fabricated in a planar structure using waveguides to split and recombine the light; these are called integrated MZIs. In such a setup, the sensing arm is functionalized and binding of the sample changes the RI within the evanescent field of the waveguide, thus modulating the phase of the propagating light and leading to interference upon recombining with light from the reference arm (Figure 3.8). The first biosensing demonstration of integrated MZIs detected human chorionic gonadotropin down to 50 pM using immobilized capture antibodies [149]. This device had an RI LOD of 5×10^{-6} RIU, but improvements in MZI fabrication and analysis have led to demonstrations of LODs down to 10^{-7} RIU [150], which is on par with most SPR instruments. Other demonstrations of MZI biosensing include detection of IgG down to 1 ng/mL [151], and the ability to distinguish the wild-type DNA (58-mer) from a mutated sequence down to 10 pM concentrations [152]. Very few reports of biosensing with MZIs have emerged following the initial interest in the 1990s. This could be due to difficulties in multiplexing and the requirement of a relatively long sensing region to generate a detectable signal. Long sensing regions not only require a larger footprint on the device, but they also work against sensitivity because of increased loss. A more recent publication addressed both of these issues by demonstrating a multiplexed device that used coiled waveguides as sensors [153]. The device had six sensors, four of which were functionalized with two different antibodies (two sensors for each antibody) and the remaining two were used as reference sensors. The sensor response corresponded to a surface coverage of just 0.3 pg/mm². However, it remains to be seen whether integrated MZIs for biosensing applications will have a significant impact in biomedical research.

Young's Interferometers

Young's interferometer (YI) can be integrated onto a chip surface in much the same way as MZIs and can be used for biosensing. Similar to MZIs, YIs split light from a single waveguide into multiple arms including one reference arm. However, instead of recombining the light back into a single waveguide, a CCD is used to record the interference fringes that result from the optical output (Figure 3.9), permitting multiplexed sensing with just one reference. An integrated YI for sensing was first demonstrated in 1994 [154], and the technique has an established RI LOD of 10^{-7} RIU [155]. YIs have been used subsequently in a number of proof-of-concept applications. For instance, a multiplexed device containing three sample arms and one reference arm enabled biosensing of herpes simplex virus type 1 (HSV-1) [156]. The authors were able to detect as few as 10^5 HSV-1 particles in serum and 10^3 particles in buffer, highlighting the potential for the device to be miniaturized and integrated for POC applications [157]. Hoffman and colleagues [158] developed a planar waveguide YI that they used to determine the binding kinetics for protein G capture of IgG and demonstrated its compatibility with biotin–streptavidin functionalization techniques. The authors reported an RI LOD of 10^{-9} RIU that corresponds to a surface coverage of just 13 fg/mm^2, one of the lowest reported values of any optical biosensor. However, it should be noted that, while there have been a number of reports demonstrating MZI and YI multiplexed sensing, interferometric biosensing has not proven to be readily amenable to high-throughput multiplexing due to the large sensing regions required and because the complexity of analysis increases significantly with each additional sensing arm.

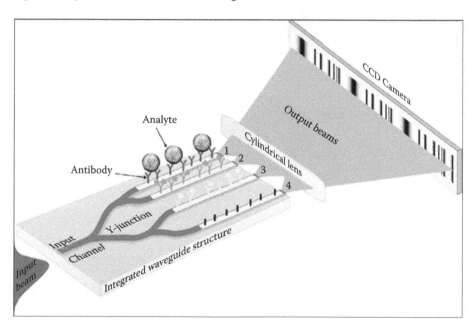

FIGURE 3.9 Schematic representation of a multiplexed integrated Young's interferometer. Antibodies with different specificities are functionalized on the waveguide arms such that specific recognition and binding of analyte alters the local refractive index, leading to changes in the interference pattern detected by the CCD camera. (Ymeti, A. et al., Fast, ultrasensitive virus detection using a Young interferometer sensor. *Nano Letters*, 2007, 7(2): 394–397. With permission of Optical Society of America.)

Resonant Cavity Sensors

Resonant cavities represent one of the most rapidly expanding and promising label-free optical biosensing techniques, due largely to their high sensitivity and their potential to be integrated into multiplexed chip-based devices. In resonant cavity sensors, which include microspheres (Figure 3.10a), microtoroids (Figure 3.10b), microrings (Figure 3.2a and b), and microcapillaries, light is coupled into an optical cavity where certain wavelengths are confined; this confinement generates a narrow dip in the transmission spectrum. The wavelengths of light that travel around the outside of the cavity and return in phase are the resonant wavelengths and can be described by the equation $\lambda = 2\pi r n_{eff}/m$, where λ is the wavelength of light, r is the radius of the cavity, n_{eff} is the effective RI of the waveguide mode, and m is an integer. A fiber or integrated bus

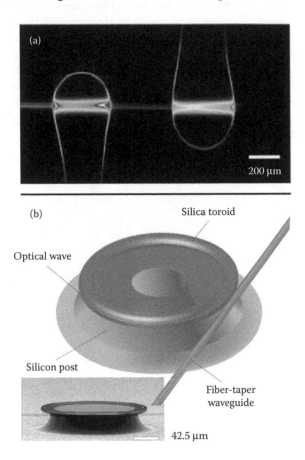

FIGURE 3.10 Resonant cavity biosensors confine the wavelengths of light which, after circumnavigating the cavity, constructively interfere with itself. Biosensing is possible because the resonant frequency is sensitive to perturbations in the surrounding RI. Examples include (a) silica microspheres (From Vollmer, F. et al., *Biophysical Journal*, 2003, 85(3): 1974–1979. With permission from the Biophysical Society) and (b) silica microtoroids fabricated on a silicon support. (With permission from Macmillan Publishers Ltd., *Nature*, Vahala, K.J., Optical microcavities, 424(6950): 839–846, copyright 2003.)

waveguide in close proximity to the resonant cavity delivers light to the cavity for coupling and away from the cavity so that the transmission spectrum can be tracked. Resonant wavelengths appear as dips in the transmission spectrum because the resonant condition extracts power from the light in the fiber/waveguide that reaches the detector [159,160]. The dependence of the resonant wavelength on n_{eff} is due to the evanescent field that extends and decays exponentially away from the surface of the cavity, and, like the other optical biosensors discussed in this chapter, it is this relationship that creates the sensing mechanism. By changing n_{eff}, biomolecules binding to the resonant cavity shift the resonant wavelengths supported by the structure. In contrast to the other evanescent sensing techniques already described (SPR, grating-coupled devices, and interferometers), where each photon only interacts with the biomolecules one time, a photon coupled into a resonant cavity interacts with biomolecules each time it travels around the cavity, which can reach into the thousands for some resonant cavities [160]. This feature bestows high sensitivities to small devices, which is not possible with other optical biosensors (e.g., interferometric sensors). The number of revolutions a photon makes around a resonant cavity before dissipating is related to the quality factor (Q) of the resonator and determines the sensitivity of the device [160]. Q is determined by the full-width at half-maximum ($\delta\lambda_r$) of the resonant dip at the resonant wavelength (λ_r), according to the equation $Q = \lambda_r/\delta\lambda_r$. Thus, a higher Q corresponds to a more narrow dip in the transmission spectrum which facilitates sensitive tracking of the resonant wavelength. Demonstrations of optical cavities implemented as biosensors include microspheres, microcapillaries, and microfabricated chip-based structures such as microtoroids and microrings.

Microspheres and Microtoroids

Microspheres exhibit Q factors over 10^6 and they have had RI LODs reported as low as 10^{-7} RIU [161,162] Resonant cavity microspheres are generally constructed by melting the tip of an optical fiber or a glass rod [119], which must then be brought into close proximity to and aligned with a tapered fiber. Demonstrations of resonant microsphere biosensors include the detection of protease activity with an LOD of trypsin at 10^{-4} units/mL [163], detection and mass determination of single influenza A virus particles [160], and the detection of single nucleotide mismatch of DNA with an LOD of 6 pg/mm^2 [164]. The device used for DNA detection used two microspheres of different sizes brought into proximity of a single tapered fiber. Because of their different size, each microsphere had a unique resonant wavelength and they could be interrogated simultaneously. Despite this proof-of-concept multiplexed device, microsphere-based resonant cavity biosensors are resistant to large-scale multiplexing because of the sensitive alignment required between the microspheres and the tapered fiber and because they are incompatible with planar fabrication techniques [25]. Armani et al. [165] developed microtoroid resonant cavities with extremely high Q ($>10^8$) which were fabricated using planar lithography. The authors report remarkable single-molecule label-free detection of interleukin-2 (IL-2) via capture by immobilized IL-2 antibody in 10-fold diluted fetal bovine serum [165]. This, while impressive, also raises a number of questions related to the reported sensing mechanism [166] and the observed mass transport [167], suggesting that we still have much to learn about ultrasensitive optical biosensing. Armani et al.'s [165] results have set a high standard for optical sensing, but they do not address our need for high-throughput multiplexed sensing. While the microtoroids were fabricated on-chip using photolithography, the technique requires alignment of a tapered optical fiber waveguide for coupling. Further, the inherent fragility of both microsphere and microtoroid systems make them sensitive to flow, particularly with viscous fluids, such as blood plasma.

Microrings

Planar microrings have arguably become the most popular form of resonant cavity biosensors, owing to their small size, high sensitivity, ease of manufacture, and multiplexing potential. Ring size can vary, but nearly all are on the order of tens of microns in diameter, which is favorable as compared to interferometric devices that require sensing lengths on the order of a centimeter [61]. Microring resonators do not have a decreased sensitivity on account of their small size because of the increased light interaction imparted by the resonance, as previously discussed. They do have a lower Q (10^4–4×10^4) and slightly higher (worse) reported RI LODs (10^5–10^7 RIU) [31,168,169] than microspheres and microtoroids, but their simple and scalable fabrication, multiplexing capability, and potential for integration with other components make them attractive for biosensing applications. Microring resonators can be fabricated using standard silicon wafer processes, enabling passive alignment of multiple microrings with on-chip bus waveguides, which is a significant advantage over the microsphere and microtoroid devices. While they are almost universally fabricated on a silicon substrate, the waveguides and rings can be made out of polymers [169,170], silicon oxide [171,172], silicon nitride [168,173,174], and SOI [175–177]. Sensitive multiplexed detection and binding assays using microring resonators are demonstrative of the advantages of this biosensing platform. Using a device containing five independent microrings, Ramachandran et al. [172] showed specific binding of *E. coli* O157:H7 to microrings functionalized with antibodies, detection of complementary DNA probes, and quantitative detection of IgG [172]. Although not unique to this device, three disadvantages become apparent: (1) relatively low acquisition rates, (2) the lack of integrated fluidics, and (3) a paucity of high-throughput functionalization techniques. In the device reported by Ramachandran et al. [172], the scan rate was 15 s per microring, limiting measurement frequency to 75 s per device, if all rings were interrogated. Faster scan rates are required to extract binding kinetics and for truly high-throughput multiplexed measurements. An instrument containing integrated fluidics and peripheral instrumentation for using disposable chips was reported by Carlborg et al. [168]. Device characterization showed an RI LOD of 5×10^{-6} RIU and a mass density detection limit of 0.9 pg/mm^2. The authors have since published on characterization of the temperature sensitivity of this device [178], but they have yet to report on its implementation in a biosensing experiment. Another instrument used extensively by ourselves and the Bailey group at the University of Illinois Urbana Champagne directly addresses the issues of scanning speed and fluidic integration, and both of our groups have devised improved techniques for differential functionalization of the microrings [179,180]. The platform has a detachable microfluidic chamber and uses high-speed scanning instrumentation that interrogates all 32 rings on the device in fewer than 10 s [31]. Bailey's group has reported on detection of carcinoembryonic antigen in undiluted serum down to 2 ng/mL [177], detection of Jurkat T lymphocyte secretions of IL-2 and IL-8 [181], detection of multiple micro RNAs with the ability to distinguish between single nucleotide polymorphisms [182], and quantitative detection of five protein biomarkers in mixed samples [183]. This group also did a thorough theoretical and empirical analysis to characterize the mass sensitivity and the evanescent sensing field of the microrings, finding a mass sensitivity of 1.5 pg/mm^2 and a 1/e evanescent decay distance of 63 nm [60]. Such characterization is rare within the field of biosensors, yet this information is critical for experimental design and interpretation of results. For multiplexing functionalization, Washburn et al. [183] have used a six-channel microfluidic device to differentially functionalize groups of microrings. Using the same microring-based biosensor, our group has implemented a piezoelectric spotter to differentially functionalize microrings on multiple chips in a single run, thus demonstrating a rapid and scalable approach [180]. An alternative method for addressing the issue of the scan speed mentioned earlier was demonstrated by Xu et al. [175]. Instead of increasing the scan speed, the investigators used a single waveguide to interrogate five rings with different radii. Since rings of varying diameter support resonances of different wavelengths, Xu et al. [175] were able distinguish shifts in the resonant frequency due to binding of species-specific IgG capture on each microring. In addition to demonstrating specific and simultaneous detection of two different IgG antibodies, the researchers deduced an impressive mass density sensitivity of 0.3 pg/mm^2. It is clear that devices implementing microring resonators have made significant advances toward realizing applications beyond the laboratory bench and into the clinic. The combination of high sensitivities, ease of manufacture, multiplexibility, and potential for integration has positioned the microring resonator-based device as one of the most promising optical sensing technologies to emerge from the biosensing community.

OUTLOOK AND CONCLUSIONS

Frustratingly, biosensor technology remains largely confined to the research setting, and very few technologies have made it to the clinic, to the general public, or to the POC setting—where the need is great. In a survey of the biosensing literature, and even within this chapter, it becomes apparent that this is not for want of new sensing techniques or increased sensitivity. Instead, the biosensing community continues to produce new devices with new or improved approaches for accomplishing similar goals. All too often, promising new technologies are falling short of the goal of making an impact in healthcare, drug discovery, environmental monitoring, defense, and so on. Clearly, increased attention needs to be directed toward realizing impactful applications of the technology.

Fully integrated devices open up many possibilities for real-world applications, but in order to gain traction and establish biosensors as an effective tool, research must focus on a few strategic areas where biosensors can make the most immediate and meaningful impact. More focused, application-driven, and collaborative research and development efforts would increase the likelihood of overcoming the hurdles that are currently preventing biosensors from being implemented in POC settings.

For instance, targeting specific applications that demonstrate the most need will attract the funding that will be needed to fully develop the biosensor and get it through clinical trials. Simply put, technology is no longer the limiting factor to more fully incorporating biosensors into healthcare—increasingly, it has become a problem of systems integration and design of application-centric biosensors.

Over the past several decades, significant effort has been invested with the aim of developing sensing technologies that will impact the practice of biomedical research and healthcare. This investment has yielded a plethora of sensing technologies built upon a host of sensing modalities (i.e., electrochemical, mechanical, optical, etc.). Ultimately, there is no one-size-fits-all solution for biosensing, and, in this chapter, we have argued for a few important design considerations for developing application-based sensors. (1) A biosensor should be *sensitive* and *selective* for the intended analyte(s) within complex samples, such as saliva, blood, or urine. (2) *Label-free* detection can decrease assay time, costs, and complexity, and is generally more flexible than its label-based counterpart. (3) *Multiplexing* confers enhanced reliability by allowing in-line controls and increased assay information density, thereby reducing costs associated with multiple tests. Finally, (4) a *fully integrated* platform, including peripheral instrumentation (e.g., a light source, a detector, and a microprocessor) and sample handling capabilities (e.g., pumps, microfluidic channels) in addition to the sensor, is essential for these devices to expand beyond the laboratory and to the POC.

Silicon photonic optical biosensors are the most promising candidate technology with the potential to integrate all of these design features. As an optical biosensing technique, these devices are label-free because they rely on the inherent refractive indices of the analyte to generate the signal. Their limited size and high sensitivity will enable massively parallelized multiplexed sensing using wafer-scale processing, dramatically reducing the cost and complexity of fabricating thousands of devices onto a single chip. In addition, by leveraging microelectronic fabrication techniques, silicon photonic biosensors can be integrated with planar on-chip light sources, detectors, and microprocessors. Microfluidics, including pumps, sample preparation strategies, and optical components, can be readily incorporated onto these planar features. Ultimately, the barriers to achieving a fully integrated biosensor using silicon photonics appear to be lower than they are for other sensing modalities.

ACKNOWLEDGMENTS

This work was supported by NSF CBET (award no. 0930411) and the Washington Research Foundation. J.W.C. wishes to thank the NSF graduate research fellowship program. The authors would also like to thank Jim Kirk and Mike Gould for their edits and valuable discussions.

REFERENCES

1. Polanski, M. and N.L. Anderson, A list of candidate cancer biomarkers for targeted proteomics. *Biomark Insights*, 2007, 1: 1–48.

2. Sun, Y.S. et al., Effect of fluorescently labeling protein probes on kinetics of protein–ligand reactions. *Langmuir*, 2008, 24(23): 13399–13405.
3. Kodadek, T., Protein microarrays: Prospects and problems. *Chemistry & Biology*, 2001, 8(2): 105–115.
4. Hood, L. et al., Systems biology and new technologies enable predictive and preventative medicine. *Science*, 2004, 306(5296): 640–643.
5. Soper, S.A. et al., Point-of-care biosensor systems for cancer diagnostics/prognostics. *Biosensors & Bioelectronics*, 2006, 21(10): 1932–1942.
6. Sidransky, D., Emerging molecular markers of cancer. *Nature Reviews Cancer*, 2002, 2(3): 210–219.
7. Wulfkuhle, J.D., L.A. Liotta, and E.F. Petricoin, Proteomic applications for the early detection of cancer. *Nature Reviews Cancer*, 2003, 3(4): 267–275.
8. Hernandez, J. and I.M. Thompson, Prostate-specific antigen: A review of the validation of the most commonly used cancer biomarker. *Cancer*, 2004, 101(5): 894–904.
9. Karlsson, K.A., Bacterium–host protein–carbohydrate interactions and pathogenicity. *Biochemical Society Transactions*, 1999, 27(4): 471–474.
10. Nagahori, N. et al., Inhibition of adhesion of type 1 fimbriated Escherichia coli to highly mannosylated ligands. *Chembiochem*, 2002, 3(9): 836–844.
11. Autar, R. et al., Adhesion inhibition of F1C-fimbriated *Escherichia coli* and *Pseudomonas aeruginosa* PAK and PAO by multivalent carbohydrate ligands. *Chembiochem*, 2003, 4(12): 1317–1325.
12. Disney, M.D. et al., Detection of bacteria with carbohydrate-functionalized fluorescent polymers. *Journal of the American Chemical Society*, 2004, 126(41): 13343–13346.
13. Disney, M.D. and P.H. Seeberger, The use of carbohydrate microarrays to study carbohydrate–cell interactions and to detect pathogens. *Chemistry & Biology*, 2004, 11(12): 1701–1707.
14. Smith, A.E. and A. Helenius, How viruses enter animal cells. *Science*, 2004, 304(5668): 237–242.
15. Eisen, M.B. and P.O. Brown, DNA arrays for analysis of gene expression, In Sherman M. Weissman (ed.), *Methods in Enzymology, Volume 303: cDNA Preparation and Characterization*. San Diego, CA: Academic Press, 1999, pp. 179–205.
16. Heller, M.J., DNA microarray technology: Devices, systems, and applications. *Annual Review of Biomedical Engineering*, 2002, 4: 129–153.
17. Templin, M.F. et al., Protein microarray technology. *Trends in Biotechnology*, 2002, 20(4): 160–166.
18. Zhu, H. and M. Snyder, Protein chip technology. *Current Opinion in Chemical Biology*, 2003, 7(1): 55–63.
19. Ratner, D.M. et al., Probing protein–carbohydrate interactions with microarrays of synthetic oligosaccharides. *Chembiochem*, 2004, 5(3): 379–382.
20. Wang, D.N. et al., Carbohydrate microarrays for the recognition of cross-reactive molecular markers of microbes and host cells. *Nature Biotechnology*, 2002, 20(3): 275–281.
21. Feizi, T. et al., Carbohydrate microarrays—A new set of technologies at the frontiers of glycomics. *Current Opinion in Structural Biology*, 2003, 13(5): 637–645.
22. Blixt, O. et al., Printed covalent glycan array for ligand profiling of diverse glycan binding proteins. *Proceedings of the National Academy of Sciences of the United States of America*, 2004, 101(49): 17033–17038.
23. Bally, M. et al., Optical microarray biosensing techniques. *Surface and Interface Analysis*, 2006, 38(11): 1442–1458.
24. Campbell, C.T. and G. Kim, SPR microscopy and its applications to high-throughput analyses of biomolecular binding events and their kinetics. *Biomaterials*, 2007, 28(15): 2380–2392.

25. Jokerst, N. et al., Chip scale integrated microresonator sensing systems. *Journal of Biophotonics*, 2009, 2(4): 212–226.
26. Myers, F.B. and L.P. Lee, Innovations in optical microfluidic technologies for point-of-care diagnostics. *Lab Chip*, 2008, 8(12): 2015–2031.
27. Monat, C., P. Domachuk, and B.J. Eggleton, Integrated optofluidics: A new river of light. *Nature Photonics*, 2007, 1(2): 106–114.
28. Momeni, B. et al., Silicon nanophotonic devices for integrated sensing. *Journal of Nanophotonics*, 2009, 3(1): 031001; doi: 10.1117/1.3122986.
29. Balslev, S. et al., Lab-on-a-chip with integrated optical transducers. *Lab Chip*, 2006, 6(2): 213–217.
30. Seo, S.W., S.Y. Cho, and N.M. Jokerst, A thin-film laser, polymer waveguide, and thin-film photodetector cointegrated onto a silicon substrate. *IEEE Photonics Technology Letters*, 2005, 17(10): 2197–2199.
31. Iqbal, M. et al., Label-free biosensor arrays based on silicon ring resonators and high-speed optical scanning instrumentation. IEEE *Journal of Selected Topics in Quantum Electronics*, 2010, 16(3): 654–661.
32. Seo, S.W. et al., High-speed large-area inverted InGaAs thin-film metal-semiconductor-metal photodetectors. *IEEE Journal of Selected Topics in Quantum Electronics*, 2004, 10(4): 686–693.
33. Cho, S.Y. and N.M. Jokerst, Integrated thin film photodetectors with vertically coupled microring resonators for chip scale spectral analysis. *Applied Physics Letters*, 2007, 90(10): 101105; doi: 10.1063/1.2711524.
34. Cho, S.Y. and N.M. Jokerst, A polymer microdisk photonic sensor integrated onto silicon. *IEEE Photonics Technology Letters*, 2006. 18(17–20): 2096–2098.
35. Seo, S.W., S.Y. Cho, and N.M. Jokerst, Integrated thin film InGaAsP laser and I X 4 polymer multimode interference splitter on silicon. *Optics Letters*, 2007, 32(5): 548–550.
36. Balslev, S. et al., Micro-fabricated single mode polymer dye laser. *Opt Express*, 2006, 14(6): 2170–2177.
37. Seo, J. and L.P. Lee, Disposable integrated microfluidics with self-aligned planar microlenses. *Sensors and Actuators B: Chemical*, 2004, 99(2–3): 615–622.
38. Llobera, A. et al., Multiple internal reflection poly(dimethylsiloxane) systems for optical sensing. *Lab on a Chip*, 2007, 7(11): 1560–1566.
39. Chediak, J.A. et al., Heterogeneous integration of CdS filters with GaN LEDs for fluorescence detection microsystems. *Sensors and Actuators A: Physical*, 2004, 111(1): 1–7.
40. Llobera, A. et al., Monolithic PDMS passband filters for fluorescence detection. *Lab on a Chip*, 2010, 10(15): 1987–1992.
41. Ligler, F.S., Perspective on optical biosensors and integrated sensor systems. *Analytical Chemistry*, 2009, 81(2): 519–526.
42. Zhang, C.S., D. Xing, and Y.Y. Li, Micropumps, microvalves, and micromixers within PCR microfluidic chips: Advances and trends. *Biotechnology Advances*, 2007, 25(5): 483–514.
43. Oh, K.W. and C.H. Ahn, A review of microvalves. *Journal of Micromechanics and Microengineering*, 2006, 16(5): R13–R39.
44. Beebe, D.J., G.A. Mensing, and G.M. Walker, Physics and applications of microfluidics in biology. *Annual Review of Biomedical Engineering*, 2002, 4: 261–286.
45. Iverson, B.D. and S.V. Garimella, Recent advances in microscale pumping technologies: A review and evaluation. *Microfluidics and Nanofluidics*, 2008, 5(2): 145–174.
46. Wang, X.Y. et al., Electroosmotic pumps and their applications in microfluidic systems. *Microfluidics and Nanofluidics*, 2009, 6(2): 145–162.
47. Mansur, E.A. et al., A state-of-the-art review of mixing in microfluidic mixers. *Chinese Journal of Chemical Engineering*, 2008, 16(4): 503–516.
48. Chang, C.C. and R.J. Yang, Electrokinetic mixing in microfluidic systems. *Microfluidics and Nanofluidics*, 2007, 3(5): 501–525.

49. Wang, Y.C. and J.Y. Han, Pre-binding dynamic range and sensitivity enhancement for immuno-sensors using nanofluidic preconcentrator. *Lab on a Chip*, 2008, 8(3): 392–394.

50. Yu, H. et al., A simple, disposable microfluidic device for rapid protein concentration and purification via direct-printing. *Lab on a Chip*, 2008, 8(9): 1496–1501.

51. Weigl, B.H. and P. Yager, Microfluidics—Microfluidic diffusion-based separation and detection. *Science*, 1999, 283(5400): 346–347.

52. Gossett, D.R. et al., Label-free cell separation and sorting in microfluidic systems. *Analytical and Bioanalytical Chemistry*, 2010, 397(8): 3249–3267.

53. Di Carlo, D. et al., Continuous inertial focusing, ordering, and separation of particles in microchannels. *Proceedings of the National Academy of Sciences of the United States of Amercia*, 2007, 104(48): 18892–18897.

54. Reed, G.T. and A.P. Knights, *Silicon Photonics—An Introduction*. West Sussex, England: John Wiley & Sons Ltd, 2004.

55. Intel Corporation, Silicon Photonics Research [cited 2011, 4/8/2011]; available from: http://techresearch.intel.com/ResearchAreaDetails.aspx?Id=26.

56. University of Washington, OpSIS (Optoelectronic Systems Integration in Silicon) [cited 2011, 4/8/2011]; available from: http://depts.washington.edu/uwopsis/

57. Hewitt, P.D. and G.T. Reed, Improving the response of optical phase modulators in SOI by computer simulation. *Journal of Lightwave Technology*, 2000, 18(3): 443–450.

58. Luxtera [cited 2011, 4/8/2011]; available from: http://www.luxtera.com/

59. Soref, R.A., J. Schmidtchen, and K. Petermann, Large single-mode rib wave-guides in GeSi-Si and Si-on-SiO$_2$. *IEEE Journal of Quantum Electronics*, 1991, 27(8): 1971–1974.

60. Luchansky, M.S. et al., Characterization of the evanescent field profile and bound mass sensitivity of a label-free silicon photonic microring resonator biosensing platform. *Biosens Bioelectron*, 2010, 26(4): 1283–1291.

61. Erickson, D. et al., Nanobiosensors: Optofluidic, electrical and mechanical approaches to biomolecular detection at the nanoscale. *Microfluidics and Nanofluidics*, 2008, 4(1–2): 33–52.

62. Newman, J.D. and A.P.F. Turner, Home blood glucose biosensors: A commercial perspective. *Biosensors & Bioelectronics*, 2005, 20(12): 2435–2453.

63. Pejcic, B. and R. De Marco, Impedance spectroscopy: Over 35 years of electrochemical sensor optimization. *Electrochimica Acta*, 2006, 51(28): 6217–6229.

64. Daniels, J.S. and N. Pourmand, Label-free impedance biosensors: Opportunities and challenges. *Electroanalysis*, 2007, 19(12): 1239–1257.

65. Patolsky, F., G.F. Zheng, and C.M. Lieber, Nanowire-based biosensors. *Analytical Chemistry*, 2006, 78(13): 4260–4269.

66. Sadik, O.A., A.O. Aluoch, and A.L. Zhou, Status of biomolecular recognition using electrochemical techniques. *Biosensors & Bioelectronics*, 2009, 24(9): 2749–2765.

67. Bellan, L.M., D. Wu, and R.S. Langer, Current trends in nanobiosensor technology. *Wiley Interdisciplinary Reviews: Nanomedicine and Nanobiotechnology*, 2011, 3: 229–246; doi: 10.1002/wnan.136.

68. Yu, X.B. et al., An impedance array biosensor for detection of multiple antibody–antigen interactions. *Analyst*, 2006, 131(6): 745–750.

69. Lodes, M.J. et al., Use of semiconductor-based oligonucleotide microarrays for influenza A virus subtype identification and sequencing. *Journal of Clinical Microbiology*, 2006, 44(4): 1209–1218.

70. Maurer, K. et al., The removal of the t-BOC group by electrochemically generated acid and use of an addressable electrode array for peptide synthesis. *Journal of Combinatorial Chemistry*, 2005, 7(5): 637–640.

71. Maurer, K. et al., Use of a multiplexed CMOS microarray to optimize and compare oligonucleotide binding to DNA probes synthesized or immobilized on individual electrodes. *Sensors*, 2010, 10(8): 7371–7385.

72. Cooper, J. et al., Targeted deposition of antibodies on a multiplex CMOS microarray and optimization of a sensitive immunoassay using electrochemical detection. *Plos One*, 2010, 5(3): e9781; doi: 10.1371/journal.pone.0009781.

73. Patolsky, F. et al., Electrical detection of single viruses. *Proceedings of the National Academy of Sciences of the United States of America*, 2004, 101(39): 14017–14022.

74. Li, C. et al., Complementary detection of prostate-specific antigen using ln(2)O(3) nanowires and carbon nanotubes. *Journal of the American Chemical Society*, 2005, 127(36): 12484–12485.

75. Cui, Y. et al., Nanowire nanosensors for highly sensitive and selective detection of biological and chemical species. *Science*, 2001, 293(5533): 1289–1292.

76. Hahm, J. and C.M. Lieber, Direct ultrasensitive electrical detection of DNA and DNA sequence variations using nanowire nanosensors. *Nano Letters*, 2004, 4(1): 51–54.

77. Zhang, G.J. et al., Highly sensitive measurements of PNA-DNA hybridization using oxide-etched silicon nanowire biosensors. *Biosensors & Bioelectronics*, 2008. 23(11): 1701–1707.

78. Malhotra, B.D., A. Chaubey, and S.P. Singh, Prospects of conducting polymers in biosensors. *Anal Chim Acta*, 2006, 578(1): 59–74.

79. Hu, P. et al., Self-assembled nanotube field-effect transistors for label-free protein biosensors. *Journal of Applied Physics*, 2008, 104(7): 074310; doi: 10.1063/1.2988274.

80. Star, A. et al., Label-free detection of DNA hybridization using carbon nanotube network field-effect transistors. *Proceedings of the National Academy of Sciences of the United States of America*, 2006, 103(4): 921–926.

81. Zheng, G.F. et al., Multiplexed electrical detection of cancer markers with nanowire sensor arrays. *Nature Biotechnology*, 2005, 23(10): 1294–1301.

82. Stern, E. et al., Label-free immunodetection with CMOS-compatible semiconducting nanowires. *Nature*, 2007, 445(7127): 519–522.

83. Yang, Y.T. et al., Zeptogram-scale nanomechanical mass sensing. *Nano Letters*, 2006, 6(4): 583–586.

84. Marx, K.A., Quartz crystal microbalance: A useful tool for studying thin polymer films and complex biomolecular systems at the solution-surface interface. *Biomacromolecules*, 2003, 4(5): 1099–1120.

85. Cooper, M.A. and V.T. Singleton, A survey of the 2001 to 2005 quartz crystal microbalance biosensor literature: applications of acoustic physics to the analysis of biomolecular interactions. *Journal of Molecular Recognition*, 2007, 20(3): 154–184.

86. Seker, S., Y.E. Arslan, and Y.M. Elcin, Electrospun nanofibrous PLGA/fullerene-C60 coated quartz crystal microbalance for real-time gluconic acid monitoring. *IEEE Sensors Journal*, 2010, 10(8): 1342–1348.

87. Hianik, T. et al., Detection of aptamer-protein interactions using QCM and electrochemical indicator methods. *Bioorganic & Medicinal Chemistry Letters*, 2005, 15(2): 291–295.

88. Hook, F. et al., Characterization of PNA and DNA immobilization and subsequent hybridization with DNA using acoustic-shear-wave attenuation measurements. *Langmuir*, 2001, 17(26): 8305–8312.

89. Su, X.D. et al., Detection of point mutation and insertion mutations in DNA using a quartz crystal microbalance and MutS, a mismatch binding protein. *Analytical Chemistry*, 2004, 76(2): 489–494.

90. Liebau, M., A. Hildebrand, and R.H.H. Neubert, Bioadhesion of supramolecular structures at supported planar bilayers as studied by the quartz crystal microbalance. *European Biophysics Journal with Biophysics Letters*, 2001, 30(1): 42–52.

91. Shen, Z.H. et al., Nonlabeled quartz crystal microbalance biosensor for bacterial detection using carbohydrate and lectin recognitions. *Analytical Chemistry*, 2007, 79(6): 2312–2319.

92. Mahon, E., T. Aastrup, and M. Barboiu, Dynamic glycovesicle systems for amplified QCM detection of carbohydrate-lectin multivalent biorecognition. *Chemical Communications*, 2010, 46(14): 2441–2443.

93. Briand, E. et al., Combined QCM-D and EIS study of supported lipid bilayer formation and interaction with pore-forming peptides. *Analyst*, 2010, 135(2): 343–350.

94. Linden, M.V. et al., Characterization of phosphatidylcholine/polyethylene glycol-lipid aggregates and their use as coatings and carriers in capillary electrophoresis. *Electrophoresis*, 2008, 29(4): 852–862.

95. Cooper, M.A. et al., Direct and sensitive detection of a human virus by rupture event scanning. *Nature Biotechnology*, 2001, 19(9): 833–837.

96. Dickert, F.L. et al., Bioimprinted QCM sensors for virus detection-screening of plant sap. *Analytical and Bioanalytical Chemistry*, 2004, 378(8): 1929–1934.

97. Su, X.L. and Y.B. Li, A self-assembled monolayer-based piezoelectric immunosensor for rapid detection of *Escherichia coli* O157:H7. *Biosensors & Bioelectronics*, 2004, 19(6): 563–574.

98. Su, X.L. and Y.B. Li, A QCM immunosensor for Salmonella detection with simultaneous measurements of resonant frequency and motional resistance. *Biosensors & Bioelectronics*, 2005, 21(6): 840–848.

99. Ma, Z.W., Z.W. Mao, and C.Y. Gao, Surface modification and property analysis of biomedical polymers used for tissue engineering. *Colloids and Surfaces B-Biointerfaces*, 2007, 60(2): 137–157.

100. Fawcett, N.C. et al., QCM response to solvated, tethered macromolecules. *Analytical Chemistry*, 1998, 70(14): 2876–2880.

101. Rabe, J. et al., Monolithic miniaturized quartz microbalance array and its application to chemical sensor systems for liquids. *IEEE Sensors Journal*, 2003, 3(4): 361–368.

102. Fritz, J. et al., Translating biomolecular recognition into nanomechanics. *Science*, 2000, 288(5464): 316–318.

103. Backmann, N. et al., A label-free immunosensor array using single-chain antibody fragments. *Proceedings of the National Academy of Sciences of the United States of America*, 2005, 102(41): 14587–14592.

104. Wu, G.H. et al., Bioassay of prostate-specific antigen (PSA) using microcantilevers. *Nature Biotechnology*, 2001, 19(9): 856–860.

105. Gupta, A.K. et al., Anomalous resonance in a nanomechanical biosensor. *Proceedings of the National Academy of Sciences of the United States of America*, 2006, 103(36): 13362–13367.

106. Ilic, B., Y. Yang, and H.G. Craighead, Virus detection using nanoelectromechanical devices. *Applied Physics Letters*, 2004, 85(13): 2604–2606.

107. Gupta, A., D. Akin, and R. Bashir, Single virus particle mass detection using microresonators with nanoscale thickness. *Applied Physics Letters*, 2004, 84(11): 1976–1978.

108. Johnson, L. et al., Characterization of vaccinia virus particles using microscale silicon cantilever resonators and atomic force microscopy. *Sensors and Actuators B: Chemical*, 2006, 115(1): 189–197.

109. Ilic, B. et al., Single cell detection with micromechanical oscillators. *Journal of Vacuum Science & Technology B*, 2001, 19(6): 2825–2828.

110. Park, K. et al., "Living cantilever arrays" for characterization of mass of single live cells in fluids. *Lab on a Chip*, 2008, 8(7): 1034–1041.

111. Ilic, B. et al., Enumeration of DNA molecules bound to a nanomechanical oscillator. *Nano Letters*, 2005, 5(5): 925–929.

112. Lee, J.H. et al., Immunoassay of prostate-specific antigen (PSA) using resonant frequency shift of piezoelectric nanomechanical microcantilever. *Biosensors & Bioelectronics*, 2005, 20(10): 2157–2162.

113. Burg, T.P. et al., Weighing of biomolecules, single cells and single nanoparticles in fluid. *Nature*, 2007, 446(7139): 1066–1069.

114. Barton, R.A. et al., Fabrication of a nanomechanical mass sensor containing a nanofluidic channel. *Nano Letters*, 2010, 10(6): 2058–2063.

115. Godin, M. et al., Using buoyant mass to measure the growth of single cells. *Nature Methods*, 2010, 7(5): 387–390.

116. Bryan, A.K. et al., Measurement of mass, density, and volume during the cell cycle of yeast. *Proceedings of the National Academy of Sciences of the United States of America*, 2010, 107(3): 999–1004.

117. von Muhlen, M.G. et al., Label-free biomarker sensing in undiluted serum with suspended microchannel resonators. *Analytical Chemistry*, 2010, 82(5): 1905–1910.

118. Rich, R.L. and D.G. Myszka, Grading the commercial optical biosensor literature-Class of 2008: "The Mighty Binders". *Journal of Molecular Recognition*, 2010, 23(1): 1–64.

119. Fan, X.D. et al., Sensitive optical biosensors for unlabeled targets: A review. *Analytica Chimica Acta*, 2008, 620(1–2): 8–26.

120. Liedberg, B., C. Nylander, and I. Lundstrom, Surface-plasmon resonance for gas-detection and biosensing. *Sensors and Actuators*, 1983, 4(2): 299–304.

121. Homola, J., Surface plasmon resonance sensors for detection of chemical and biological species. *Chemical Reviews*, 2008, 108(2): 462–493.

122. Boozer, C. et al., Looking towards label-free biomolecular interaction analysis in a high-throughput format: A review of new surface plasmon resonance technologies. *Current Opinion in Biotechnology*, 2006, 17(4): 400–405.

123. Qavi, A.J. et al., Label-free technologies for quantitative multiparameter biological analysis. *Analytical and Bioanalytical Chemistry*, 2009, 394(1): 121–135.

124. Wassaf, D. et al., High-throughput affinity ranking of antibodies using surface plasmon resonance microarrays. *Analytical Biochemistry*, 2006, 351(2): 241–253.

125. Usui-Aoki, K. et al., A novel approach to protein expression profiling using antibody microarrays combined with surface plasmon resonance technology. *Proteomics*, 2005, 5(9): 2396–2401.

126. Dhayal, M. and D.A. Ratner, XPS and SPR analysis of glycoarray surface density. *Langmuir*, 2009, 25(4): 2181–2187.

127. Corn, R.M. et al., Fabrication of DNA microarrays with poly(L-glutamic acid) monolayers on gold substrates for SPR imaging measurements. *Langmuir*, 2009, 25(9): 5054–5060.

128. Spangler, B.D. et al., Comparison of the Spreeta (R) surface plasmon resonance sensor and a quartz crystal microbalance for detection of *Escherichia coli* heat-labile enterotoxin. *Analytica Chimica Acta*, 2001, 444(1): 149–161.

129. Chinowsky, T.M. et al., Performance of the Spreeta 2000 integrated surface plasmon resonance affinity sensor. *Sensors and Actuators B: Chemical*, 2003, 91(1–3): 266–274.

130. Codner, E.P. and R.M. Corn, *Portable Surface Plasmon Resonance Imaging Instrument*. US: Wisconsin Alumni Research, 2006.

131. Fu, E. et al., SPR imaging-based salivary diagnostics system for the detection of small molecule analytes. *Oral-Based Diagnostics*, 2007, 1098: 335–344.

132. Chinowsky, T.M. et al., Portable 24-analyte surface plasmon resonance instruments for rapid, versatile biodetection. *Biosensors & Bioelectronics*, 2007, 22(9–10): 2268–2275

133. Lee, C.Y. et al., Surface coverage and structure of mixed DNA/alkylthiol monolayers on gold: Characterization by XPS, NEXAFS, and fluorescence intensity measurements. *Analytical Chemistry*, 2006, 78(10): 3316–3325.

134. Laibinis, P.E. et al., Comparison of the structures and wetting properties of self-assembled monolayers of normal-alkanethiols on the coinage metal-surfaces, Cu, Ag, Au. *Journal of the American Chemical Society*, 1991, 113(19): 7152–7167.

135. Ulman, A., Formation and structure of self-assembled monolayers. *Chemical Reviews*, 1996, 96(4): 1533–1554.

136. Rich, R.L. and D.G. Myszka, Survey of the year 2006 commercial optical biosensor literature. *Journal of Molecular Recognition*, 2007, 20(5): 300–366.
137. Rich, R.L. and D.G. Myszka, Survey of the year 2007 commercial optical biosensor literature. *Journal of Molecular Recognition*, 2008, 21(6): 355–400.
138. Kersey, A.D. et al., Fiber grating sensors. *Journal of Lightwave Technology*, 1997, 15(8): 1442–1463.
139. Chryssis, A.N. et al., Detecting hybridization of DNA by highly sensitive evanescent field etched core fiber Bragg grating sensors. *IEEE Journal of Selected Topics in Quantum Electronics*, 2005, 11(4): 864–872.
140. DeLisa, M.P. et al., Evanescent wave long period fiber Bragg grating as an immobilized antibody biosensor. *Analytical Chemistry*, 2000, 72(13): 2895–2900.
141. Chen, X. et al., Dual-peak long-period fiber gratings with enhanced refractive index sensitivity by finely tailored mode dispersion that uses the light cladding etching technique. *Applied Optics*, 2007, 46(4): 451–455.
142. Washburn, A.L. and R.C. Bailey, Photonics-on-a-chip: Recent advances in integrated waveguides as enabling detection elements for real-world, lab-on-a-chip biosensing applications. *Analyst*, 2011, 136(2): 227–236.
143. Szekacs, A. et al., Development of a non-labeled immunosensor for the herbicide trifluralin via optical waveguide lightmode spectroscopic detection. *Analytica Chimica Acta*, 2003, 487(1): 31–42.
144. Adanyi, N. et al., Development of immunosensor based on OWLS technique for determining Aflatoxin B1 and Ochratoxin A. *Biosensors & Bioelectronics*, 2007, 22(6): 797–802.
145. Wittmer, C.R. and P.R. Van Tassel, Probing adsorbed fibronectin layer structure by kinetic analysis of monoclonal antibody binding. *Colloids and Surfaces B: Biointerfaces*, 2005, 41(2–3): 103–109.
146. Blattler, T.M. et al., High salt stability and protein resistance of poly(L-lysine)-g-poly(ethylene glycol) copolymers covalently immobilized via aldehyde plasma polymer interlayers on inorganic and polymeric substrates. *Langmuir*, 2006, 22(13): 5760–5769.
147. Horvath, R. et al., Structural hysteresis and hierarchy in adsorbed glycoproteins. *Journal of Chemical Physics*, 2008, 129(7): 071102; doi:10.1063/1.2968127.
148. Adrian, J. et al., Wavelength-interrogated optical biosensor for multi-analyte screening of sulfonamide, fluoroquinolone, beta-lactam and tetracycline antibiotics in milk. *TRAC: Trends in Analytical Chemistry*, 2009, 28(6): 769–777.
149. Heideman, R.G., R.P.H. Kooyman, and J. Greve, Performance of a highly sensitive optical wave-guide Mach-Zehnder interferometer immunosensor. *Sensors and Actuators B: Chemical*, 1993, 10(3): 209–217.
150. Heideman, R.G. and P.V. Lambeck, Remote opto-chemical sensing with extreme sensitivity: Design, fabrication and performance of a pigtailed integrated optical phase-modulated Mach–Zehnder interferometer system. *Sensors and Actuators B: Chemical*, 1999, 61(1–3): 100–127.
151. Shew, B.Y., Y.C. Cheng, and Y.H. Tsai, Monolithic SU-8 micro-interferometer for biochemical detections. *Sensors and Actuators A: Physical*, 2008, 141(2): 299–306.
152. Sánchez del Río, J., L.G. Carrascosa, F.J. Blanco, M. Moreno, J. Berganzo, A. Calle, C. Domínguez, and L.M. Lechuga, Lab-on-a-chip platforms based on highly sensitive nanophotonic Si biosensors for single nucleotide DNA testing (Proceedings Paper in Joel A. Kubby and Graham T. Reed, eds, *Silicon Photonics II*). *Proceedings of the SPIE*, 6477: 64771B; doi: 10.1117/12.713977.
153. Densmore, A. et al., Silicon photonic wire biosensor array for multiplexed real-time and label-free molecular detection. *Optics Letters*, 2009, 34(23): 3598–3600.
154. Brandenburg, A. and R. Henninger, Integrated optical Young interferometer. *Applied Optics*, 1994, 33(25): 5941–5947.

155. Brandenburg, A., Differential refractometry by an integrated-optical Young interferometer. *Sensors and Actuators B: Chemical*, 1997, 39(1–3): 266–271.

156. Ymeti, A. et al., Realization of a multichannel integrated Young interferometer chemical sensor. *Applied Optics*, 2003, 42(28): p. 5649–5660.

157. Ymeti, A. et al., An ultrasensitive Young interferometer handheld sensor for rapid virus detection. *Expert Review of Medical Devices*, 2007, 4(4): 447–454.

158. Hoffmann, C. et al., Interferometric biosensor based on planar optical waveguide sensor chips for label-free detection of surface bound bioreactions. *Biosensors & Bioelectronics*, 2007, 22(11): 2591–2597.

159. Griffel, G. et al., Morphology-dependent resonances of a microsphere-optical fiber system. *Optics Letters*, 1996, 21(10): 695–697.

160. Vollmer, F., S. Arnold, and D. Keng, Single virus detection from the reactive shift of a whispering-gallery mode. *Proceedings of the National Academy of Sciences of the United States of America*, 2008, 105(52): 20701–20704.

161. Vollmer, F. et al., Protein detection by optical shift of a resonant microcavity. *Applied Physics Letters*, 2002, 80(21): 4057–4059.

162. Hanumegowda, N.M. et al., Refractometric sensors based on microsphere resonators. *Applied Physics Letters*, 2005, 87(20): 201107; doi: 10.1063/1.2132076.

163. Hanumegowda, N.M. et al., Label-free protease sensors based on optical microsphere resonators. *Sensor Letters*, 2005, **3**(4): 315–319.

164. Vollmer, F. et al., Multiplexed DNA quantification by spectroscopic shift of two microsphere cavities. *Biophysical Journal*, 2003, 85(3): 1974–1979.

165. Armani, A.M. et al., Label-free, single-molecule detection with optical microcavities. *Science*, 2007, 317(5839): 783–787.

166. Arnold, S., S.I. Shopova, and S. Holler, Whispering gallery mode bio-sensor for label-free detection of single molecules: Thermo-optic vs. reactive mechanism. *Optics Express*, 2010, 18(1): 281–287.

167. Squires, T.M., R.J. Messinger, and S.R. Manalis, Making it stick: Convection, reaction and diffusion in surface-based biosensors. *Nature Biotechnology*, 2008, 26(4): 417–426.

168. Carlborg, C.F. et al., A packaged optical slot-waveguide ring resonator sensor array for multiplex label-free assays in labs-on-chips. *Lab on a Chip*, 2010, 10(3): 281–290.

169. Chao, C.Y., W. Fung, and L.J. Guo, Polymer microring resonators for biochemical sensing applications. *IEEE Journal of Selected Topics in Quantum Electronics*, 2006, 12(1): 134–142.

170. Huang, Y.Y. et al., Fabrication and replication of polymer integrated optical devices using electron-beam lithography and soft lithography. *Journal of Physical Chemistry B*, 2004, 108(25): 8606–8613.

171. Yalcin, A. et al., Optical sensing of biomolecules using microring resonators. *IEEE Journal of Selected Topics in Quantum Electronics*, 2006, 12(1): 148–155.

172. Ramachandran, A. et al., A universal biosensing platform based on optical micro-ring resonators. *Biosensors & Bioelectronics*, 2008, 23(7): 939–944.

173. Barrios, C.A. et al., Label-free optical biosensing with slot-waveguides. *Optics Letters*, 2008, 33(7): 708–710.

174. Hosseini, E.S. et al., Systematic design and fabrication of high-Q single-mode pulley-coupled planar silicon nitride microdisk resonators at visible wavelengths. *Optics Express*, 2010, 18(3): 2127–2136.

175. Xu, D.X. et al., Label-free biosensor array based on silicon-on-insulator ring resonators addressed using a WDM approach. *Optics Letters*, 2010, 35(16): 2771–2773.

176. De Vos, K. et al., SOI optical microring resonator with poly(ethylene glycol) polymer brush for label-free biosensor applications. *Biosensors & Bioelectronics*, 2009, 24(8): 2528–2533.

177. Washburn, A.L., L.C. Gunn, and R.C. Bailey, Label-free quantitation of a cancer bio-marker in complex media using silicon photonic microring resonators. *Analytical Chemistry*, 2009, 81(22): 9499–9506.
178. Gylfason, K.B. et al., On-chip temperature compensation in an integrated slot-waveguide ring resonator refractive index sensor array. *Optics Express*, 2010, 18(4): 3226–3237.
179. Bailey, R.C., A robust silicon photonic platform for multiparameter biological analysis (Proceedings Paper in in Joel A. Kubby and Graham T. Reed, eds, *Silicon Photonics IV*). *Proceedings of the SPIE*, 2009, 7220: 72200N-1–72200N-6; doi: 10.1117/12.809819.
180. Kirk, J.T. et al., Multiplexed inkjet functionalization of silicon photonic biosensors. *Lab Chip*, 2011, 11(7): 1372–1377.
181. Luchansky, M.S. and R.C. Bailey, Silicon photonic microring resonators for quantitative cytokine detection and T-cell secretion analysis. *Analytical Chemistry*, 2010, 82(5): 1975–1981.
182. Qavi, A.J. and R.C. Bailey, Multiplexed detection and label-free quantitation of microRNAs using arrays of silicon photonic microring resonators. *Angewandte Chemie-International Edition*, 2010, 49(27): 4608–4611.
183. Washburn, A.L. et al., Quantitative, label-free detection of five protein biomarkers using multiplexed arrays of silicon photonic microring resonators. *Analytical Chemistry*, 2010, 82(1): 69–72.
184. Cooper, M.A., Optical biosensors in drug discovery. *Nature Reviews Drug Discovery*, 2002, 1(7): 515–528.
185. Smith, K.H. et al., Surface-relief fiber Bragg gratings for sensing applications. *Applied Optics*, 2006, 45(8): 1669–1675.
186. Ymeti, A. et al., Fast, ultrasensitive virus detection using a Young interferometer sensor. *Nano Letter*, 2007, 7(2): 394–397.
187. Vahala, K.J., Optical Microcavities. *Nature*, 2003, 424(6950): 839–846.

4 Luminescent Thermometry for Sensing Rapid Thermal Profiles in Fires and Explosions

Joseph J. Talghader and Merlin L. Mah

CONTENTS

INTRODUCTION

The interior of an explosion is one of the harshest environments on earth. The violent chemical reactions and rapid changes in pressure and temperature make sensing with conventional electronics virtually impossible. One can obtain significant information using remote optical techniques, and indeed the spectroscopy of explosions is a major field of research [1,2]; however, the presence of debris or opaque chemical reactions can make the fireball impossible to probe, even with the most advanced equipment. In this chapter, we describe temperature and thermal history sensors based on microparticle or nanoparticle luminescence. These particles are embedded in explosive material and disperse with it but are undamaged by the explosion since they have no mechanical or electronic "parts." The luminescence of the particles can be examined in the debris or measured in a laboratory and can ascertain the distribution of temperatures seen in the periphery of an explosive fireball. This method is currently under development at the University of Minnesota and Oklahoma State University and shows good promise as a thermal diagnostic where common sensing methods fail.

Any sensor that is placed inside a fire or explosion will be subject to conditions so violent that they are likely to destroy any mechanical components or conventional electronics. A sensor that could work within these limitations is a block of material with some nonvolatile, temperature-dependent property. Many oxides, fluorides, and other materials have embedded traps with a distribution of trap energies. If the material has traps filled with electrons or holes that are significantly deeper than an electron volt, the traps will usually be stable under normal conditions over many years or even centuries. High temperatures can excite and empty these traps, a process that is described more fully in the next section. Generally, higher temperatures empty deeper traps. The thermoluminescence (TL) of these particles is an indication of trap population, and thus gives one insight into the thermal history of the particle. These particles could be placed near explosive material or embedded within it and could thus probe temperature distributions around the explosive area, or post detonation environment, without fear of sensor damage. Since the particles can be developed with a variety of sizes and shapes they can mimic the aerodynamic behavior of components of an explosive (e.g., metal nanoparticles) or its target (e.g., a bioweapon containing viruses or bacteria). Figure 4.1 shows a conceptual diagram of how such a process might work. A group of particles, numbering in the thousands to millions, is exposed to deep ultraviolet (UV) light or some other ionizing radiation source to fill the particle traps. After this, the particles are embedded in a flammable or explosive material (or placed around it) so that they will be dispersed throughout during the reaction. Finally, the debris or ashes are examined in the field using a handheld luminescence reader or taken to a laboratory to assess the temperature and/or thermal history distribution.

This approach has several advantages:

1. The particles themselves have no parts that can be damaged by a fire or explosion; they are merely doped glasses, crystals, or other materials that can withstand harsh environments and procedures.
2. The materials are usually very cheap, and millions of particles can be fabricated using existing processes.
3. The particles can be micromachined to take on just about any shape or size, and thus will have essentially identical aerodynamics and thermal experience to any nanoparticles used in the explosion itself.
4. Fluctuations in the thermal history can be seen from the differences in luminescence from region to region or even particle to particle.
5. Luminescence data are exponentially weighted toward the highest temperatures experienced, which are the most critical in understanding postdetonation effects.
6. Extreme thermal profiles can be simulated using micromachined heaters that increase temperature to hundreds of degrees and cool on timescales on the order of a millisecond, or longer if desired. This also allows one to isolate temperature effects from pressure effects.

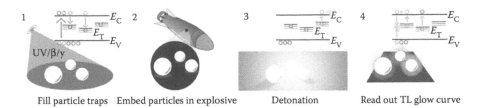

Fill particle traps   Embed particles in explosive   Detonation   Read out TL glow curve

FIGURE 4.1 Conceptual diagram of measuring thermal history using microparticles. Initially, the traps inside thermoluminescent particles are filled using ionizing radiation. The particles are embedded in an explosive, which detonates and empties the shallower traps. By examining the luminescence, one can determine the trap population distribution and therefore the temperature-time relationship of the explosive event.

TL AS A THERMOMETER

A luminescent material is one that emits light in response to some external stimulus such as heat or optical pumping. Many applications of luminescent materials depend on the presence of very long-lived deep traps that can store charge for years or even millennia, as shown in Figure 4.2. Perhaps the most widely known application of this type is radiation dosimetry, where luminescent particles are embedded in an identification badge or other object. If this object is exposed to ionizing radiation, then electrons from the valence band are excited into the conduction band and the resulting

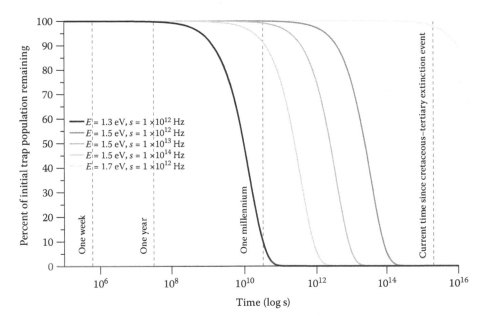

FIGURE 4.2 Trap population versus time for a simple cluster of traps at room temperature for a variety of trap energies and frequency factors. Deep traps can remain filled for thousands of years and thermoluminescent dating is commonly used in archaeology, albeit for time periods less than that since the Age of Dinosaurs.

electrons or holes can fall into traps within the gap. These trapped charges can be released upon stimulation by heat in TL or by light in optically stimulated luminescence (OSL), after which they recombine and emit light. The number of filled traps is, for many materials, directly proportional to the dose of ionizing radiation received; therefore, the amount of light emitted upon heating or optical excitation is a direct indication of the amount of radiation to which the object has been exposed.

The difference in using luminescent materials for dosimetry and temperature sensing is that in radiation dosimetry one is interested in how the trapping states are filled by exposure to radiation, whereas in temperature sensing one is interested in how the trapping states are *emptied* by a certain temperature profile. Therefore, for temperature sensing the trapping states need to be filled *prior* to use of the materials as a temperature sensor. This can be easily accomplished using ionizing radiation or, more conveniently for some materials, deep UV light.

Basics of Trap Luminescence

The simplest model of the population statistics of a trap is governed by an Arrhenius-type equation. Assuming first-order kinetics [3], the probability per unit time of an electron being released from an electron trap via thermal energy is

$$p = s \ \exp\left(-\frac{E_t}{kT}\right) \tag{4.1}$$

where p is the probability of emission in s^{-1}, E_t is the depth of the trap below the conduction band (or above the valence band), k is Boltzmann's constant, T is the temperature, and s is a frequency factor that can be roughly understood to be related to the thermal phonon interactions that an electron undergoes with the surrounding lattice. The parameter s often has a value between 10^{12} and 10^{14} s^{-1}, but in the literature, the parameter is frequently used for curve-fitting and highly nonphysical values of s can be reported. This does not mean the values are *wrong*; it just indicates that the simple equation above is insufficient to describe the complexity of most traps. Traps may interact with one another, requiring a kinetic model that includes multiparticle interactions. Traps may also have tunneling-assisted transitions with surface or boundary states that require explicitly quantum mechanical treatments. Finally, defects in many materials have behaviors yet to be fully explained. So while the simple Arrhenius expression is very useful for understanding concepts and developing simple models, we should recognize that it rigorously applies to only a very limited number of traps and materials. A discussion on more complex models is beyond the scope of this review but is covered in other excellent texts [3–5].

The energy of the trap can be understood using a bandgap model for the charge carrier states of the solid, as shown in Figure 4.3. A valence band filled with electrons is separated from an empty conduction band by an energy gap of several electron volts. The size of the gap for the thermoluminescent materials under discussion is large enough that the number of thermally excited valence electrons that reach the conduction band is effectively zero. However, there can be a number of traps distributed throughout the bandgap. These traps are not necessarily isolated levels but

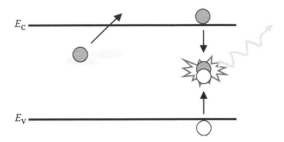

FIGURE 4.3 Bandgap model of charge traps. An electron (or hole) is excited to the conduction (valence) band where it travels until recombining at a luminescence center, emitting a photon.

generally have a density of states that clusters at one or more energies. Some materials, such as carbon-doped aluminum oxide (Al_2O_3:C), have a single dominant trap cluster while others, such as LiF:Mg,Ti, have a number of trap clusters that contribute to the luminescence.

It is not enough for a material to have traps—it must also luminesce. Visible radiation from nonvolatile traps usually does not arise from direct transitions from conduction band to valence band (which would be deep in the UV) but rather through a luminescence defect in the material. This is typically an impurity such as Cr^{3+}, Eu^{2+}, or Tb^{3+} or a vacancy in the lattice that attracts and recombines oppositely charged carriers, emitting a photon of characteristic energy. The terms "F-center" or "color center" are often used to describe luminescence centers that derive from vacancies in a crystal lattice [3]. The recombination process can create an excited luminescence center that relaxes to the ground state by photon emission.

In TL, the trap characteristics are obtained from a plot of emitted light intensity versus temperature. The temperature is usually ramped linearly with time, say at a heating rate of 1°C/s, but there are certain applications where other profiles such as short thermal [6] or laser [7] pulses are used. The resultant TL glow curve usually has one or more peaks, each corresponding to a different trap distribution. An example with a single peak is shown in Figure 4.4. These trap distributions can be

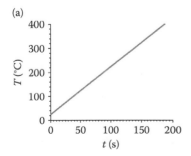
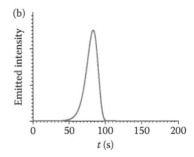

FIGURE 4.4 Heating profile (a) and thermoluminescent glow curve (b) of a material with a classic single trap system.

assigned an activation energy E_t and a frequency factor s as described above. To measure temperature, one fills the traps using ionizing radiation prior to measurement. In most applications, from radiation dosimetry to archeological dating, one fills the traps using radioactive decay from gamma or beta sources, since radiation dose is the parameter to be measured or calibrated against. However, deep UV light will also work with many materials such as $Al_2O_3{:}C$ and $Mg_2SiO_4{:}Tb,Co$. This may seem to be somewhat of a contradiction since it is well known that daylight, which contains UV light, removes carriers from certain types of traps, and the last exposure to daylight is considered the "zeroing" point from which a date is reckoned in archeological TL. The difference involves the wavelength of the UV light. If a photon is extremely energetic, it will be able to excite an electron from the valence band all the way to the conduction band, giving the conduction band a very large number of carriers. Excited carriers then fall into the traps, filling them. Shallow UV photons will be unable to bridge the bandgap in TL materials. Occasionally, there are other absorptions, such as the F-center absorption of $Al_2O_3{:}C$ at 205 nm [8], that also fill the traps effectively. In both cases, however, near-UV and visible photons will empty the traps.

In addition to TL, there are other methods to probe the trap characteristics of a particle. Two of these are OSL and electron spin resonance (ESR). In OSL, the foremost information is the intensity versus time curve, obtained as the sample is illuminated with light of constant wavelength. The total light emitted is proportional to the trapped charge population. Therefore, OSL is an alternative way to probe trap population that does not involve heating in the measurement process. Provided this difference is understood, the concept of using OSL for temperature sensing is similar to the case of TL. Due to the all-optical nature of the OSL process, one has some flexibility in choosing how the stimulation is performed in terms of intensity and wavelength. Most OSL systems are designed to stimulate one particular trapping level; therefore, the light stimulation spectrum is fixed at an appropriate wavelength. Systems capable of wavelength scanning are used to measure the stimulation spectrum and determine the optical depth of the trapping levels [6,7]. For the application being proposed here, one wants to obtain the relative populations of traps. Therefore, one of the best ways to extract this information is to use OSL measurements in series at different wavelengths, starting with low-energy photons and scanning to higher energies. This way, traps will be probed sequentially in order of optical depth. If the optical depth in the particular material is proportional to the thermal depth (activation energy E_t), successive measurements would give information on progressively deeper traps.

OSL, however, has an important flaw that makes it less desirable than TL for temperature measurement: inherent light sensitivity. Many or most traps will empty in response to visible or near-UV light, which is a process called "photobleaching." All particles in an explosion are exposed to some amount of UV light due to the extreme combustion. After the explosion or during/after a fire, they are exposed to daylight or other visible and UV light. The traps probed by OSL are, by their nature, light-sensitive and so loss of data will occur unless strict precautions are taken, such as coating the particles with an opaque material prior to use. Such a coating would have to be removed before analysis, making the entire process more difficult. Most TL materials utilize traps that are light sensitive, but many, such as LiF:Mg,Ti, do not, so a proper choice of TL materials with traps that are not light sensitive or only weakly sensitive can eliminate this issue.

ESR is another technique that can be used to probe trap characteristics, where a magnetic field is applied across a system and the microwave absorption between magnetic sublevels of unpaired electrons is measured. ESR requires highly specialized equipment for analysis and is therefore less suitable for a rapid diagnostic, so it is not treated further in this review.

TL FOR TEMPERATURE AND THERMAL HISTORY MEASUREMENT

The use of mineral or ceramic luminescence to determine temperature has been utilized previously in space sciences and archeology but not extensively because of complications in assessing initial (prior to the thermal event) trap populations among other issues. In the early 1980s [3,9–11], there was interest in determining the former orbital characteristics of meteorites that had fallen on the earth. One way to determine this was to look at the TL of mineral particles within the meteorite. Since the TL would be characteristic of the highest temperature commonly reached by the meteorite while it was in orbit, it was assumed that the luminescence would indicate, via the temperature, the approximate distance of the perihelion of its orbit around the sun. This particular application was much more complex in many ways than what is proposed here because one must measure temperature during entry into the earth's atmosphere, determine cosmic ray exposure that would give a steady-state trap population, and predict the albedo (reflectivity) of the overall meteor. Despite this, the spread in the perihelia of known meteorite sources could be roughly correlated to the spread in TL measurements [11]. Trap luminescence has also been used to obtain information about the firing temperatures of pottery [12]. However, in this case, it is not the change in trap population that is measured, but rather how the firing temperature changes the sensitivity of the materials to radiation dose.

While, in theory, it is possible to obtain temperature measurements in rapid events using only the absolute intensity of the luminescence (as used in radiation dosimetry, archeological dating, and the meteorite temperature measurements just described), in reality, the complexity of the combustion environment makes this extremely difficult. For example, temperature-sensing particles will be mixed in with debris and soot; therefore, the signal will be attenuated to an unknown and uncontrollable extent. A better method of temperature measurement relies on examining the ratios of the intensities of two or more peaks. The population of the traps corresponding to a lower-temperature peak will empty at lower temperatures than those corresponding to a higher-temperature peak. Since the higher-temperature peak can be used as a relative reference, the ratios of the intensities of the two peaks will be a more reliable indicator of temperature than the intensity of just one.

In order to understand the details of obtaining a thermal history from TL, we must develop an expression for trap population as a function of temperature, where temperature can vary with time. Consider a TL particle with a single peak that undergoes a rapid heating followed by a (relatively) slow cooling:

Rapid Heating:

$$T(t) = \beta_{\text{expl}} \cdot t + T_{\text{init}} \qquad (4.2)$$

Slow Cooling:

$$T(t) = T_{\text{max}} e^{-t/\tau} + T_{\text{init}} \qquad (4.3)$$

where β_{expl} is the (very large) slope of the temperature versus time curve during heating, T_{init} is the ambient temperature before the explosion, T_{max} is the peak temperature to which the particle is exposed, τ is a cooling time constant, t is the time in seconds, and T is the temperature of the particle at a time t. A graph of this heating profile is shown in Figure 4.5.

The population of a specific set of traps can be derived by multiplying the probability that a trap is filled (Equation 4.1) by the number of traps in that group and taking the derivative. At this point, the temperature is a single number. If we wish to analyze trap population as a function of time, the temperature must become a function of time and then the entire expression must be integrated:

$$\frac{dn}{dt} = -p \cdot n = -nse^{-E/kT(t)} \qquad (4.4)$$

$$n(t) = n_0 \exp\left(-\int_0^t se^{-E/kT(t)} \, dt\right) \qquad (4.5)$$

where n is the trap population at time t, n_0 is the initial trap population before the event, and p is the probability of a single trap state being filled as described in Equation 4.1.

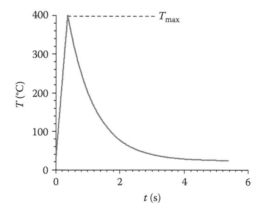

FIGURE 4.5 Temperature profile used to simulate an explosion. There is an extremely fast initial rise (usually considered to be instantaneous) followed by slow asymptotic cooling.

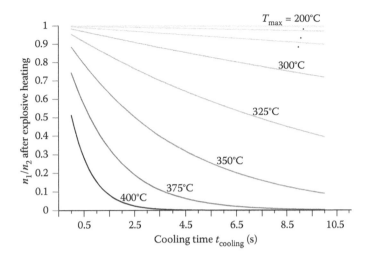

FIGURE 4.6 Trap population ratio as a function of cooling time for a variety of maximum temperatures. Note the cooling time only has a large impact once the temperature has reached a high value.

Now, consider two sets of traps whose energies E_{t1} and E_{t2} cluster about 1.277 and 1.30 eV, respectively. For simplicity, we will assume that the frequency factor, s, of both traps is the same at $s = 10^{12}$ s^{-1}. If we make a plot of the population ratio, n_1/n_2, as a function of cooling time for a variety of maximum temperatures, we will obtain Figure 4.6. (Note that the cooling time is related to the thermal time constant, τ defined earlier.) Here we can see that even small changes in maximum temperature can have a dramatic effect on the populations of two sets of traps, even when those trap sets are separated by only relatively small energies. On the other hand, Figure 4.7

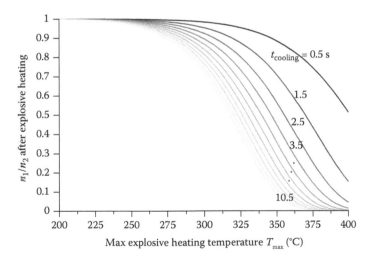

FIGURE 4.7 Trap population ratio versus maximum heating temperature for a variety of cooling times. The transitions from high ratio to low ratio all cluster near similar temperatures.

shows a plot of n_1/n_2 as a function of T_{max} for a variety of cooling times. The cooling time is much less important in making quantitative changes to the population ratio. This can be understood to be a result of the exponential dependence on temperature of the detrapping probability rate of Equation 4.1. A maximum temperature increase, even if held for a small period of time, is more effective at causing carriers to escape traps than long periods of exposure to lower temperatures. It is a valid question to ask whether the heating rate plays a large role in the trap population ratios. The answer depends on the speed of the heating. In the simulations leading to Figures 4.6 and 4.7, we have assumed a more or less infinitely fast heating rate. As long as the heating taking place over time scales much shorter than the cooling, we may neglect the effect of β_{expl} on the population profile. However, if the cooling and heating times become comparable, then both must be included. Attempts to extract both temperature and cooling times will be discussed in more detail in Section "Thermal History Modeling and Reconstruction."

Now that we have established that the population ratio of two traps is intimately related to the thermal history to which the traps have been exposed, it is useful to examine how the luminescence might indicate these ratios. Ideally, a TL curve will show multiple peaks that are spaced far enough apart that their populations can be individually determined without reference to any others. However, if two traps are extremely close together in energy, such as the traps described in Figures 4.6 and 4.7, the TL from each trap will overlap and be difficult to separate, even if there are significant population differences between the two. This phenomenon is shown in Figure 4.8, where a trap with $E_{t1} = 1.4$ eV and $s_1 = 2 \times 10^{12}$ s^{-1} is plotted with another trap with $s_2 = 1 \times 10^{13}$ s^{-1} where E_{t2} changes from 1.3 to 1.7 eV in steps of 0.1 eV.

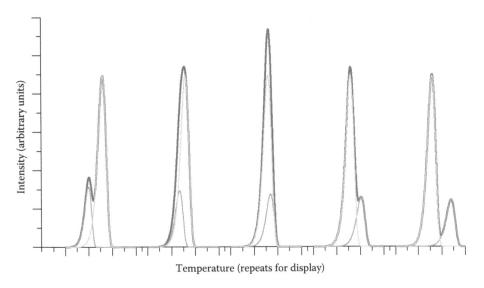

Temperature (repeats for display)

FIGURE 4.8 Overlapping thermoluminescent glow curves as the energy between two traps is changed. The parameters of the highest peak trap are $E_{t1} = 1.4$ eV and $s_1 = 2 \times 10^{12}$ s^{-1} while the parameters of the variable trap are $s_2 = 1 \times 10^{13}$ s^{-1} and E_{t2} changes from 1.3 to 1.7 eV in steps of 0.1 eV.

(The populations of both traps are identical; it is the variation in s that causes the difference in height.) The top curve is the sum of the two luminescences, which is what would be measured by a detector. It can be seen that an easy distinction between the two really only occurs at $E_{t2} = 1.3$ and 1.7 eV. This does not mean that we cannot use the information in the merged TL curve, but it does mean that our model does more than just measure peak heights (see the Section on "Thermal History Modeling and Reconstruction").

So far, we have considered that we are using a single type of particle, even if there may be millions of them in a fire or explosion. It is theoretically possible to enhance the amount of information that we can extract from the luminescence using multiple types of particles. Perhaps the most attractive feature is that each type of particle could have traps that luminesce at different wavelengths from other types of particles. Many, but not all, TL materials have fixed emission characteristics, even if they have multiple trap populations. The reason for this is that the luminescence center is often a completely separate entity from the trap itself. Carriers are excited from traps to the conduction band and from there the carriers travel to recombine at the luminescence center. This means that within a single particle, one cannot count on being able to distinguish traps by wavelength, but with multiple particles, where there are different luminescence centers with different emission wavelengths, this should be possible. However, one must be aware that the spectral width of the luminescence can be very large. For example, Al_2O_3:C has a peak emission near 420 nm with a spectral width of over 100 nm [13]. The emission of LiF:Mg,Ti [14] has peaks that overlap with Al_2O_3:C, and so it would not be straightforward to use emission wavelength selectivity to distinguish the luminescence of particles of these two materials. On the other hand, there are materials, such as MgO:Tm,Li and MgO:Eu,Li that have relatively narrow emission spectra [15], such that wavelength selectivity among multiple particles would be viable.

MICROHEATERS

Obviously, developing TL temperature measurements will be easier if one does not have to use large-scale explosions just to obtain a data point, particularly in a university environment. In order to eliminate the need to explode particles, we perform much of our thermoluminescent testing on micromachined heaters. The heating element of a traditional TL measurement is usually a hotplate or planchet heater, in which a sample-bearing platform is heated by an electrical current passing through a resistive element [16]. While these conventional devices serve their purpose well enough for TL dosimeters and the like, their size and thermal mass make them ponderous to heat and cool, vulnerable to contamination, power-hungry, and difficult to control.

Many of these problems can be solved by shrinking the heating element into the micro regime. Using very-large-scale integration microfabrication techniques, semiconductor materials can be patterned into heating elements with dimensions of tens of microns. Examples are shown in Figure 4.9. At these scales, the thermal mass of a heater is so small that temperature excursions of many hundreds of degrees can be made on timescales of milliseconds. It should be explicitly noted that an arbitrary

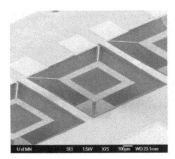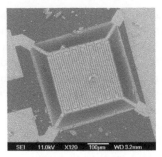

FIGURE 4.9 Scanning electron microscope images of microheaters used in luminescent particle studies.

temperature waveform can be simulated with a microheater. It is extremely easy, using a pulse generator or other high-speed current source, to drive the heater to heat within hundreds of microseconds to hundreds of degrees and then cool over several seconds. Further, heat transfer through convection is limited, thus making control and analysis relatively simple. The microheaters can often eschew the complex control schemes needed to deal with response delay and ringing in conventional hotplates. The highly concentrated application of heat greatly reduces unwanted TL signals from sample residue or any other contamination that may be nearby, as well as background light and distortions from convection and radiative heating. This selectivity of heating means that thermoluminescent films can be deposited directly onto the heater surfaces during fabrication if desired, providing for a sample with integrated heating facilities on a single chip. Finally, their miniature size, power, and processing requirements make microheaters ideal for integration into portable TL readers.

On a microheater, measurement with a commercial thermistor or thermocouple is out of the question, as the heater would be dwarfed by the measurement device. However, platinum can be easily integrated into the heater and exhibits an extremely linear thermal coefficient of resistance (TCR). Since platinum is also well suited for a heating resistor, the sensing and driving elements can even be one and the same, which greatly simplifies the device and eliminates concerns about sensor placement.

The advantages of microheaters do not come without complications. The miniscule surface area and highly focused heating of the heaters inherently translate to a smaller volume of TL material and thus lower signal intensity. If the microheaters are poorly designed, the temperature of one section of the microheater may differ enough from another so that particles on the two sections will have TL glow curves that are displaced along the temperature axis from one another. This phenomenon can be seen in Figure 4.10, where two Al_2O_3:C particles are tested on a microheater where the heating element has been placed around the edges rather than throughout the device. Also, differing film stress characteristics can cause heater platforms to take on curved shapes, and these shapes can buckle rapidly under extreme temperatures. One batch of our early microheaters would abruptly transform from concave to convex (or vice versa) and thus launch any loosely adhered particles into the air!

Ultimately, the previously listed complications are a matter of proper engineering and have seldom caused trouble after initial development. However, one issue is

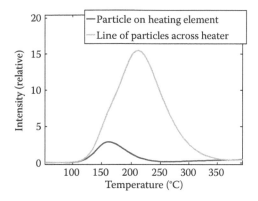

FIGURE 4.10 TL curves of particles of Al$_2$O$_3$:C on a microheater where one particle lies directly on a portion of the heater element and the another group has variable placement. Clearly, the temperature data from this microheater cannot be trusted and a better microheater with a more uniformly placed heating element is required.

more fundamental: the speed of the heat transfer from the microheater to the particles. The small time constants that allow temperature profiles that mimic explosions very accurately can be too short for large particles. This problem is magnified when dealing with TL powders with highly irregular shapes which may result in poor thermal contact between heater and material, resulting in inaccurate temperature readings and smeared-out TL curves. Whereas a conventionally sized heater never heats fast enough to make these issues relevant, a microheater makes the calculation of the time constants of the particles important to the analysis. Thus, one must estimate the thermal mass of any large particle on the microheater to ensure that the temperature at the platinum thermistor is actually the temperature of the particle. It should be noted, however, that this will also be an issue in explosions (postdetonation environments), and particles larger than an order of 10 μm or so should be avoided.

Our latest-generation microheaters use a serpentine resistor of sputtered platinum a few hundred nanometers thick, patterned by liftoff photolithography and underlain by a 10 nm titanium adhesion layer. These electrical parts are sandwiched by silicon nitride films, with a plasma-enhanced chemical vapor deposition layer deposited on top for wet etch protection and a low-stress low-pressure chemical vapor deposition layer underneath. A plasma etch through both nitride layers to the silicon wafer allows the use of anisotropic wet etching to carve a pit below the heater, resulting in a suspended membrane of silicon nitride. The sizable air gaps below and to all sides of the heater platform limit heat conduction paths to four support beams of high aspect ratio; this degree of isolation, when combined with the minuscule thickness—in the hundreds of nanometers—and mass of the nonheating support structure, ensures a rapid response to joule heating and a time constant of around 30 ms. For applications where a much faster time constant is desired, the release etch can be skipped to leave the underlying silicon substrate intact; this provides a massive heat conduction path from the heater platform's entire underside to a relatively enormous block of material, and thus drives the time constant down to the 200 μs range. To

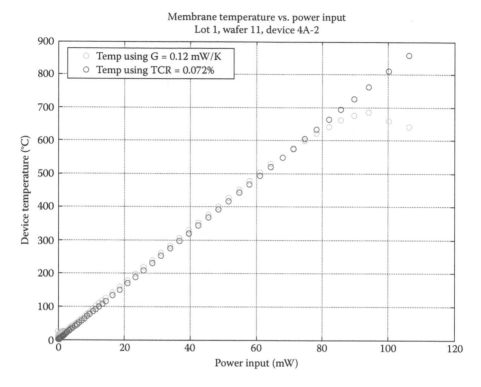

FIGURE 4.11   Microheater temperature versus input power for an early pair of microheaters.

date, we have not seen any background TL signals from any of the materials used to fabricate the heaters. Experience has shown that the released heaters are mechanically robust enough to handle powder samples being removed by pressurized nitrogen spray. Even with single-wafer processing, hundreds of devices can be derived from a single batch, giving the option of a single use per heater, which all but eliminates sample-to-sample cross-contamination.

Before operation, the heaters are placed on a conventional hotplate and their resistance is measured as the die is heated; this yields the TCR and I–V relationships that are used to measure and control their temperature in use. Stress testing has shown that these microheaters can easily and repeatedly ramp to temperatures of over 500°C without change in characteristics. The thermal and speed performance of two example microheaters is presented in Figures 4.11 and 4.12.

A SAMPLE PROCESS

Using the rapid heating technology of microheaters, we can study the effects of an explosion-type temperature profile on TL glow curves in conditions much more easily produced, measured, and controlled than actual explosions [17–19]. For an initial test, microparticles of Mg_2SiO_4:Tb,Co of the order of 10–20 μm in diameter were prepared [20] by Dr. M. Prokic, Institute of Nuclear Sciences, Vinča, Belgrade. This

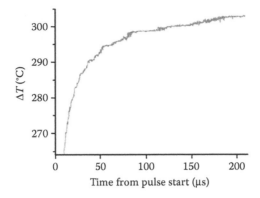

FIGURE 4.12   Time response of a microheater that has not been separated from the underlying substrate. This configuration makes the microheater very fast but prevents it from reaching high temperatures.

terbium- and cobalt-doped magnesium silicate formulation produces at least two easily distinguishable TL peaks below 350°C upon irradiation with broad-spectrum 180–400 nm UV light, which allows us to refill sample traps inside our TL setup without the need for more hazardous radiation sources or heavier shielding. The entirety of the experiment can thus be run without removing the sample from the TL chamber, eliminating the risk of photobleaching (to which the closely related Mg_2SiO_4:Tb is known to be vulnerable [21]) during handling and ensuring that all data are taken from the same sample with a consistent thermal contact.

Microheaters similar to the variety previously described, with platforms 300 μm on a side, were used for both explosive and linear (glow curve) heating. Four calibrations of the microheater resistors on hot plates produced closely matching linear TCR results that, at the highest temperature used in the experiment, produced a maximum temperature deviation of 4.4%. The microheaters were also run through a temperature ramp to record the nonlinear relationship of voltage and temperature, which would be used to target the output power during pulse heating. (The actual pulse temperature reached is still determined via resistance measurement, however.) A computer-controlled source meter was used to drive the microheater in both modes of heating; while this was observed to work very well for slow heating, later revisions of the experiment use faster control programs and dedicated pulsed generation devices to achieve better accuracy during rapid operation. TL readout is handled by a photomultiplier tube, which is protected during irradiation by a motorized metal shutter. This setup is shown in Figure 4.13.

Mg_2SiO_4:Tb,Co microparticles were sprinkled onto the heater die and adjusted where necessary with microprobe manipulators; the microheaters are thermally isolated from their substrates, so the excess particles and any other contaminants that land near the device will not be heated enough to luminesce. Once laden with samples, the microheater is placed inside the darkened TL chamber and its microparticles are irradiated with UV light from a 30 W deuterium lamp. A temperature pulse with a total duration 200 ms (±13 ms) and variable maximum temperature

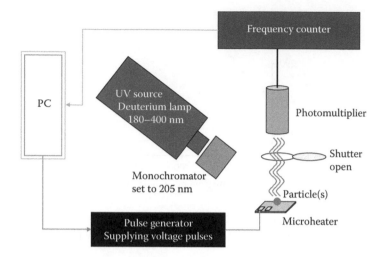

FIGURE 4.13 Experimental set-up used to test the response of thermoluminescent Al_2O_3:C microparticles to rapid thermal profiles. Included are devices to fill the traps, simulate explosive heating, and measure TL glow curves.

(i.e., variable current amplitude) is applied in noncontrol runs. A TL glow curve is then collected using a linear temperature ramp of 2°C/s up to 350°C, where the magnesium silicate glow is observed to die away while a strong background intensity—most likely thermal blackbody emission—rises to prominence. Although no ill effects from repeated rapid cooling were observed in stress testing, the microheater temperature is gradually ramped down to ambient to avoid any thermal gradient bimorph stresses that might jostle the particle. The entire experiment run is automated and is repeated as necessary without manual handling of the sample.

To reduce photomultiplier tube noise and fluctuations in overall intensity, each raw data point is averaged with its 20 closest neighbors (corresponding to approximately 10°C) and each curve multiplicatively scaled so that their heights are equal at 266°C, the apex of the second peak. The blackbody emission background, which becomes dominant at ramp temperatures above 310°C after the Mg_2SiO_4:Tb,Co TL peaks (if any is present) have died away, is modeled as an exponential function curve-fitted to the nonpulsed control curve above 310°C, and subtracted from the intensity data of each subsequent pulsed run. This is shown in Figure 4.14.

In actual use, our technique must compare the TL glow curves of different samples of the same material, some scarred and sullied by explosion, possibly using detectors and read-out systems with slightly different characteristics. To compensate for the changes in overall intensity that will follow, we must preserve one of our two distinguishable TL peaks undiminished as an intensity reference; this can be done by choosing a TL material with a high-temperature peak that will not be affected by the expected heating level, which we may set by observing the effects of slow preheating (using a conventional planchet heater) on the glow curve of Mg_2SiO_4:Tb,Co.

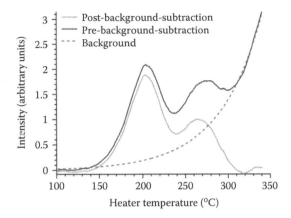

FIGURE 4.14 Background signal and subtraction performed for Mg_2SiO_4:Tb,Co thermoluminescence. (Reprinted from M. L. Mah et al., Measurement of rapid temperature profiles using thermoluminescent microparticles, *IEEE Sensors Journal*, 10(2), 311–15, 2010. © 2010 IEEE.)

Theory predicts that a pulse with a higher peak temperature will have a progressively stronger depopulation effect on the relatively shallow electron traps. Our data suggest that the deeper traps responsible for the second peak are relatively unaffected by the pulse temperatures and durations used here, but will begin to empty at higher excitation temperatures as can be seen by slow preheating. Increasing the maximum pulse-heating temperature [17] applied by 5°C at a time yields easily discernible decreases in the height of the low-temperature peak until the peak dwindles to an indistinguishable height when the maximum preheating pulse temperature reaches around 300°C, as shown in Figure 4.15.

To compare these results to theoretical prediction, we chose trap parameters to closely fit our nonpulsed control curve; numerical simulations based on first-order kinetics theory were then used to produce postpulse glow curves for these theoretical particles. Figure 4.16 compares the relation between the intensity peak height and temperature pulse maxima of the resulting simulations with that of the experimental data obtained using traditional TL heating. As expected, a clear one-to-one decrease occurs in the intensity ratio of the first peak to second peak heights as the pulse temperature climbs [17]. The slope of the curve is reasonably stable, indicating that the sensitivity of this Mg_2SiO_4:Tb,Co will be fairly steady over the usable pulse temperature range. In the range we observed, model and data differed by an average of 4.4%, with no errors greater than 9.1%. (Even better accuracy could be obtained by simply curve fitting the TL peak ratio in much the same manner that the resistance of commercial thermistors is calibrated.)

THERMAL HISTORY MODELING AND RECONSTRUCTION

We have confirmed that the TL of a particle can measure the temperatures to which it has been exposed; the question now is how this information can be recovered and

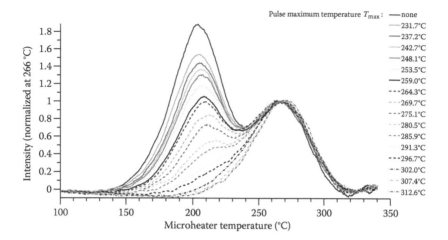

FIGURE 4.15   The thermoluminescence glow curves of Mg_2SiO_4:Tb,Co microparticles after a 190 ms explosive heating pulse. The legend at right indicates the peak temperature of the pulse corresponding to each curve. The intensities shown are normalized at 266°C, the top of the second observable peak. The ratios of the intensities of the peaks strongly correlate with the maximum temperature to which the particles had been exposed. (Reprinted from M. L. Mah et al., Measurement of rapid temperature profiles using thermoluminescent microparticles, *IEEE Sensors Journal*, 10(2), 311–15, 2010. © 2010 IEEE.)

reconstructed from a single experimental glow curve. Under a first-order kinetics model, the percentage of carriers remaining in a partially depopulated trap is rather straightforward to deduce. In Equation 4.5, the population at the end of preheating n'_0 is a coefficient in the expression for the glow curve intensity $I(t)$, and so the population that survives preheating is directly reflected in the amount that the peak

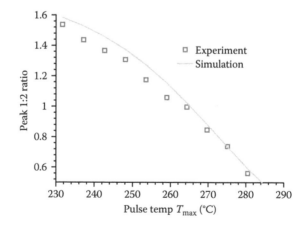

FIGURE 4.16   The ratio of the height of the first TL peak of Figure 4.15 to the height of the second as a function of pulse temperature for simulated and experimental data. (Reprinted from M. L. Mah et al., Measurement of rapid temperature profiles using thermoluminescent microparticles, *IEEE Sensors Journal*, 10(2), 311–15, 2010. © 2010 IEEE.)

intensity decreases from its fully populated maximum. If this population value is taken to be n'_0 after $T(t)$ has run its course, then the TL intensity observed from the remaining population during a linear temperature ramp $T'(t')$ is expressed as

$$
\begin{aligned}
I(t') &= C\frac{dn}{dt} \\
&= n'_0 \left(Cn_0 \exp\left(-\int_0^t se^{-E/kT(t)}dt \right) \right) se^{-E/kT'(t')}
\end{aligned}
\tag{4.6}
$$

To subsequently recover the temperature profile that was responsible for the observed depopulation, we must solve Equation 4.6 for $T(t)$. Here we run into trouble: this expression cannot be solved analytically for most functional forms of $T(t)$. Either a symbolic approximation or decomposition must be substituted, or a numerical computation method used. To maximize the generality of our technique, we choose the latter.

Perhaps the simplest and most general numerical simulation method devisable is the brute-force exhaustive search, where the space of plausible solutions for T_{max} and t_{cool} is canvassed until a combination of values that give minimal error is found. This approach is usually inadvisable except as a last resort because each additional parameter added to the temperature profile will exponentially increase the number of possible combinations. However, in this application, several mitigating factors justify this technique: first, a trap must be partially depopulated in order to give useful information, delineating a temperature range outside of which there is no need to search; second, experimental error will set a minimum temperature resolution, and finer scales will not yield greater accuracy; and finally and most crucially, presetting a limit on the number of parameters we seek will yield a predictable pattern to the solution space that transforms a random search into a guided one.

To expound on this, let us take the previously described experiment involving the microheater-driven pulsed heating of Mg_2SiO_4:Tb,Co. As we noted earlier, this material manifested two easily distinguished, fairly well-separated TL peaks under the conditions used, allowing the higher temperature peak to be used as a constant intensity reference in order to calculate normalized relative intensities for its lower-temperature sibling. If this peak derives from a single trap modeled with first-order kinetics, we will be able to easily determine the degree to which the trap has been depopulated by the preheating, idealized (for the moment) as a fixed-duration square pulse of variable maximum temperature. To find the pulse temperature responsible for a glow peak of a certain height, we can simply simulate the glow curves—or even more simply, the remaining trap population—which would result from a selection of possible temperatures and compare the magnitude of error between these and the actual result. If we have little idea of the approximate range in which our actual pulse temperature lies, many of the temperatures we simulate will leave the trap's population either completely depleted or wholly untouched and thus result in a peak height that does not change from temperature to temperature. As we near our actual pulse temperature, the magnitude of error should decrease until zero at the correct pulse temperature (ignoring experimental read error) and then increase again as we travel past and away from the answer. Therefore, we should be able to locate the correct pulse temperature much more efficiently by simulating a selection of widely separated points, examining the changes in error, and honing our search range until we have our answer to the desired accuracy.

The same strategy holds in more complex situations as well. We will assume that our preheating profile is a single explosive heating event, approximated by an instantaneous jump from the ambient temperature to a maximum followed immediately by slow cooling back to the original ambient, as described in Section "TL as a Thermometer" and shown in Figure 4.5. With these simplifications, the preheating (i.e., explosion) scenario can be expressed with two parameters: the maximum temperature, symbolized as T_{max}, and the time taken for the temperature to fall back to within 1°C of the ambient temperature, referred to as the cooling time or t_{cool}.

Our solution space is now a two-dimensional plane of possible T_{max} and t_{cool} pairings, a much larger number of precision-limited points than the base case. But different combinations of duration and temperature can produce the exact same depopulation effect on a given trap, so that for each possible value of t_{cool}, there will be a corresponding T_{max} to produce the effect seen in our experiments, leaving us with an entire line of (T_{max}, t_{cool}) points which perfectly explain the observed sample TL, as shown in Figure 4.17. A second trap of sufficiently different parameters will also generate a line of situations that produce identical depopulation effects, but this line will not, in general, be the same as that of the first trap. There should be precisely one preheating scenario that the two equal-depopulation lines have in common: the scenario that actually took place. This type of graph with intersecting lines is shown in Figure 4.18.

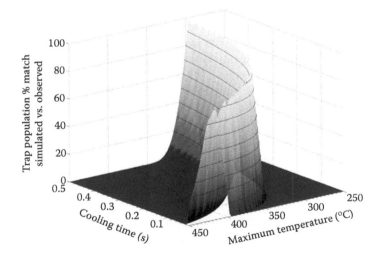

FIGURE 4.17 Shows the % correctness (1-%error) of the simulated trap population for various pre-heating scenarios compared to the actual observed trap population. The line of points at the top are caused by an insufficiently high simulation resolution, that is, the simulated points are a visible distance away from the theoretically-100%-correct points. This diagram used the ratio from one pair of traps.

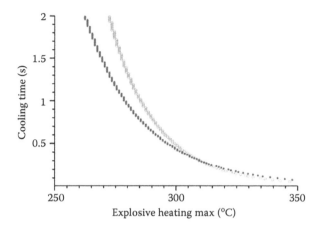

FIGURE 4.18   Each of the two lines is the result of tracing the 100% correct line seen in Figure 4.17 except for different trap pairs. There were three traps simulated total, with one chosen to not depopulate significantly and two chosen to depopulate partially in the "actual" explosive heating scenario; each of these two lines is calculated from the ratio of one of the two depopulating traps and the reference trap.

This method could, in theory, be extended to temperature profiles of greater complexity. The additional solution space dimensions would of course come at increased computational cost. A naive implementation would scale in the worst case as $O(n^x)$, where n is the number of precision-limited points possible in each preheating parameter and x is the number of parameters needed to describe the temperature profile sought, but various optimizations can be made. If the individual glow peaks of the traps are well separated or easily distinguished, the equivalent-depopulation lines can be extrapolated from a few points; locating these lines requires a much smaller number of points to be simulated than directly seeking their intersection point, reducing n. An iterative search by choosing and simulating ever-shrinking ranges of parameters can also bring down the rate of growth. On the hardware side, the simulation of each preheating scenario is independent of the results of others, allowing multiprocessor computers—or even specialized hardware such as graphics processor units and field programmable gate arrays—to effectively divide the time required by the number of simultaneously active threads sustained.

Some complications still arise. Many TL materials are known to have glow curves that can be altered depending on preparation and preirradiation treatment; unexpected inconsistencies in trap parameters can lead us to spurious matching situations. Fortunately, a random change in the parameters of multiple traps is highly unlikely to coincidentally form another unanimous intersection point, and so the damage can be easily detected and the trap parameters then recalibrated from a control sample. In our tests of this search method, correct temperature profiles have been recovered from postexplosion glow curves that were simulated using another set of trap parameters differing in E_t and s, indicating some natural robustness against parameter mismatch. It should be noted that the tolerance is likely to vary based on the temperature profile and material used, and is yet to be studied systematically.

A difficulty in comparing first-order simulations with experimental results is that the TL glow curve gives the combined TL intensity of all the peaks, which in many cases will contain substantial overlap. Although techniques exist that attempt to discern the luminescence contributions of individual peaks from their summed intensity, most are targeted at gauging material trap parameters and thus assume that bountiful supplies of sample are available for repeated heating and reirradiation [22–24]. Our thermal history reconstruction method can adapt to overlapping peaks by using the overall intensity at points along the summed glow curve for comparison instead of individual peak height, which is just as effective and requires no additional experiments, but it does add computational overhead and restricts some optimizations. Finally, it should be stressed that this thermal history reconstruction method is likely to be the most general and robust possible, but it is by nature very computation heavy. If computer speed or computational load is paramount, other methods such as computerized glow curve decomposition may be worth investigating.

SENSOR SURVIVAL

The strong overpressure wave and shock stimuli produced by an explosion are violently intense [25]: while our solid-state, low-mass particles will be able to withstand much more punishment than conventional contact sensors, a sufficient physical impulse could plausibly introduce changes to the defect and crystal structure [26] which help define the TL properties of a material. Surface contamination and reactions with the chemicals in the postdetonation environment, which tends to include components of the explosive itself in various stages of reaction or decomposition, could introduce unexpected luminescence effects or block the observation of the intended ones.

We tested the stability of the thermoluminescence of Al_2O_3:C, a common commercial dosimetry material, before and after an explosion. Particles of this aluminum oxide 15–90 μm in diameter were obtained from Landauer, Inc. (Stillwater, Oklahoma) and between 5 and 15 g poured into a paper cup. A high explosive charge consisting of either 2 or 25 g of pentaerythritol tetranitrate and a detonator was placed on top of or immersed in the Al_2O_3:C. The assembly was suspended inside a steel blast chamber and the charge detonated, as represented in Figure 4.19, after which the particles were recovered from a plastic collection cone located underneath the charge.

Our primary concern in this experiment was not to measure temperature but to assess whether or not the Al_2O_3:C would survive as a serviceable TL material. To this end, the postexplosion and control nonexploded samples were all irradiated with broad-spectrum 180–400 nm UV light and subsequently linearly heated at a rate of 0.75°C/s on a hotplate. The TL glow curves thus collected are qualitatively and quantitatively in agreement, with both the full-width at half-maximum and peak maximum location varying by 4.0% and 1.9% respectively, as shown in Figure 4.20; this is within the bounds of experimental error. Due to the extreme proximity of the explosives, it is estimated that the particles were subjected to pressures of the order of 2000 psi (Jim Lightstone, Naval Surface Warfare Center, priv. comm.).

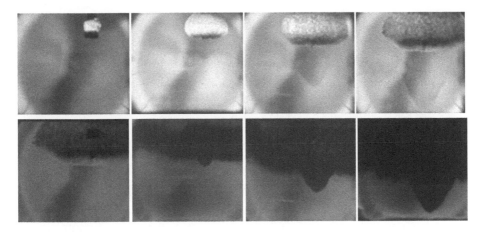

FIGURE 4.19   The detonation of the charge + assembly in which the Al$_2$O$_3$:C microparticles were embedded. These images were captured as 20 ns exposures at 5 ms intervals. (Images courtesy of Jim Lightstone of the NSWC, Indian Head, MD.)

There is no indication that this or any other detonation effects caused significant change in the TL ability of the material regardless of the charge placement or size.

With this most vital point established, we are free to satisfy our curiosity by examining the particles for any additional explosion effects. It was likely that there were surface contaminants or reactions on the particle, even though they had clearly not impacted luminescence. The elemental composition of the particle surfaces were therefore compared using x-ray photoelectron spectroscopy (XPS) as shown in Figure 4.21. In some samples, the Al 2p and O 1s peaks were found to be significantly weaker after being exposed to detonation, suggesting that some aluminum combustion had taken place. By contrast, carbon had a stronger presence through an elevated C 1s peak height. Since XPS generally does not penetrate more than a few atomic layers—no more than a few tens of Å or so—beneath the

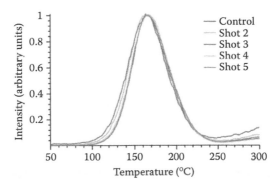

FIGURE 4.20   Thermoluminescence glow curves of Al$_2$O$_3$:C before and after explosion with PETN charges. The data were collected using a linear heating ramp of approximately 0.9°C/s.

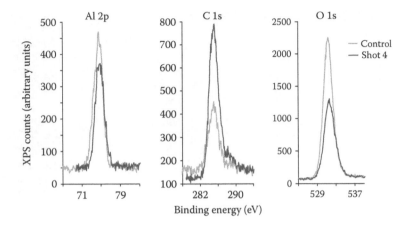

FIGURE 4.21   High-resolution XPS spectra of Al_2O_3:C, centered on the primary peaks of its elemental constituents, after detonation of a 25 g PETN charge.

surface of its sample [27], the opposing changes in content could also be simply due to an added outer garnish of carbon obscuring the view of unchanged proportions of aluminum and oxygen, or even a result of a smaller sample cross-section being observed.

Most basic of characterizations, a visual examination, was carried out by a scanning electron microscope shown in Figure 4.22. The images revealed an interesting shift in the shape of the particles: whereas the original pre-explosion specimens (Figure 4.22a) were angular bodies with sharp corners and smooth planes, the post-explosion particles (Figure 4.22b) displayed much more rounded, compact configurations with higher surface roughness. The average particle size also seemed to have grown after detonation, thanks mainly to a dearth of sub-20-μm particles. This size bias likely arises from the collection process rather than anything that happens due to detonation.

FIGURE 4.22   SEM image of Al_2O_3:C microparticles (a) before undergoing high explosive detonation and (b) after detonation with 25 g of pentaerythritol tetranitrate (PETN).

CONCLUSION AND FUTURE DIRECTIONS

We conclude that thermoluminescent microparticles and nanoparticles with multiple, light-insensitive TL peaks provide an excellent means of measuring thermal history in extreme environments such as fires and explosions, where no traditional sensors could function. These have the advantages of no separable parts, low cost, compatible aerodynamics, easy dispersal, and high sensitivity to peak temperatures. They can be simulated many times per day using microheaters instead of explosions but survive an actual detonation even when placed directly on high explosive charges. The concept has been experimentally proved using $MgSiO_4$:Tb,Co particles on microheaters with rapid heating and cooling over approximately 232–313°C.

We feel future work in the area will proceed in four directions. The first, obviously, is to test TL thermal history sensing in real explosions, and, in fact, this work is already proceeding as a collaboration between our group and Jim Lightstone at the Naval Surface Warfare Center in Indian Head, Maryland. Second, there is a definite need for materials research. Almost all TL materials have been developed and characterized for radiation dosimetry, a technology where materials with a single well-behaved peak and high radiation sensitivity are desirable. For temperature measurement, multiple light-insensitive peaks are preferred and radiation sensitivity is largely irrelevant. Many materials that were rejected for dosimetry may be viable for sensing. In addition, new materials are being developed at Oklahoma State University and Clemson University using solution combustion synthesis [15]. Third, one can develop specially engineered particles and materials to enhance the collection of thermal history beyond what we have described here. For example, one might construct a thin film stack of alternating TL materials to engineer a "particle" with a desired set of traps. One may also create thick core-shell particles where the shell will follow temperatures at the surface very rapidly while the core will only follow slower average events. By examining the luminescence as a function of depth in the particle (say by using core/shell materials with different emission wavelengths) one can extract temperature information in much the same manner that previous climate history can be obtained from deep caves and boreholes drilled into the ice in Antarctica or Greenland [28]. Finally, it may be possible to measure parameters other than temperature. It will be interesting to see whether extreme pressures in the center fireball of an explosion cause any changes in trap populations. From one point of view, one would like to ignore such a complication. On the other hand, if there is a dependence, then certain particles are likely to be more sensitive to this than others, and the development of pressure sensors could begin. In any case, it will be interesting to see whether other applications of microparticle thermoluminescent sensors arise!

ACKNOWLEDGMENTS

We would particularly like to thank the Defense Threat Reduction Agency for funding our work under grants HDTRA1-07-1-006 and HDTRA1-10-0007. We would also like to thank Su Peiris of the DTRA for ongoing guidance and support. Portions of this chapter are based on research in our group from prior publications, particularly in the section entitled: "An Example Process" using References [17–19]. We would like to thank Eduardo Yukihara of Oklahoma State University and Jim Lightstone of the Naval Surface Warfare Center for their fine research in the area and for their continuing collaboration and assistance. We would also like to acknowledge our colleagues at the University of Minnesota: Michael Manfred, Sangho Kim, and Nick Gabriel, who have greatly contributed to the work in this chapter.

REFERENCES

1. W. K. Lewis and C. G. Rumchik, Measurement of apparent temperature in post-detonation fireballs using atomic emission spectroscopy, *Journal of Applied Physics*, 105(5), 056104, 2009.
2. J. Wilkinson, J. M. Lightstone, C. J. Boswell, and J. R. Carney, Emission spectroscopy of aluminum in post-detonation combustion, *AIP Conference Proceedings*, 955(1), 1271–1274, 2007.
3. S. W. S. McKeever, *Thermoluminescence of Solids*, Cambridge University Press, Cambridge, 1985.
4. Y. S. Horowitz, *Thermoluminescence and Thermoluminescent Dosimetry*, CRC Press, Boca Raton, FL, 1984.
5. L. Botter-Jensen, S. W. S. McKeever, and A. G. Wintle, *Optically Stimulated Luminescence Dosimetry*, Elsevier, Amsterdam, 2003.
6. M. E. Manfred, N. Gabriel, E. Yukihara, and J. J. Talghader, Thermoluminescence measurement technique using millisecond temperature pulses, *Radiation Protection Dosimetry*, 139(4), 560–564, 2010.
7. J. Gasiot, P. Braunlich, and J. P. Fillard, Laser heating in thermoluminescence dosimetry, *Journal of Applied Physics*, 53, 5200, 1982.
8. K. H. Lee and J. H. Crawford, Jr., Luminescence of the F center in sapphire. *Physical Review B*, 19(6), 3217–3221, 1979.
9. G. Valladas and C. Lalou, Thermoluminescence of the Saint-Severin meteorite, *Earth and Planetary Science Letters*, 18(1), 168–171, 1973.
10. S. W. S. McKeever and D. W. Sears, Thermoluminescence and terrestrial age of the Estacado meteorite, *Nature*, 275(5681), 629–630, 1978.
11. S. W. S. McKeever and D. W. Sears, The natural thermoluminescence of meteorites: A pointer to meteorite orbits? *Modern Geology*, 7(3), 137–145, 1980.
12. G. S. Polymeris, A. Sakalis, D. Papadopoulou, G. Dallas, G. Kitis, and N. C. Tsirliganis, Firing temperature of pottery using TL and OSL techniques, *Nuclear Instruments and Methods in Physics Research A*, 580, 747–750, 2007.
13. E. G. Yukihara, V. H. Whitley, J. C. Polf, D. M. Klein, S. W. S. McKeever, A. E. Akselrod, and M. S. Akselrod, The effects of deep trap population on the thermoluminescence of Al_2O_3:C, *Radiation Measurements*, 37, 627–638, 2003.
14. P. D. Townsend, Analysis of TL emission spectra, *Radiation Measurements*, 23(2/3), 341–348, 1994.
15. V. R. Orante-Barrn, L. C. Oliveira, J. B. Kelly, E. D. Milliken, G. Denis, L. G. Jacobsohn, J. Puckette, and E. G. Yukihara, Luminescence properties of MgO produced by solution combustion synthesis and doped with lanthanides and Li, *Journal of Luminescence*, 131(5), 1058–1065, 2011.
16. I. Bibicu and S. Calogero, High-efficiency heater for a thermoluminescence apparatus, *Journal de Physique III*, 6, 475–480, 1996.
17. M. L. Mah, M. E. Manfred, S. S. Kim, M. Prokić, E. G. Yukihara, and J. J. Talghader, Measurement of rapid temperature profiles using thermoluminescent microparticles, *IEEE Sensors Journal*, 10(2), 311–15, 2010.

18. M. L. Mah, M. E. Manfred, S. S. Kim, M. Prokić, E. G. Yukihara, and J. J. Talghader, Sensing of thermal history using thermoluminescent microparticles, *IEEE/LEOS International Conference on Optical MEMS and Nanophotonics 2009*, Clearwater, Florida, August 2009, pp. 23–24.
19. J. J. Talghader, Micro- and nano-particles for the distributed sensing of thermal history, *Seventh International Conference on Networked Sensing Systems*, Kassel, Germany, June 15–18, 2010, pp. 169–170.
20. J. C. Mittani, M. Prokić, and E. G. Yukihara, Optically stimulated luminescence and thermoluminescence of terbium-activated silicates and aluminates, *Radiation Measurements*, 43(2), 323–326, 2008.
21. M. Prokić and E. G. Yukihara, Dosimetric characteristics of high sensitive Mg2SiO4:Tb solid TL detector, *Radiation Measurements*, 43(2), 463–466, 2008.
22. K. H. Nicholas and J. Woods, The evaluation of electron trapping parameters from conductivity glow curves in cadmium sulphide, *British Journal of Applied Physics*, 15, 83–795, 1964.
23. A. Halperin and A. A. Braner, Evaluation of thermal activation energies from glow curves, *Physical Review*, 117(2), 408–415, 1960.
24. V. Pagonis, G. Kitis, and C. Furetta, *Numerical and Practical Exercises in Thermoluminescence*, Springer Science + Business Media, New York, 2006.
25. M. M. Swisdak, *Explosion Effects and Properties. Part I. Explosion Effects in Air*. Naval Surface Warfare Center, Maryland, 1975.
26. R. A. Graham, B. Morosin, E. L. Venturini, and M. J. Carr, Materials modification and synthesis under high-Pressure shock compression, *Annual Review of Materials Science*, 16, 315–341, 1986.
27. C. R. Brundle, C. A. Evans, and S. Wilson (eds), *Encyclopedia of Materials Characterization*, Butterworth-Heinemann, Boston, 1992.
28. W. S. B. Patterson, *The Physics of Glaciers*, 3rd edn, Butterworth-Heinemann, New York, 1994.

5 Solar Cell Analyses with Ultraviolet–Visible–Near-Infrared Spectroscopy and I–V Measurements

Andreas Stadler

CONTENTS

ULTRAVIOLET–VISIBLE–NEAR-INFRARED (UV–VIS–NIR) SPECTROSCOPY

Solar cell development demands exact and efficient analyzses of transparent conducting oxide (TCO) and absorber materials. Optical analysis is usually done by means of ultraviolet–visible–near-infrared (UV–Vis–NIR) spectroscopy provides transmission spectra, $T(\lambda)$, as a function of wavelength λ from about 200 to 3300 nm. Transmission, $T(\lambda)$, and reflection spectra, $R(\lambda)$, can be measured up to $\lambda = 2500$ nm by using a spectralon-coated integrating sphere. A novel, nonnumerical model, without approximations, is provided for analyzsing a single solid-state layer. Approximations are shown for multilayered systems and measurements without an integrating sphere. Extracted values are compared with those of the well-known Keradec–Swanepoel model (KSM). Quantum mechanical potential barrier models are used to reevaluate spectra out of these physical values, in order to compare them with the originally measured spectra.

CORRECT AND EFFICIENT THEORY FOR THIN-FILM INVESTIGATIONS: SINGLE-LAYER MODEL

SURFACES, INTERFACES, AND BULKS OF MATERIALS: REFLECTIONS, TRANSMISSIONS, AND ABSORPTIONS

Combining Maxwell's equations with the Poynting Theorem results in the vertical and parallel components and therefore gives the total amount of the incident, e, reflected, r, and transmitted, t, beam power [1,2]

$$P_{i,j} = \frac{1}{2}\sqrt{\frac{\varepsilon_i}{\mu_i}}\vec{E}_{i,j,0}^2 A \cos\theta_i, \quad i \in \{e,r,t\}, \quad j \in \{\perp,\|\}, \tag{5.1}$$

where $\vec{E}$ is the electrical field vector, ε_i is the permittivity, and μ_i is the permeability. The investigated area is denoted as A and the corresponding angle versus the normal to the interface with θ. The *reflection, $R_{/j}$, at an interface between two media* is then defined by $(\theta_r = \theta_e)$

$$R_{/,j} = \frac{P_{r,j}}{P_{e,j}} = \frac{\vec{E}_{r,j,0}^2}{\vec{E}_{e,j,0}^2} = r_j^2, \quad j \in \{\perp,\|\}. \tag{5.2}$$

The *transmission, $T_{/j}$,* can be written as $(\theta_t \neq \theta_e)$

$$T_{/,j} = \frac{P_{t,j}}{P_{e,j}} = \frac{c_t \cdot \varepsilon_t \cdot \vec{E}_{t,j,0}^2 \cdot \cos\theta_t}{c_e \cdot \varepsilon_e \cdot \vec{E}_{e,j,0}^2 \cdot \cos\theta_e} = \frac{(n_t/\mu_t)}{(n_e/\mu_e)}\frac{\sqrt{1-(n_e/n_t)^2 \cdot \sin^2\theta_e}}{\cos\theta_e} t_j^2, \quad j \in \{\perp,\|\}. \tag{5.3}$$

Herein the reflection, r_j, and transmission coefficients, t_j, are defined by Fresnel's equations.

In the case of *normal incidence* $\theta_e, \theta_r \in [0°, \dots, 8°]$ Fresnel's equations are identical for vertical and parallel components of the electrical field, resulting in

$$R_{/,\perp} = R_{/,\|} = r_\perp^2 = r_\|^2 = \left(\frac{n_e/\mu_e - n_t/\mu_t}{n_e/\mu_e + n_t/\mu_t}\right)^2,$$

$$T_{/,\perp} = T_{/,\|} = \frac{n_t/\mu_t}{n_e/\mu_e}t_\perp^2 = \frac{n_t/\mu_t}{n_e/\mu_e}t_\|^2 = 4\frac{n_e n_t}{\mu_e \mu_t}\bigg/\left(\frac{n_e}{\mu_e} + \frac{n_t}{\mu_t}\right)^2. \tag{5.4}$$

For electromagnetic waves within media, besides Maxwell's equations, three material equations have to be taken into account: $\vec{D} = \varepsilon_0\vec{E} + \vec{P} = (1+\chi_P)\varepsilon_0\vec{E} = \varepsilon_r\varepsilon_0\vec{E} = \varepsilon\vec{E}$ for *dielectric effects*, $\vec{B} = \mu_0\vec{H} + \vec{M} = (1+\chi_M)\mu_0\vec{H} = \mu_r\mu_0\vec{H} = \mu\vec{H}$ for *magnetic effects*, and *Ohm's law* $\vec{j} = \sigma_L\vec{E}$, where $\sigma_L = 1/\rho_L$. Using these material equations, the Maxwell equations can be transferred into a 2×2 differential equation system for the electrical and magnetic field [3]. In detail, on the one hand differentiation, $\partial/\partial t$, and on the other rotation, $\vec{\nabla}\times$, of $\vec{\nabla}\times\vec{B} = \mu\sigma_L\vec{E} + \mu\varepsilon\,\partial\vec{E}/\partial t$, leads to two equations. Using $\vec{\nabla}\times\vec{E} = -\partial\vec{B}/\partial t$, the operator identity $\vec{\nabla}\times\vec{\nabla} = \vec{\nabla}(\vec{\nabla}\cdot) - \vec{\nabla}^2$ and $E = B/\sqrt{\mu\varepsilon}$ result in

$$
\left[\vec{\nabla}^2 - \mu\varepsilon\frac{\partial^2}{\partial t^2} - \mu\sigma_L\left(\vec{r},t\right)\frac{\partial}{\partial t} - \mu\left(\frac{\partial}{\partial t}\sigma_L\left(\vec{r},t\right)\right)\right]\vec{E}\left(\vec{r},t\right) = \vec{\nabla}\frac{\rho_L\left(\vec{r},t\right)}{\varepsilon},
$$

$$
\left[\vec{\nabla}^2 - \mu\varepsilon\frac{\partial^2}{\partial t^2} - \mu\sigma_L\left(\vec{r},t\right)\frac{\partial}{\partial t} - \sqrt{\frac{\mu}{\varepsilon}}\left(\vec{\nabla}\times\sigma_L\left(\vec{r},t\right)\right)\right]\vec{B}\left(\vec{r},t\right) = 0.
$$

(5.5)

Solving this equation system with $\vec{E}(\vec{r},t) = \vec{E}_0 e^{-\vec{k}_{L,I}\cdot\vec{r}} \times e^{i(\vec{k}_{L,R}\cdot\vec{r}\mu\omega t)}$, $|\vec{B}| = |\vec{E}|\sqrt{\mu\varepsilon}$, $c = 1/\sqrt{\varepsilon\mu}$, $\omega/c = 2\pi/\lambda$, and $\sigma_L = 1/\rho_L$ leads for space, $\vec{r}$, and time, t, independent conductivities, σ_L, permittivities, ⊠, and permeabilities, μ, and finally to the wave vector [3]

$$
\vec{k} = k\cdot\vec{e}_k = \left(k_{L,R} - ik_{L,I}\right)\cdot\vec{e}_k,
$$

$$
k_{L,R} = \frac{2\pi}{\lambda}\frac{1}{\sqrt{2}}\sqrt{\sqrt{1 + \frac{\mu}{\varepsilon}\left(\frac{\lambda\sigma_L}{2\pi}\right)^2} + 1} \xrightarrow{\sigma_L\to 0} \frac{2\pi}{\lambda},
$$

$$
k_{L,I} = \frac{2\pi}{\lambda}\frac{1}{\sqrt{2}}\sqrt{\sqrt{1 + \frac{\mu}{\varepsilon}\left(\frac{\lambda\sigma_L}{2\pi}\right)^2} - 1} \xrightarrow{\sigma_L\to 0} 0.
$$

(5.6)

Therefore, the beam power for electromagnetic waves within volumes of materials can be written as

$$
P_{t,j}\left(\vec{r}\right) = \frac{1}{2}\sqrt{\frac{\varepsilon_t}{\mu_t}}\vec{E}_{t,j,0}^2 e^{-2\vec{k}_{L,I}\cdot\vec{r}} A\cos\theta_t, \quad j \in \{\perp, \|\}.
$$

(5.7)

The quotient of the beam power at the positions $\vec{r} \neq 0$ and $\vec{r} = 0$ is equal to the transmission, $T_{\#,j}(\vec{r})$, of an electromagnetic wave *through the bulk of a material*

$$
T_{\#,j}\left(\vec{r}\right) = \frac{P_{t,j}\left(\vec{r}\right)}{P_{t,j}\left(0\right)} = e^{-2\vec{k}_{L,I}\cdot\vec{r}} = e^{-\vec{\alpha}_L\cdot\vec{r}}, \quad j \in \{\perp, \|\}.
$$

(5.8)

Every photon which is not transmitted will be absorbed—independent of its path. Thus, the *adequate absorption* results in

$$A_{\#,j}(\vec{r}) = 1 - T_{\#,t,j}(\vec{r}) = 1 - e^{-2\vec{k}_{L,I}\cdot\vec{r}} = 1 - e^{-\bar{\alpha}_L\cdot\vec{r}}, \quad j \in \{\perp,\parallel,\}, \tag{5.9}$$

(Lambert's law) where $\boxtimes_L = 2k_{L,I}$ is the *absorption coefficient*.

EXACT AND COMPLEX PARAMETER EXTRACTION FOR A SINGLE LAYER

Refractive Indices and Absorption Coefficients

In the case of a *single-layer system (SLM)*, an incident beam is repeatedly reflected and transmitted at/through both surfaces of the medium. Integration over all parts of the beam, released on the reflection side of the sample, results in the total reflection, $R(\lambda)$; integration over all transmitted parts, results in the total transmission, $T(\lambda)$. Measured UV–Vis–NIR spectra provide the total reflections, $R(\lambda)$, and transmissions, $T(\lambda)$, as functions of the wavelengths, λ (Alternatives: [4, Ch. 2.2], [5, 6, Ch. AII], [7, Ch. 2.4]).

The mathematical formulation, taking into account Equations 5.2, 5.3, 5.8, and 5.9, results in Equation 5.10. Moreover the measured *reflections, R*, and *transmissions, T*, satisfy the balance equation $100\% = 1 = A + R + T$ [8,9], see Figure 5.1, which allows to compute the according *absorptions, A*,

$$\begin{aligned}
R &= R_{l,eL} + T_{l,eL}T_{l,Le}T_{\#,L}^2 R_{l,Lt} \cdot \Im, \\
A &= 1 - (R + T), \\
T &= T_{l,eL}T_{\#,L}T_{l,Lt} \cdot \Im.
\end{aligned} \tag{5.10}$$

The sum over all reflections within the layer $\Im = \sum_{p=0}^{P_0}(T_{\#,L}^2 R_{l,Lt}R_{l,Le})^p$ reaches a well-defined limit with increasing p. For $p \to \infty$ the theoretical solution of the physical problem is exact. Practically, p should be high such that the theoretical failure of Equation 5.10 is less than the measurement error.

Using Equations 5.4, 5.8, and 5.9, within the measurable reflections, R, and transmissions, T, in Equation 5.10, leads to a transcendent 2×2 equation system for the refractive index quotient n_l/n_e and the absorption coefficient α_L.

Transmissions $T_{l,Le} = T_{l,Lt}$ and reflections $R_{l,eL} = R_{l,Lt}$ are identical for *normal light incidence* ($\theta_e \leq 8°$; see Equation 5.4). Therefore, the total transmissions, T, and reflections, R, can be simplified as

$$R = R_{l,eL}(1 + T_{\#,L}T) = r_{eL,j}^2(1 + Te^{-\alpha_L d_L \cos\theta_e}) = \left(\frac{n_L\mu_e/n_e\mu_L - 1}{n_L\mu_e/n_e\mu_L + 1}\right)^2 (1 + Te^{-\alpha_L d_L}). \tag{5.11}$$

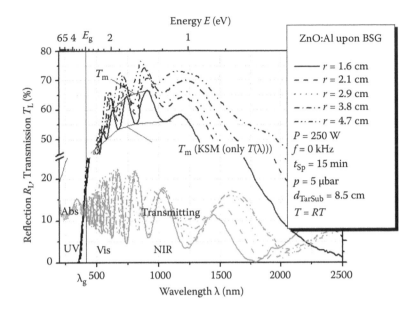

FIGURE 5.1   Reflection, R_L, and transmission, T_L, spectra for sputtered aluminum-doped zinc-oxide (ZnO:Al) thin films upon a boron silicate glass (BSG) substrate (Leybold Optics CLUSTEX 100M sputtering cluster tool, Perkin-Elmer Lambda 750 UV–Vis–NIR spectrometer). The UV, Vis, and NIR region of the wavelength and energy–dependent spectrum as well as the band gap energy, E_g (wavelength λ_g), which separates absorbing from the transmitting region of the spectrum are shown. The KSM uses exclusively the envelope functions T_m and T_M of the transmission spectra for parameter extraction.

Equation 5.11 includes no approximations and is independent of the beam path within the layer, that is, from $\Im = \sum_{p=0}^{P_0}(T_{\#,L}^2 R_{l,Li} R_{l,Le})^p$.

On the one hand, solving Equation 5.11 for the *refractive index quotient* n_L/n_e and the *absorption coefficient* α_L results in

$$\frac{n_L\mu_e}{n_e\mu_L} = \frac{\sqrt{1+T\,e^{-\alpha_L d_L}}+\sqrt{R}}{\sqrt{1+T\,e^{-\alpha_L d_L}}-\sqrt{R}},$$

$$\alpha_L = -\frac{1}{d_L}\ln\left(\left(\frac{n_L\mu_e/n_e\mu_L+1}{n_L\mu_e/n_e\mu_L-1}\right)^2\frac{R}{T}-\frac{1}{T}\right). \tag{5.12}$$

On the other hand, the substitution of the absorption $A = 1 - (R + T)$ into Lambert's Law $1 - A = e^{-\alpha_L d_L}$ results in

$$\alpha_L = \frac{1}{d_L}\ln\left(\frac{1}{1-A}\right) = \frac{1}{d_L}\ln\left(\frac{1}{R+T}\right)\xrightarrow[R+T\to 1]{}0. \tag{5.13}$$

Elimination of the absorption coefficient, α_L, from this equation system leads to, see Figure 5.2a,

$$\frac{n_L \mu_e}{n_e \mu_L} = \frac{\sqrt{1+T(R+T)}+\sqrt{R}}{\sqrt{1+T(R+T)}-\sqrt{R}} \xrightarrow{T\to 0} \frac{1+\sqrt{R}}{1-\sqrt{R}}. \tag{5.14}$$

UV–Vis–NIR spectra measured with an integrating sphere result in lossless reflections, $R(\lambda)$, and transmissions, $T(\lambda)$. Using Equation 5.14 leads to an approximation free, wavelength-dependent quotient $n_L \mu_e / n_e \mu_L$ (λ). Finally, Equation 5.13 provides wavelength-dependent absorption coefficients, α_L, if the layer thickness d_L is known, see Equation 5.19.

An approximation can be made *for a UV–Vis–NIR spectrometer without an integrating sphere* (e.g., [10–13]). Taking Equation 5.4, 5.8, and $\mu_e = \mu_L = \mu_0$ into account, then for $p_0 = 0$ the transmission, T, from Equation 5.10 directly leads to

$$T = \left(\frac{4 n_L / n_e}{(n_L / n_e + 1)^2}\right)^2 e^{-\alpha_L d_L} \xrightarrow[d_L \to 0]{\alpha_L \to 0} \left(\frac{4 n_L / n_e}{(n_L / n_e + 1)^2}\right)^2. \tag{5.15}$$

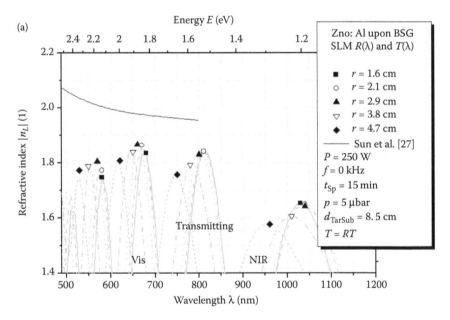

FIGURE 5.2   Wavelength-, λ, and energy-, $E = hc/\lambda$ (h = Planck's constant, c = light velocity), dependent refractive indices, $|n_L|$, extracted with (a) the SLM, using reflection and transmission spectra (integrating sphere, Equation 5.14), (b) the SLM, using exclusively transmission spectra (Equation 5.16), and (c) the KSM, using solely the envelope functions T_m, T_M of the transmission spectra. For comparison, the refractive index after Sun et al. [15] is shown. Slightly elevated refractive indices occur when reflection spectra are neglected for parameter extraction. The distance, r, from the deposition center is proportional to the layer thickness, d_L.

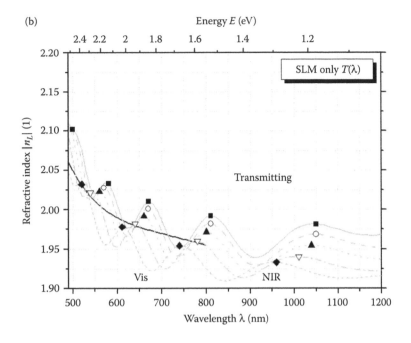

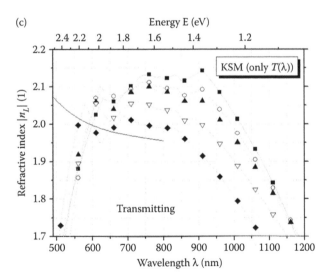

FIGURE 5.2 Continued.

According to Equation 5.15 and Figure 5.2b, the refractive index quotient can be determined for *transparent layers* $\alpha_L \rightarrow 0$ to

$$\frac{n_L}{n_e} \approx 2\left(-\sqrt{\frac{1}{T} + \frac{1}{\sqrt{T}}} - \frac{1}{\sqrt{T}} - \frac{1}{2}\right). \tag{5.16}$$

Moreover, for *opaque layers* (i.e., for $R = R_{l,eL}$ in Equation 5.10, for $\alpha_L \to \infty$ in Equation 5.12 or for $T \to 0$ in Equation 5.14, the refractive index can be evaluated with [14]

$$\frac{n_L}{n_e} \approx \frac{1+\sqrt{R}}{1-\sqrt{R}}. \tag{5.17}$$

Refractive indices, calculated with the KSM are shown in Figure 5.2c.

Light Velocities, Permittivities, Wavelengths, Wave Numbers, Layer Thicknesses, and Deposition Rates

The quotient of the refractive indices can be written as

$$\frac{n_L}{n_e} = \frac{c_e}{c_L} = \sqrt{\frac{\varepsilon_L \mu_L}{\varepsilon_e \mu_e}} = \frac{\lambda_e}{\lambda_L} = \frac{k_L}{k_e} \neq 1. \tag{5.18}$$

This is, because the light velocity is equal to the reciprocal square root of the according permittivity and permeability, $c_i = 1/\sqrt{\varepsilon_i \mu_i}$, $i \in \{e, L\}$, and it is the product of wavelength and frequency $c_i = \lambda_i \nu$, $\nu = \text{const.}$, $i \in \{e, L\}$, where the wavelength $\lambda_i = 2\pi/k_i$, $i \in \{e, L\}$ depends on the wave number.

Herein, the quotient of the wave numbers k_L/k_e is of special importance, as the wave number of the layer material is complex, $k_L = k_{L,R} - jk_{L,I}$; see Equation 5.6. The real and imaginary parts of the complex wave numbers are only dependent on parameters of the medium (wavelength λ_e, permittivity ε_e, and permeability μ_0) and on the conductivity σ_L of the layer. n_L/n_e can be taken from the spectrum using Equations 5.12, 5.14, 5.16, or 5.17. As a result, the *refractive index* n_L, the *light velocity* c_L, the *square root of the permittivity* ε_L and the *permeability* μ_L, and the *wavelength* λ_L can be treated as *complex numbers*, see Figure 5.3a,b.

If the optical thickness, $n_L d_L$, of a layer is larger than the wavelength, $n_L d_L > \lambda_e$, of an incident beam, and then the layer thickness, d_L, can be easily calculated. Every consecutive maximum of a reflection spectrum (resp. minimum of a transmission spectrum) at a wavelength $\lambda_{e,i}$ shall be enumerated with m_i, where m_i is a natural number ($m_i \in IN$). Then for normal incidence, the *layer thickness,* d_L, can be estimated using

$$d_L = \frac{m_i - m_j}{2\left(\dfrac{n_L(\lambda_{e,i})}{\lambda_{e,i}} - \dfrac{n_L(\lambda_{e,j})}{\lambda_{e,j}} \right)}, \quad m_i, m_j \in IN. \tag{5.19}$$

Herein, $n_L(\lambda_{e,i})$ is the wavelength-dependent refractive index, see Figure 5.3c. For a well-known sputter duration, t_{Sp}, the *deposition rate,* v_L, can be computed using $v_L = d_L/t_{Sp}$.

Band Gap Energies and Conductivities

Optical stimulation of semiconductor materials with a direct band gap can be described with the relativistic Compton effect, Fermi–Dirac statistics for

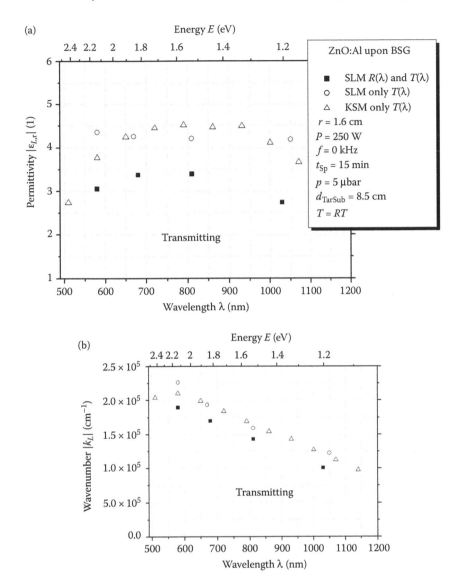

FIGURE 5.3 (a) Permittivity, $\varepsilon_{L,r}$, and (b) wave number, k_L, of a ZnO:Al thin film upon BSG at a distance $r = 1.6$ cm from the sputter deposition center.

spontaneous and stimulated transitions and the Klein–Nishina cross section. It can be shown that the absorption coefficient, α_L, is a function only of the incident light energy, E, the band gap, E_g, and a physical constant

$$\alpha_L E = C_{dir} \sqrt{E - E_g} \tag{5.20}$$

The constant C_{dir} is primarily a function of the electron charge, q_e, the electric constant, ε_0, the light velocity, c_0, Planck's constant, $\hbar$, and the combined mass,

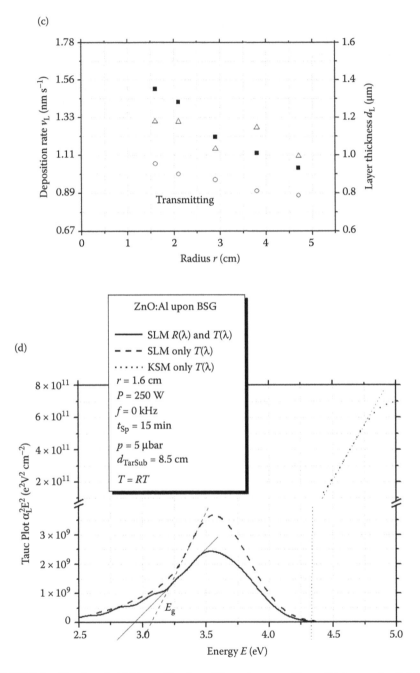

FIGURE 5.3 Continued. (c) Gauss-like deposition rate, v_L, and layer thickness, d_L, distribution (center of deposition at $r = 0$). (d) Tauc Plot for band gap energy, E_g, extraction.

TABLE 5.1

Direct Band Gap Energies, E_g, of ZnO:Al Thin Films Upon a BSG for Different Distances, r, from the Sputter Deposition Center. Values have been Extracted with the SLM (with and without Reflection Spectra) and with the KSM (Neglecting Reflection Spectra)

r/cm	1.6	2.1	2.9	3.8	4.7
$E_{g,SLM}(R,T)$	2.93	2.96	2.99	3.00	3.01
$E_{g,SLM}(T)$	3.09	3.08	3.07	3.05	3.03
$E_{g,KSM}(T)$	4.35	4.36	4.36	4.36	—

$m_{comb} = (1/m_e + 1/m_h)^{-1}$ (m_e = electron mass, m_h = defect electron mass). Therefore, the linear approximation of $\alpha_L^2 E^2$ versus E (*Tauc Plot*) results in a straight line that intersects the abscissa at the *band gap energy E_g*; see Figure 5.3d, 5.4a and Table 5.1.

Similarly, for semiconductor materials with an indirect band gap $\alpha_L^{1/2}E^{1/2}$ has to be plotted versus E.

Absorption coefficients, α_L, are a function of conductivity, σ_L, according to Equation 5.9, $\alpha_L = 2k_{L,I}$, and Equation 5.6, $k_{L,I}(\sigma_L)$. Moreover, $\alpha_L = -\ln (R_L + T_L)/d_L$ (see Equation 5.13, a better conformity with four-tip measurements is reached with $\alpha_L = -\ln (R_{LS} + T_{LS})/d_L$). Therefore, taking Equation 5.14 into account, the *conductivity, σ_L*, is a function of measured spectra and well-known constants, see Figure 5.4b,

$$\sigma_L = \frac{2\pi \, n_L^2}{\lambda_e \, n_e^2} \sqrt{\frac{\varepsilon_e}{\mu_0} \left(\left(1 + \left(\frac{n_e}{n_L} \frac{\alpha_L \lambda_e}{4\pi} \right)^2 \right)^2 - 1 \right)}. \tag{5.21}$$

Effective Dopant Concentrations, Mobilities, and Lifetimes

For electrons in the conduction band of a semiconductor with a *parabolic band structure*, the wave number is given with $k = \sqrt{2m_C (\hbar\omega - E_C)}/\hbar$. Herein, E_C is the minimum energy within the conduction band and m_C is the according effective mass of the electrons. The volume $V_k = 4\pi k^3/3 = (4\pi/3\hbar^3)[2m_C (\hbar\omega - E_C)]^{3/2}$ within the reciprocal space and the Pauli principle lead to the density of states, $D_C(\hbar\omega) = 2dV_k/Vdh$ $\omega = (2m_C)^{3/2} (\hbar\omega - E_C)^{1/2}/2\pi^2\hbar^3$. Electrons are Fermi particles, so that the Fermi–Dirac function, $f_c(\hbar\omega) = (e^{(\hbar\omega - E_F)/kT} + 1)^{-1}$, E_F = Fermi energy, in combination with the density of states, results in the *electron density within the conduction band*

$$n_e = \int_{E_C}^{\hbar\omega} f_C(\hbar\omega) D_C(\hbar\omega) \, d\hbar\omega$$

$$= \frac{(2m_C kT)^{3/2}}{2\pi^2\hbar^3} \int_0^\infty \frac{\sqrt{x}}{e^{x-(E_F-E_C)/kT} + 1} \, dx, \tag{5.22}$$

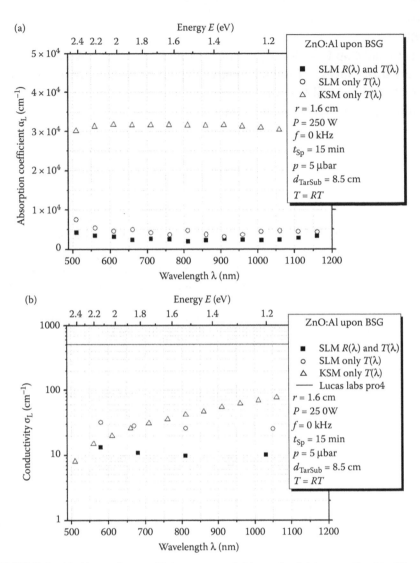

FIGURE 5.4 (a) Absorption coefficients, α_L, and (b) conductivities, σ_L, for thin films on BSG, a comparison. The noncontact, optically measured conductivities are compared with that of a four-tip measurement system from Lucas Labs. (Adapted from V.A. Sokolov, *Russian Phys. J.*, DOI 10.1007BF00890479, 1158–1159.)

where $x = (\hbar\omega - E_C)/kT$ and $F_{1/2}\left((E_F - E_C)/kT\right) = \int_0^\infty (\sqrt{x}/e^{x-(E_F-E_C)/kT} + 1)\,dx = (\sqrt{\pi}/2)e^{(E_F-E_C)/kT}$. For an *intrinsic semiconductor*, the Fermi level, E_F, is exactly in the middle of the band gap, E_g, so that $E_F - E_C = -E_g/2$. As a result, the electron density within the conduction band can be written as

$$n_{e,i} = \frac{1}{\sqrt{2}}\left(\frac{kT}{\pi\hbar^2}m_C\right)^{3/2} e^{-E_g/2kT}. \qquad (5.23)$$

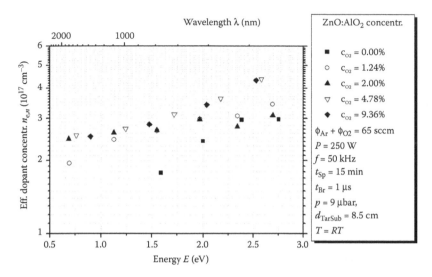

FIGURE 5.5 Effective electron density, $n_{e,n}$, of a semiconducting ZnO:Al thin film as a function of the incident optical energy, E. The influence of different reactive oxygen amounts within the inert argon gas is shown.

For an *n-doped semiconductor*, electron activation occurs from a donator level, E_D, into the conduction band, E_C. Thus, the energy difference $E_d = E_C - E_D$ has to be passed—instead of the whole band gap, E_g, as for intrinsic semiconductors, see Figure 5.5,

$$n_{e,n} = \frac{1}{\sqrt{2}} \left(\frac{kT}{\pi \hbar^2} m_C \right)^{3/2} e^{-E_d/2kT}. \tag{5.24}$$

Hydrogen model: The effective band gap, E_d, shall be determine by the means of the hydrogen model. Here, the balance between centripetal force and Coulomb force leads to the kinetic energy, $E_{kin,n} = m_0 v^2 /2 = q^2 /8\pi \varepsilon_0 r_n$. Taking the quantification of the angular momentum, $L_n = m_0 v r_n = n\hbar$, into account, leads to Bohrs radii $r_n = (4\pi\varepsilon_0 \hbar^2/m_0 q^2)n^2$.

For a hydrogen atom in the ground state, there is $n = 1$ and the Bohr radius accounts to $r_1 = 52.9$ pm. For $m \to \infty$, the electron is released. Thus, the ionization energy can be written $E_{ion} = E_{kin,n} - E_{kin,m} = m_0 q^4/32\pi^2\varepsilon_0^2\hbar^2$. Inserting all the common constants results in $E_{ion} = 13.6$ eV.

Solid-state approximation: The ionization energy is adequate for isolated hydrogen atoms, but not for semiconductor atoms, bound into a semiconductor lattice. In order to get the energy E_d from E_{ion}, the rest mass, m_0, has to be replaced by the effective mass of electrons within the conduction band, m_C, and the dielectric constant of the vacuum, ε_0, by that of the semiconductor, ε. This results in $E_d = m_C q^4/32\pi^2\varepsilon\hbar^2 = (\varepsilon_0/\varepsilon)^2 (m_C/m_0) E_{ion}$. Finally, the ionization energy, E_{ion}, is to be substituted by the work function, E_{WF}, of the semiconductor, leading to

$$E_d = \left(\frac{\varepsilon_0}{\varepsilon}\right)^2 \left(\frac{m_C}{m_0}\right) E_{WF,S},$$

(5.25)

$$E_{WF,S} \approx E_{WF,M} - \left(E_{F,S} - E_{F,M} + E_g/2\right).$$

The work function of a semiconductor may be estimated by means of a Schottky contact. The Fermi levels of the semiconductor, $E_{F,S}$, and the metal, $E_{F,M}$, as well as the work function of the metal, $E_{WF,M}$, shall be known or are easy to calculate. The dielectric constant, ε, the band gap energy, E_g, and the effective mass, m_C, can be evaluated; see above.

According to Ohms law, $j = \sigma_L E$, an electrical field, E, causes a current density, j, in a material with the conductivity, σ_L. The conductivity, $\sigma_L = \rho\mu$, is directly proportional to the mobility, μ, and the carrier density, $\rho = qn_e$, where $q = 1.602 \times 10^{-19}\, As$ is the elementary charge and $n_e = N/V$ the electron density, which is the number of electrons, N, per volume, V. The electrical field, E, moves the electrons with the velocity, $v_D = \mu E$, through the crystal, according to the mobility, μ. Therefore, Ohm's law can be expressed as $j = \sigma_L E = \rho\mu E = qn\mu E = qnv_D$, and the *mobility* can be written as

$$\mu = \sigma/q;$$

(5.26)

(see Figure 5.6.) Herein, the conductivity can be calculated with Equation 5.21 and the electron density with Equations 5.23 or 5.24, dependent on whether the semiconductor is intrinsic or n-doped.

According to the *Drude Model*, an electrical field, E, accelerates charges, q, until they are slowed down by interactions with crystal atoms or other electrons. This motion can be expressed by electrical and mechanical forces, that is

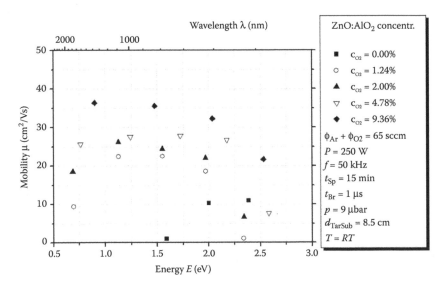

FIGURE 5.6 Effective mobilities, μ, of electrons within a semiconducting ZnO:Al thin film as a function of the incident optical energy, E. The influence of different reactive oxygen amounts within the inert argon gas is shown.

$F_e = qe = F_m = m_C \dot{v} + m_C v_D / \tau$. Herein, m_C is the effective mass of an electron, $\dot{v}$ its acceleration, τ the time between two interactions with crystal atoms, and v_D the average drift velocity. After acceleration from the resting position, an approximately steady motion can be assumed ($\dot{v} = 0$) with the *drift velocity* $v_D = qE\tau/m_C$. Taking into account Ohm's law $j = \sigma_L E = qnv_D$, again, leads to the *lifetime*,

$$\tau = m_C \sigma_L / n_e q^2. \tag{5.27}$$

The *mean free path* can be calculated with $l = v_D \tau = qE\tau^2/m_C$, if the electrical field, E, is known.

APPROXIMATE PARAMETER EXTRACTION FOR A MULTILAYERED SYSTEM

An electromagnetic wave with normal incidence upon a substrate will be reflected R_S, absorbed A_S, and transmitted T_S according to the *balance equation*

$$A_i = 1 - (R_i + T_i), \quad i \in \{S, SL, L\}. \tag{5.28}$$

This equation also holds for a layer upon a substrate *SL*, or a single layer *L*.

Reflections R_i and transmissions T_i can be measured, absorptions A_i have to be calculated. Regarding the absorption of a layer stack ($i = SL$), incident photons may be absorbed either by the layer ($i = L$) or by the substrate ($i = S$) [8,9]. Therefore, a second *balance equation for the absorptions*, $A_{SL} = A_S + A_L$, can be written as

$$A_L = A_{SL} - A_S = (R_S - R_{SL}) + (T_S - T_{SL}), \tag{5.29}$$

taking Equation 5.28 for a previously measured substrate and a total layer stack, $i \in \{S, SL\}$, into account. Combining the remaining balance equation, $A_L = 1 - (R_L + T_L)$, for an assumptive single layer and the balance equation for absorptions results in

$$R_L + T_L = 1 + (R_{SL} - R_S) + (T_{SL} - T_S) \tag{5.30}$$

The right-hand sides of Equations 5.29 and 5.30 can be measured directly, while the left-hand sides cannot. As all reflections within Equation 5.30 refer to the incident/reflection side of the sample and all transmissions belong to the transmission side of the sample, Equation 5.30 can be separated into two equations—one for the reflections and the other for the transmissions. For practical use in photovoltaic, transmissions through the substrate are high, that is, $T_S \to 100\% = 1$, and reflections low, $R_S \to 0$, so that Equation 5.30 results in

$$R_L = R_{SL} - R_S \xrightarrow[R_S \to 0]{} R_{SL},$$
$$T_L = T_{SL} + (1 - T_S) \xrightarrow[T_S \to 1]{} T_{SL}. \tag{5.31}$$

Glass substrates usually provide $T_S \geq 92\%$ and $R_S \leq 8\%$ over a wavelength region from about 300 to at least 2500 nm. The approximation with Equation 5.31 is negligible for almost all practical applications, as will be shown below. An iterative application of Equation 5.31 may be used to investigate *more-than-two-layer systems*. Therefore, a UV–Vis–NIR spectroscopic analysis has to be lone after adding a further layer. The arithmetic errors for parameter extraction increase with every added layer.

In the case of a thin film upon a substrate, reflection and transmission spectra of the substrate, R_S, T_S, and the layer/substrate stack, R_{LS}, T_{LS} have to be measured. Application of Equation 5.31 provides the (approximate) reflections and transmissions for the layer, R_L, T_L. In the case of just a single layer, R_L and T_L can be measured directly.

Now, using R_L and T_L, the refractive index quotient n_L/n_e can be evaluated exactly with Equation 5.14 ($\mu_e = \mu_L = \mu_0$), for transparent layers (TCOs) approximately with Equation 5.12, Equation 5.16 ($\alpha_L = 0$), for opaque layers with Equation 5.17, and for measurements without integrating spheres with Equation 5.16.

Once the refractive index quotient n_L/n_e is known, Equations 5.18 and 5.6 provide further and complex parameter extraction. Absorptions, A_L, and absorption coefficients, α_L, can be calculated directly with Equation 5.10 and 5.13. In the case of the absorption coefficients, the layer thicknesses, d_L, have to be calculated with Equation 5.19. Effective dopant concentrations can be calculated with Equations 5.23 through 5.25, mobilities and lifetimes with Equations 5.26 and 5.27.

QUANTUM MECHANICAL POTENTIAL BARRIER MODELS

SINGLE-LAYER MODEL

Incident photons with the wave functions $\psi_e = e^{-ik_{e,r}r}$, $r \leq -r_0$, will be partly reflected $\psi_r = \psi_{r,0}e^{+ik_{e,r}r}$, $r \leq -r_0$, by the potential barrier of a layer or transmitted $\psi_t = \psi_{t,0}e^{-ik_tr}$, $r \geq r_0$, through it; see Figure 5.7. The photon–current densities can be calculated with

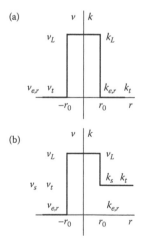

FIGURE 5.7   Potential barriers for (a) the SLM after the correction of the measured spectra with Equation 5.31 and (b) the KSM.

$j_i = i\hbar\left(\psi_i\vec{\nabla}\psi_i - \psi_i\vec{\nabla}\psi_i\right)\big/2m$ to $|j_i| = \hbar k_i \psi_{i,0}^2/m$, $i \in \{e, r, t\}$. Reflections and transmissions are then defined as

$$R = \left|\frac{j_r}{j_e}\right| = \psi_{r,0}^2, \quad T = \left|\frac{j_t}{j_e}\right| = \frac{k_t}{k_{e,r}}\psi_{t,0}^2, \tag{5.32}$$

where wave numbers within the same medium are identical, $k_{e,r} = k_t = k$.

Using the above-mentioned wave functions outside and $\psi_L = \psi_{L,+}e^{+ik_Lr} + \psi_{L,-}e^{-ik_Lr}$, $r < |r_0|$, within the layer, the boundary conditions for both surfaces of a single layer lead to a 4×4 equation system for the unknown amplitudes $\psi_{r,0}$, $\psi_{L,+}$, $\psi_{L,-}$ and $\psi_{t,0}$. In the case of transparent layers (TCO), this equation system can be solved using Equation 5.32, giving

$$R = \frac{\left[1 - \left(k_L/k\right)^2\right]^2 \sin^2\left(k_L d_L\right)}{4\left(k_L/k\right)^2 + \left[1 - \left(k_L/k\right)^2\right]^2 \sin^2\left(k_L d_L\right)},$$

$$T = \frac{4\left(k_L/k\right)^2}{4\left(k_L/k\right)^2 + \left[1 - \left(k_L/k\right)^2\right]^2 \sin^2\left(k_L d_L\right)}, \tag{5.33}$$

where $d_L = 2r_0$ is the layer thickness.

KERADEC–SWANEPOEL MODEL

The Keradec–Swanepoel Model (KSM) itself is given in [11,17]; the according quantum mechanical model is shown below. Here, the wave numbers within the transmission region are identical to those of the substrate, $k_t = k_S$. Thus, the wave functions of the photons can be written as

$$\psi = \begin{cases} \psi_e + \psi_r = e^{-ikr} + \sqrt{R}\,e^{ikr} & \forall -\infty < r \le -r_0, \\ \psi_L = \psi_{L,+}e^{ik_Lr} + \psi_{L,-}e^{-ik_Lr} & \forall -r_0 < r < +r_0, \\ \psi_S = \sqrt{kT/k_S}\,e^{-ik_Sr} & \forall +r_0 \le r < +\infty, \end{cases} \tag{5.34}$$

before the layer stack k, within the layer k_L and within the substrate k_S. Inserting these wave functions into the boundary conditions for the surface and the interface of this quasi-single-layer model (SLM)

$$\begin{aligned}
e^{ikr_0} + \sqrt{R}\,e^{-ikr_0} &= \psi_{L,+}e^{-ik_Lr_0} + \psi_{L,-}e^{ik_Lr_0} \\
\sqrt{kT/k_S}\,e^{-ik_Sr_0} &= \psi_{L,+}e^{ik_Lr_0} + \psi_{L,-}e^{-ik_Lr_0} \\
-ike^{ikr_0} + ik\sqrt{R}\,e^{-ikr_0} &= ik_L\psi_{L,+}e^{-ik_Lr_0} - ik_L\psi_{L,-}e^{ik_Lr_0} \\
-ik_S\sqrt{kT/k_S}\,e^{-ik_Sr_0} &= ik_L\psi_{L,+}e^{ik_Lr_0} - ik_L\psi_{L,-}e^{-ik_Lr_0}
\end{aligned} \tag{5.35}$$

leads to a 4×4 equation system with the four unknown amplitudes $\psi_{L,+}$, $\psi_{L,-}$, $\sqrt{R}$, and $\sqrt{kT/k_S}$. This equation system can be solved for TCO giving

$$R = \frac{2\gamma_3 \left(\dfrac{\gamma_2}{\gamma_1} - \dfrac{\gamma_1}{\gamma_2}\right)\left(\cos(2k_L d_L) + \cos(k\, d_L)\right)}{\gamma_2^2 + \dfrac{1}{\gamma_2^2} - 2\cos(4k_L d_L)}$$
$$+ \frac{\gamma_3^2 + \dfrac{\gamma_2^2}{\gamma_1^2} + \dfrac{\gamma_1^2}{\gamma_2^2} - 2\cos(4k_L d_L)}{\gamma_2^2 + \dfrac{1}{\gamma_2^2} - 2\cos(4k_L d_L)},$$

$$T = \frac{\gamma_6 \left(\gamma_4^2 + \gamma_5^2 + 2\gamma_4\gamma_5 \cos(2k_L d_L)\right)}{\gamma_2^2 + \dfrac{1}{\gamma_2^2} - 2\cos(4k_L d_L)},$$
(5.36)

where, again, the layer thickness can be written as $d_L = 2r_0$ and

$$\gamma_1 = \frac{k + k_L}{k - k_L}, \qquad\qquad \gamma_4 = \frac{2k\left(k_S^2 + k_L^2\right)}{k_S\left(k_S - k_L\right)\left(k - k_L\right)},$$

$$\gamma_2 = \gamma_1\left(\frac{k_S + k_L}{k_S - k_L}\right), \qquad\qquad \gamma_5 = \frac{4kk_L}{\left(k_S + k_L\right)\left(k + k_L\right)},$$
(5.37)

$$\gamma_3 = \gamma_1\left(1 - \frac{1}{\gamma_2}\right) + \frac{1}{\gamma_1}(1 + \gamma_2), \quad \gamma_6 = \frac{k_S}{k}.$$

The trigonometric functions in Equations 5.33 and 5.36 lead to the already mentioned *Fabry–Perot extrema*; see Figure 5.8.

CURRENT–VOLTAGE (I–V) MEASUREMENTS

IDEAL AND REAL *I–V* CHARACTERISTIC

Principally, solar cells are bipolar semiconductor diodes. Here, the optical sensitivity of the *pn*-junction is used to convert optical energy into electrical energy. The current–voltage characteristic (*I–V* characteristic) of conventional diodes is given by the Schockley equation [3,18,19,20]. For solar cells, the Schockley equation is to be supplemented with the luminescence current, I_{L0} (light current), as additional conducting electrons are generated by incident photons. Only photons, with natural multiples, x_{el}, of the band gap energies, E_g, contribute to this luminescence current. Energy excesses, $\Delta E_{th} = \hbar\omega - x_{el}E_g$, are converted in to heat. In accordance with the equivalent circuit of an ideal solar cell, the *ideal I–V equation* can be written as

$$I_0(V_0) = I_{s0}\left(e^{qV_0/kT} - 1\right) - I_{L0}$$
(5.38)

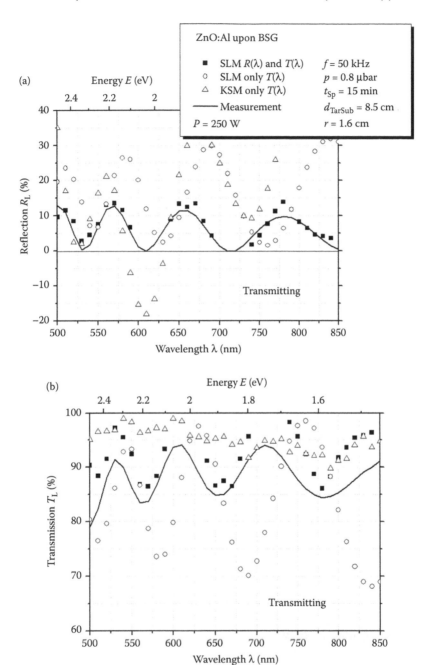

FIGURE 5.8 Measured and quantum mechanically evaluated (a) reflections, R_L, and (b) transmissions, T_L. Applied were the SLM and the KSM. The best conformance is given for the SLM using reflection and transmission spectra for parameter extraction.

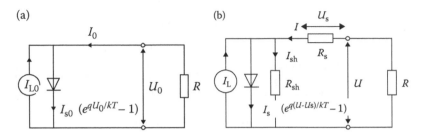

FIGURE 5.9 Equivalent circuits for (a) an ideal and (b) a real solar cell.

Herein, I_{s0} is the saturation current, q the charge of an electron, k the Boltzmann constant, and T the temperature.

For an equivalent circuit of a real solar cell, a serial R_s and a shunt R_{sh} resistance have to be added; the first one for ohmic losses within the used materials and at the electrical contacts and the second one for insulating losses and tunneling currents. Using the according mesh rules, compare Figure 5.9, results in $I(V) = I_{s0}\,(e^{q(V-V_s)/kT} - 1) - I_{L0} + I_{sh}$ with $V_s = R_s\,I$, $I_{sh} = (V - V_s)/R_{sh} = (V - R_s I)/R_{sh}$ and therefore in a *transcendent, real I–V equation*

$$I(V) = I_s\left(e^{q(V-R_s I(V))/kT} - 1\right) - I_L + \frac{V}{R_{sh} + R_s} \quad \underset{\substack{R_{sh}\to\infty \\ R_s\to 0}}{\longrightarrow} \quad \text{Equation 5.38,} \quad (5.39)$$

where $I_s = \left(R_{sh}/(R_{sh} + R_s)\right) I_{s0} \underset{\substack{R_{sh}\to\infty \\ R_s\to 0}}{\longrightarrow} I_{s0}$, $I_L = \left(R_{sh}/(R_{sh} + R_s)\right) I_{L0} \underset{\substack{R_{sh}\to\infty \\ R_s\to 0}}{\longrightarrow} I_{L0}$.

In practice, all currents, I, are related to the measured area, A, resulting in current densities, $j = I/A$.

The intersection point between the I–V curve, of an illuminated solar cell, and the current axis is called *short-circuit current*, I_{sc}, and is approximately equal to the *luminescence current*, I_L, see Figure 5.11. Setting $V = 0$ ($R \to 0$) results in a transcendent

equation for the current, $I(V = 0) = I_{sc} = -(I_L + I_s) + \dfrac{kT}{qR_s} W_{\text{Lambert}}\left(\dfrac{qI_s R_s}{kT}\, e^{q(I_L + I_s)R_s/kT}\right)$.

The according intersection point with the voltage axis is called the *open-circuit voltage*, V_{oc}. Setting $I = 0$ ($R \to \infty$) results in $V(I = 0) = V_{oc} = (I_L + I_s)(R_{sh} + R_s)$ $-(kT/q)W_{\text{Lambert}}((qI_s(R_{sh} + R_s)/kT)\,(e^{q(I_L + I_s)(R_{sh} + R_s)/kT}))$. Herein, the *Lambert W-function*, W_{Lambert}, is defined as the inverse function of $z = W_{\text{Lambert}}(z)\,e^{W_{\text{Lambert}}(z)}$; see Figure 5.10. Nevertheless, these both equations for the short-circuit current, I_{sc}, and the open-circuit voltage, V_{oc}, are just of *theoretical interest*.

Practically, they are taken from the intersection points between the measured or theoretical I–V curve, of an illuminated solar cell, and the current, respectively, on the voltage axis.

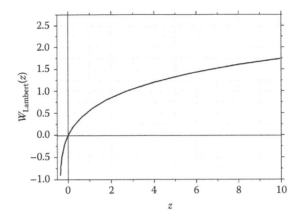

FIGURE 5.10 Lamberts W-function.

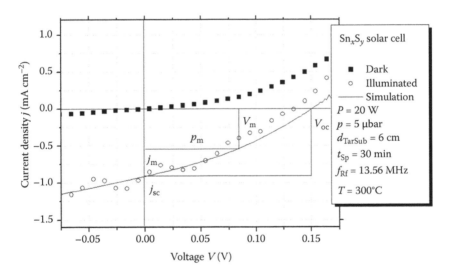

FIGURE 5.11 Comparison of measured and simulated j–V curves for a sputtered solar cell with a molybdenum ground contact, a tin-sulfide absorber layer, and a zinc-oxide TCO-layer (Keithley 2601 System Source Meter, LabTracer 2.0). The measurement was made with sunlight at an illuminance of $E = 91$ klux, an area of $A = 0.25$ cm^2, a *short-circuit current density* of $j_{sc} = 0.924$ mAcm^{-2}, and an *open-circuit voltage* of $U_{oc} = 129$ mV. The slope of the curve is $a_{sc} = 3.30$ mA/Vcm^2, at the position of the short-circuit current density, and $a_{oc} = 10.8$ mA/Vcm^2, at the open-circuit voltage. The charge of an electron is $q = 1.602 \times 10^{-19}$ As, the Boltzmann constant accounts $k = 1.38 \times 10^{-23}$ Ws/K, and the temperature is set to $T = 300$ K. For a given luminescence current density of $j_L = 0.925$ mAcm^{-2}, the saturation current density can be computed to $j_s = 1.34$ µA/cm^2, the serial resistance to $R_s = 1.04$ kΩ/cm^2 ($R_s = 260$ Ω), and the shunt resistance to $R_{sh} = 4.31$ kΩ/cm^2 ($R_{sh} = 1.08$ kΩ). The oscillation of the illuminated curve is due to oscillations of the illuminance (according to the frequency of the electricity network, taking the measuring velocity into account) or to the interference of electron wave functions within a layer of the solar cell. The legend contains the sputter parameters for the absorber thin film (see Table 5.2).

TABLE 5.2
Process Parameters for a Sputtered Solar Cell (Leybold Optics CLUSTEX 100M Sputtering Cluster Tool), with a Molybdenum (Mo) Ground Contact, a Tin-Sulfide (SnS) Absorber Layer, a Cadmium Sulfide (CdS) Buffer Layer, an Intrinsic ZnO:Al Intermediate Layer, and an Aluminum, *n*-Doped Zinc-Oxide (ZnO:Al) TCO Thin Film

Layer	d_{TarSub} (cm)	t_{Sp} (min)	p (μbar)
Mo	7.6	10	5
SnS	6	30	5
CdS	×	5	×
i-ZnOAl	8.5	3	5
n-ZnO:Al	8.5	10	3

T (°C)	P (W)	f (Hz)	t_{Br} (μs)	Deposition-Method
20	250	0	×	DC-sputtering
300	20	13.56 M	×	RF-sputtering
75	×	×	×	Wet-chemical
20	250	350 k	1	PDC-sputtering
20	250	50 k	1	PDC-sputtering

In order to calculate the *serial resistance* R_s and the *shunt resistance* R_{sh} from a measured *I–V* curve, the slopes a_{sc} and a_{oc} at the short-circuit current position and the open-circuit voltage position are to be measured. They are equal to

$$a_{sc} = \left(\frac{dI}{dV}\right)_{V=0} = \frac{qI_s}{kT}e^{-qR_sI_{sc}/kT} + \frac{1}{R_{sh}+R_s},$$

$$a_{oc} = \left(\frac{dI}{dV}\right)_{I=0} = \frac{qI_s}{kT}e^{qV_{oc}/kT} + \frac{1}{R_{sh}+R_s}.$$

(5.40)

This equation system includes no approximations and can be solved for R_s and R_{sh}

$$R_s = \frac{kT/qI_{sc}}{\ln\left[e^{qV_{oc}/kT} - \frac{kT}{qI_s}\left(a_{oc}-a_{sc}\right)\right]},$$

$$R_{sh} = \frac{2}{a_{oc}+a_{sc}-\frac{qI_s}{kT}\left(e^{qV_{oc}/kT}+e^{-qR_sI_{sc}/kT}\right)} - R_s.$$

(5.41)

For calculating these resistances, we just need the *saturation current* I_s. Therefore, we keep the light on (illuminated cell: $I_L \neq O$) and switch the voltage off (short-circuit condition: $V = 0$). Then, Equation 5.39, $I_{sc} = I_s (e^{-qR_s I_{sc}/kT} - 1) - I_L$, and Equation 5.41 result in

$$I_s = \frac{I_{sc} + I_L + \frac{kT}{q}(a_{oc} - a_{sc})}{e^{qV_{oc}/kT} - 1} \qquad (5.42)$$

without approximations. For a given electron charge, q, and temperature, T, the saturation current, I_s—and further the serial resistance, R_s, and the shunt resistance, R_{sh}—is only dependent on measurable values (V_{oc}, I_{sc}, a_{oc}, a_{sc}) and on the luminescence current, I_L. So, with a given luminescence current, we have got all physical values to calculate the real $I–V$ curve with Equation 5.39; see Figure 5.11.

MAXIMUM POWER RECTANGLE

According to Equation 5.39 the electrical power for a solar cell is

$$P = IV = I_s V \left(e^{(-R_s I)/kT} - 1 \right) V + \frac{V^2}{R_{sh} + R_s} \qquad (5.43)$$

Increasing the voltage, V (and the current, I), during a measurement of an illuminated cell, causes a switch of the algebraic sign from positive, via negative, to positive.

The cell consumes power, for positive signs—and it generates power, for negative signs. In order to calculate the maximum of the generated power, we have to differentiate the power with respect to the voltage, $dP/dV = 0$, and set the result to zero. This leads to the first equation, of a 2×2 equation system, for the current I_m and the voltage U_m; the second equation is given by Equation 5.39

$$I_m = \frac{U_m}{R_s} + \frac{kT}{qR_s} \ln\left(\frac{I_L + I_s - 2U_m/(R_{sh} + R_s)}{I_s(1 + qU_m/kT)} \right),$$

$$I_m = I_s\left(e^{q(U_m - R_s I_m)/kT} - 1 \right) - I_L + \frac{U_m}{R_{sh} + R_s}. \qquad (5.44)$$

Solving this equation system allows us to calculate a *just theoretical value for the maximum power*, $p_m = I_m U_m$. Practically, the measured or calculated $I–V$ curve is used to calculate an according power–voltage curve ($P–V$ curve). The minimum of this $P–V$ curve, a negative value, provides the *maximum power* P_m.

With an adequate load resistance, R, that is, with a fitting snubber circuit, the operating point of the cell can be placed at the maximum power point (U_m, I_m), what maximizes the energy generation.

The better the maximum power rectangle $P_m = I_m U_m$ fits into the power rectangle $P = I_{sc} U_{oc}$, which is built by the open-circuit voltage, U_{oc}, and the short-circuit current, I_{sc}, the higher is the *fill factor*

$$FF = \frac{P_m}{P} = \frac{I_m U_m}{I_{sc} U_{oc}}, \qquad (5.45)$$

TABLE 5.3

Optical Power Density, p_{solar}, of the Sunlight upon the Earth's Surface, Dependent on the Weather

p_{solar} (Wm^{-2})	Sunny	Cloudy	Overclouded
Summer	600–1000	300–600	100–300
Winter	300–500	150–300	50–150

Note: It also strongly depends on season and daytime—all listed values apply for noontime.

see Figure 5.11. Realistic fill factors are between 25% and about 94%. Very low fill factors are typical for nearly linear I–V curves and according to Equation 5.39, due to high serial resistances, R_s, or low shunt resistances, R_{sh}. High serial resistances may be provoked by inadequate materials or material combinations (Schottky contacts), which provide very low—charge mobilities, μ. On the contrary, low shunt resistances are due to insulation losses or tunneling currents.

In order to reach high *conversion efficiencies*,

$$= \frac{P_m}{P_{solar}} = \frac{I_m U_m}{P_{solar}} = \frac{FF I_{sc} U_{oc}}{P_{solar}}, \tag{5.46}$$

the three physical values of the numerator (fill factor, short-circuit current, and open-circuit voltage) have to be optimized, technologically. The solar power density can be assumed to be $p_{solar} \approx 1000$ Wm^{-2} for AM1.5g—which means at noontime on a bright summer day in Asia, Europe, North-America, or Australia—higher toward the equator, lower away from it. Moreover, the solar power density is dependent on the weather; see Table 5.3, Figure 5.12.

Solar Spectrum and *I–V* Curve

The earth's atmosphere is a complex optical filter for the solar spectrum. The smaller the incident angle, γ_S, between the sunbeam and the surface of the earth, the thicker is the effective atmosphere—the air mass, $AM = 1/\sin\gamma_s$—which an incident photon has to pass. For Asia, Europe, North America, or Australia at noontime in summer, the air mass accounts to about AM1.5, that is $\gamma_S \approx 41.8°$; see Table 5.4.

Sunlight beyond the earth's atmosphere possesses AM0. AM0 spectra are applied for solar cells on satellites. *Sunlight* on the earth's surface, *filtered by the atmosphere*, provides by definition AM1.5g. The spectrum is called AM1.5d, when reflections within the atmosphere are hidden from the detector. Figure 5.13 shows these three spectra as functions of the wavelength, λ, and the energy, E.

Generally, incident photons may be reflected by or transmitted through an absorber layer of a solar cell. These photons do not contribute to electron generation. Only absorbed photons do, according to *the absorption spectrum*,

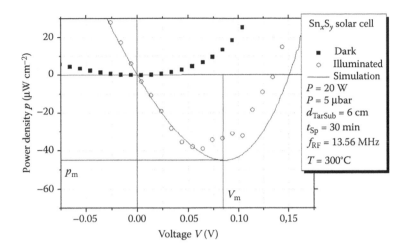

FIGURE 5.12   Power density as a function of the voltage; the minimum of the (simulated) curve provides a negative value, which indicates a maximum power density generation, p_m, of about 44.9 μW cm^{-2} at a voltage of $U_m = 84.1$ mV. Then, the maximum power rectangle leads to a maximum current density of $j_m = 0.543$ mA cm^{-2}, compare Figure 5.3. Taking a short-circuit current density of $j_{sc} = 0.924$ mA cm^{-2} and an open-circuit voltage of $U_{oc} = 129$ mV into account, results in a fill factor of about $FF = 38.3\%$. Relating the maximum power density, p_m, to the solar constant, $p_{solar} \approx 1000$ W m$^{-2} = 0.1$ W cm^{-2}, leads to a conversion efficiency of, $\eta = 45.7 \times 10^{-3}\%$.

TABLE 5.4

Solar Power Densities, p_{solar}, Depend on the Air Masses

Air Mass	AM0	AM1	AM1.5		AM2
γ_s (deg.)	×	0	41.8	≈45	60
[a]p_{solar} (W m^{-2})	1353	925	×	844	691
[b]p_{solar} (W m^{-2})	1367	×	1000	(943)	(770)

Air Mass	AM1.15	AM1.5	AM2	AM3	AM4
γ_s (deg.)	60.8	41.8	30	19.5	14.1
[c]Date	Jun. 21	Apr. 1	Mar. 2	Jan. 30	
		Sept. 12	Oct. 12	Nov. 13	Dec. 22
Season	Summer	Spring/Autumn			Winter

[a] Sze, physics of semiconductor devices [18].

[b] Irradiance-standard curve (ASTM 891/87), see Figure 5.1. The values within paranthesis were calculated using the formula, $AM = 1/\sin_s$, where γ_s is the angle between the sunbeam and the earth's surface.

[c] The air masses are strongly dependent on season and daytime. Season values for central Europe are shown.

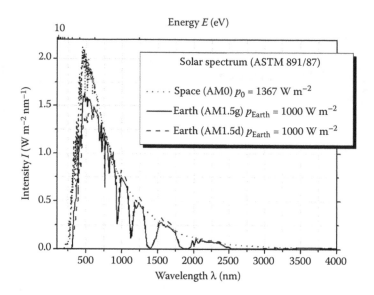

FIGURE 5.13 The air mass, AM, where sun rays have to pass on their way down to earth, influences the solar spectrum. Typically, the air mass values are calculated with AM = 1/sinγ_s; γ_s is the angle between the sunbeam and the earth's surface. AM0 represents the spectrum from outside the atmosphere.

$A(\lambda) = 1 - (R(\lambda) + T(\lambda)) < 1$, for wavelengths below λ_g or energies above the band gap energy, $E_g = hc / \lambda_g$, see [3,18,19,20]. Therefore, the *solar power per area on earth*, p_{earth}, accounts to

$$p_{earth} = \int_{\infty}^{0} I(\lambda)\, d\lambda,$$

$$p_{abs} = \int_{\infty}^{0} A(\lambda) I(\lambda)\, d\lambda,$$

(5.47)

where only the *power density*, p_{abs}, *is absorbed;* see Figure 5.14a. The difference, $p_{earth} - p_{abs}$, cannot be converted into electrical energy, due to reflections or transmissions. Generally, these formulas hold for any kind of spectrum, $I(\lambda)$.

Typical absorber materials provide band gap energies of about $E_g \approx 1.37$ eV (see Figure 5.14b). Only photons, with natural multiples, $x_{el} \in IN$, of these band gap energies, $E \approx x_{el} E_g$, contribute—ideally x_{el} electrons—to the luminescence current. Excess of energy , $\Delta E_{th} = E - x_{el} E_g$, is converted in to heat. Using the integer function, which sets all internal decimal places to zero, results in

$$x_{el}(E) = int(E / E_g),$$

$$x_{th}(E) = (E / E_g) - int(E / E_g).$$

$$x(E) = x_{el}(E) + x_{th}(E).$$

(5.48)

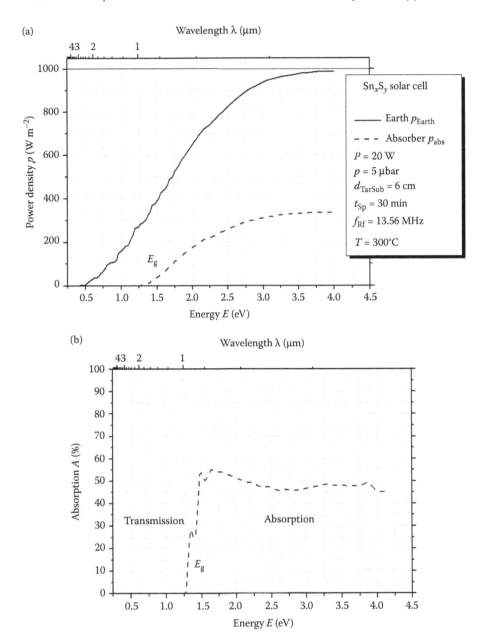

FIGURE 5.14 (a) Optical power density, p, as a function of the energy, E. The absorbed power density, p_{abs}, is proportional to the amount of energy, which is converted by the solar cell. (b) The absorption spectrum of the absorber layer, used within this solar cell. The legend contains the sputter parameters for the absorber thin film.

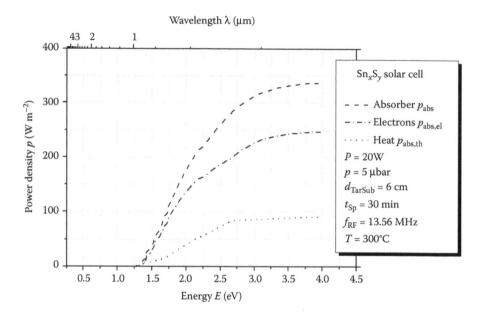

FIGURE 5.15 Optical power density, p, as a function of the energy, E. Shown are the totally absorbed power density, p_{abs}, the part of it, $p_{abs,el}$, which is responsible for electrical energy generation, and the part, $p_{abs,th}$, which is for sure transferred into heat.

So, the absorbed power density, p_{abs}, can be split into two parts: The part with the index el, which is *responsible for generating free electrons*, and the part with the index th, which is *exclusively transferred into thermal energy*,

$$p_{abs,el} = \int_{\infty}^{0} A(\lambda) I(\lambda) g_{el}(\lambda) \, d\lambda,$$

$$p_{abs,th} = \int_{\infty}^{0} A(\lambda) I(\lambda) g_{th}(\lambda) \, d\lambda,$$

(5.49)

where $g_{el}(E) = x_{el}(E)/x(E)$, $g_{th}(E) = x_{th}(E)/x(E)$, see Figure 5.15.

QUANTUM EFFICIENCY

The *number of photons*, $n_{ph,earth}$, *which reaches* in a certain time, t, a well-defined area, A, upon *the earth*, is given by differentiation of the according power density with respect to the energy of the incident light,

$$n_{ph,earth} = \frac{dp_{earth}}{dE},$$

$$n_{ph,abs} = \frac{dp_{abs}}{dE},$$

(5.50)

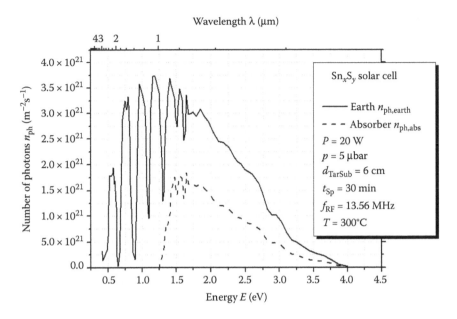

FIGURE 5.16   Number of photons per area and time; shown are those on the earth's surface, $n_{ph,earth}$, and those absorbed by the absorber thin film, $n_{ph,abs}$, as a function of energy, E, and wavelength, λ.

See Figure 5.16. An equal expression is given for the *number of absorbed photons*, $n_{ph,abs}$, per time and area. Again, $n_{ph,abs}$, can be split into a part, which is *responsible for free electron generation*, $n_{ph,abs,el}$, and a part, which *just produces thermal energy*, $n_{ph,abs,th}$.

$$n_{ph,abs,el} = \frac{dp_{abs,el}}{dE},$$

$$n_{ph,abs,th} = \frac{dp_{abs,th}}{dE}. \tag{5.51}$$

Integration over $n_{ph,Abs,el}$ leads—toward the high-energy end of the solar spectrum—to the number of photons, which potentially may generate free electrons; see Figure 5.17.

$$n_{ph,q_e}(E) = \sum_{E=0}^{\infty} n_{ph,abs,el}(E) \tag{5.52}$$

But not every photon, which is potentially able, generates a free electron, contributing to the charge transport. The quotient of the *number of actually*, per time and

(a)

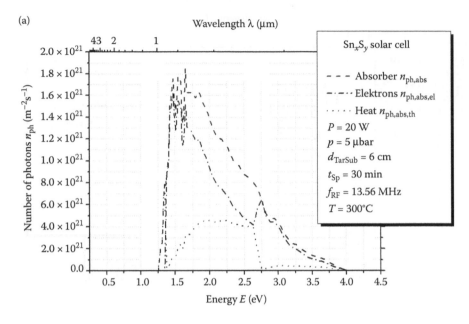

(b)

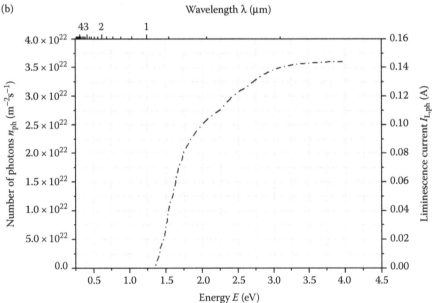

FIGURE 5.17 (a) Number of photons per area and time; shown are those absorbed by the absorber thin film, $n_{ph,abs}$, and those potentially able to generate electrical energy, $n_{ph,abs,el}$, and those, transferred into thermal energy, $n_{ph,abs,th}$, as a function of energy, $E = hc/\lambda$. (b) Integration over $n_{ph,abs,el}$ provides, at the high-energy end, the number of ideally generated electrons. The correspondent luminescence current, $I_{L,ph}$, is shown too.

area, *generated charges*, n_{q_e}, to the *number of incident photons*, n_{ph,q_e}, per time and area, is called the *quantum efficiency*,

$$Y = \frac{n_{q_e}}{n_{ph,q_e}} = \frac{q_e n_{q_e}}{q_e n_{ph,q_e}} = \frac{j_L}{j_{L,ph}}$$

$$= \frac{j_L A}{j_{L,ph} A} = \frac{I_L}{I_{L,ph}}. \tag{5.53}$$

Consecutively, the quotients of the *luminescence current densities*, j_L, $j_{L,ph}$, and the *luminescence currents*, I_L, $I_{L,ph}$, can be calculated. As the luminescence current, I_L, can be measured with *I–V* measurements from illuminated solar cells (intersection point with the current axis) and the luminescence current, $I_{L,ph}$, can be calculated using the solar spectrum and the absorption spectrum of the absorber material (see above), the quantum efficiency, Y, can be evaluated directly using Equation 5.53.

Moreover, the wavelength- or energy-dependent quantum efficiencies, $Y(\lambda)$ or $Y(E)$, can be calculated, using $n_{ph,abs,el}$ is the wavelength- respectively energy-dependent "number of absorbed photons per time and area that may generate free electrons."

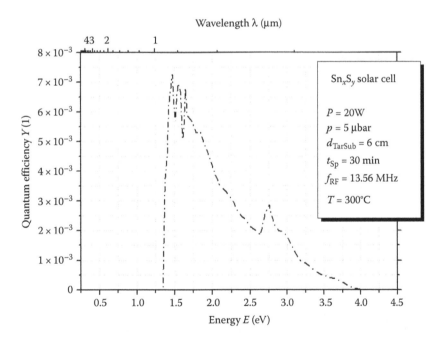

FIGURE 5.18 Energy-dependent quantum efficiency $Y(E)$ for a solar cell with sputtered Sn_xS_y absorber thin film (sputter parameters are shown within the legend). The measurements were made with the solar spectrum (AM1.5g). According to the I–V curve, the luminescence current, I_L, can be evaluated to $I_L = j_L A = 0.144\,\text{mA}$ and the luminescence current, $I_{L,ph}$, accounts to 0.145 A. So, the quantum efficiency can be calculated $Y = I_L/I_{L,ph} = 0.0993\%$, and the energy-dependent quantum efficiency $Y(E) = (n_{ph,Abs,el}(E)/n_{ph,q_e})Y = (n_{ph,Abs,el}(E)/3.60 \times 10^{22}\,m^{-2}s^{-1}) \times 99.3\,\text{m}\%$.

So, the *energy-dependent quantum efficiency*, see Figure 5.18, can be estimated using

$$Y(E) = \frac{n_{ph,abs,el}(E)}{n_{ph,q_e}} Y \tag{5.54}$$

REFERENCES

1. M. Born, *Optik*, 3. Aufl., ISBN 3-540-05954-7, Springer-Verlag, Berlin, 1985.
2. E. Hecht, *Optik*, 4. Aufl., ISBN 3-486-27359-0, Oldenbourg Verlag, München, 2005.
3. A. Stadler, *Analysen für Chalkogenid-Dünnschicht-Solarzellen*, 1. Aufl., ISBN 978-3-8348-0993-3, Vieweg + Teubner, 2010.
4. A. Schütz, *Einsatz der reflektometrischen Interferenzsektroskopie (RIfS) zur markierungsfreien Affinitätsdetektion für das Hochdurchsatzscreening*, Dissertation, Eberhard-Karls-Universität Tübingen, 2000.
5. S.A. Scheer, The design of a cost effective multi wavelength development rate monitoring tool, Thicknessdetection.com, 2005.
6. H.J. Muffler, *Umsetzung und Funktionsprinzip eines alternativen Material- und Abscheidekonzepts für Pufferschichten von Solarzellen*, Dissertation, Freie Universität Berlin, 2001.
7. J. Rostalski, *Der photovoltaisch aktive Bereich molekularer organischer Solarzellen*, Dissertation, Rheinisch-Westfälische Technische Hochschule Aachen, 1999.
8. H.H. Perkampus, *UV–VIS Spectroscopy and its Application*, ISBN 3-540-55421-1, Springer-Verlag, Berlin, 1992.
9. W. Gottwald, K.H. Heinrich, *UV/VIS-Spektroskopie für Anwender*, ISBN 3-527-28760-4, Wiley-VCH, Weinheim, 1998.
10. J. Szczyrbowski, Determination of optical constants of real thin films, *J. Phys. D: Appl. Phys.*, 11, 583, 1978.
11. R. Swanepoel, Determination of the thickness and optical constants of amorphous silicon, *J. Phys. E: Sci. Instrum.*, 16, 1214, 1983.
12. M. Kubinyi et al., Determination of the thickness and optical constants of thin films from transmission spectra, *Thin Solid Films*, 286, 164–169, 1996.
13. D. Poelman et al. Methods for the determination of the optical constants of thin films from single transmission measurements: A critical view, *J. Phys. D: Appl. Phys.* 36, 1850–1857, 2003.
14. A.N. Banerjee et al., Electro-optical characteristics and field-emission properties of reactive DC-sputtered p-CuAlO$_2$ + x thin films, *Elsevier Phys. B*, 370, 264–276, 2005.
15. X.W. Sun et al., Optical properties of epitaxially grown zinc oxide films on sapphire by pulsed laser deposition, *J. Appl. Phys.*, 86(1), 408–411, 1999.
16. V.A. Sokolov, Electrical conductivity of zinc oxide in atomic nitrogen, *Russian Phys. J.*, DOI 10.1007BF00890479, 1158–1159.
17. J. Keradec, Contribution a l'étude expérimental des alliages amorphes Ge$_x$Te$_{18x}$ mesures electriques e photoélectriques, Thesis L'Université Scientifique et Médicale de Grenoble, 1973.
18. N. Sze, *Physics of Semiconductor Devices*, ISBN 0-471-05661-8, John Wiley & Sons, Inc., 1981.
19. N. Sze, *Physics of Semiconductor Devices*, ISBN-13 978-0-471-14323-9, John Wiley & Sons, Inc., 2007.
20. D.K. Schroder, *Semiconductor Material and Device Characterization*, ISBN-13 978-0-471-73906-7, John Wiley & Sons, Inc., 2006.

6 Sensing Applications Using Photoacoustic Spectroscopy

Ellen L. Holthoff and Paul M. Pellegrino

CONTENTS

INTRODUCTION

In recent years, photoacoustic spectroscopy (PAS) has emerged as an attractive and powerful technique well suited for sensing applications. The development of high-power radiation sources and more sophisticated electronics, including sensitive microphones and digital lock-in amplifiers, has allowed for significant advances in PAS. Recent research suggests that PAS is capable of trace gas detection at parts-per-trillion (ppt) levels [1,2]. Furthermore, photoacoustic (PA) detection of infrared absorption spectra using modern tunable lasers offers several advantages, including simultaneous detection and discrimination of numerous molecules of interest.

Successful applications of PAS in gases and condensed matter have made this a notable technique and it is now studied and applied by scientists and engineers in a variety of disciplines. Therefore, a substantial body of literature on PAS exists today. The following discussion summarizes PAS and the experimental components and arrangements for PA detection, as well as its use in the past 15 years for sensing applications. PA sensing, specifically laser-based PA sensing schemes, of gas, liquid, and solid samples is reviewed, along with standoff detection methods. It is not our intention to discuss PA theory in great detail, as this has been presented in numerous publications on the topic. Instead, we hope to provide the reader with a general understanding of PAS as it applies to sensing.

FUNDAMENTALS OF PHOTOACOUSTICS

PHOTOTHERMAL PHENOMENA

One should always begin discussion on photoacoustic or optoacoustic spectroscopy with a more general discussion on the phenomenon of spectroscopy, namely photothermal spectroscopy. Photothermal spectroscopy encompasses a group of highly sensitive methods that can be used to detect trace levels of optical absorption and subsequent thermal perturbations of the sample in gas, liquid, or solid phases. The underlying principle that connects these various spectroscopic methods is the measurement of physical changes (i.e., temperature, density, or pressure) as a result of a photo-induced change in the thermal state of the sample. In general, most scientists classify photothermal methods as indirect methods for detection of trace optical absorbance, because the transmission of the light used to excite the sample is not measured directly. On closer inspection, a counter position asserting that these techniques may be a more direct measure of optical absorption could be appropriate due to its sole dependence on optical absorption and its immunity to optical scattering and reflections. Examples of photothermal techniques include photothermal interferometry, photothermal lensing, photothermal deflection, and photoacoustic spectroscopy. All photothermal processes consist of several linked steps that result in a change of the state of the sample. In general, the sample undergoes an optical excitation, which can take various forms of radiation, including laser radiation. This radiation is absorbed by the sample placing it in an excited state (i.e., increased internal energy). Some portion of this energy decays from the excited state in a nonradiative fashion. This increase in local energy results in a temperature change in the sample or the coupling fluid (e.g., air). The increase in temperature can result in a density change and, if it occurs at a faster rate than the sample or coupling fluid can expand or contract, the temperature change will result in a pressure change. Figure 6.1 pictorially shows the process associated with photothermal phenomena. As mentioned above, all photothermal methods attempt to key in on the changes in the thermal state of the sample by measuring the index of refraction change as with photothermal interferometry, photothermal lensing, and, photothermal deflection; temperature change as with photothermal calorimetry and photothermal radiometry; or pressure change as with photoacoustic spectroscopy [3].

PHOTOACOUSTIC SPECTROSCOPY

In order to generate acoustic waves in a sample, periodic heating and cooling of the sample is required to produce pressure fluctuations. This is accomplished using modulated or pulsed excitation sources [4–6]. The pressure waves detected in PAS are generated directly by the absorbed fraction of the modulated or pulsed excitation

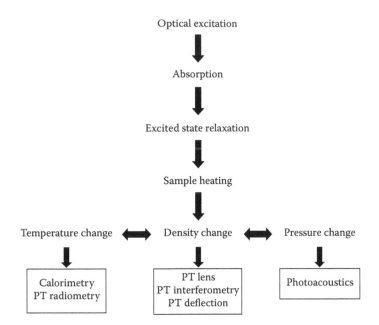

FIGURE 6.1 The basic process for signal generation with photothermal spectroscopy. Following absorption of radiation of the appropriate wavelength, the sample undergoes nonradiative excited-state relaxation resulting in temperature, density, and pressure changes. (Adapted from Bialkowski, S. E., *Photothermal Spectroscopy Methods for Chemical Analysis*, Vol. 134, John Wiley & Sons, New York, 1996.)

beam. Therefore, the signal generated from a PA experiment is directly proportional to the absorbed incident power. However, depending on the type of excitation source (i.e., modulated or pulsed), the relationship between the generated acoustic signal and the absorbed power at a given wavelength will differ [7]. The theory of PA signal generation and detection has been extensively outlined in the literature [3,8–11] and will not be discussed here.

The oldest application of photoacoustics dates back to Alexander Graham Bell's photophone ca. 1880 [12]. This communication device used Bell's discovery of the PA effect when he filled a transparent tube with various materials and modulated the light impinging on this tube. The effect saw little activity during this period until the advent of the laser with a notable exception of Viengerov's study of absorption in gases in the late 1930s, which represents the first example of PAS [13]. The first studies involving lasers were performed by Kerr and Atwood [14], but the technique gained more popularity when Kreuzer detected methane (CH_4) and ammonia (NH_3) by laser excitation at parts-per-billion (ppb) and sub-ppb levels, respectively [15].

Early work by Kreuzer [8,15–18] demonstrated the analytic power of PAS in gases, but the technique can also be applied readily to liquids and solids. PAS in these phases can be accomplished using both a direct and indirect coupling method. Direct coupling is the most straightforward and, as the name implies, the acoustic wave generated in the sample is detected by a transducer in direct contact with the solid or liquid sample. Since the acoustic wave generated in the sample never crosses a high-impedance interface, it is easily detected by the transducer. In comparison, indirect coupling methods are not as straightforward. In fact, this method was demonstrated at the onset of photoacoustics by Bell, but was not rediscovered until Parker noted increased signal contributions from his cell windows in his gaseous PAS experiments [19]. One possible explanation for its elusiveness may have been the fact that the original acoustic wave is not the origin of the main signal in an indirect PAS method. Usually, there is a large acoustical impedance at the sample–fluid interface such that most acoustic energy will be reflected back into the sample. The indirect PA detection of liquids or solids relies on the gas-coupling method and is explained clearly by Rosencwaig as the gas–piston model [10]. In this model the periodic heating of the sample surface occurs within a diffusion length of the surface and this thermal wave is responsible for the subsequent heating of the layer of gas directly above the surface (diffusion length in the gas). The periodic expansion in this gas layer produces an acoustic wave that can be detected using standard gas-phase microphones.

EXPERIMENTAL ARRANGEMENTS FOR PA DETECTION

In comparison with other photothermal techniques, which measure the changes in refractive index or temperature using combinations of probe sources and detectors, PAS measures the pressure wave produced by sample heating. Although the basic experimental embodiment of PAS can take many forms, several key elements are constants. This typical setup has been discussed in the literature [6,11,20–22]. Figure 6.2 shows the main components of a PAS apparatus. An excitation source, usually a laser or filtered lamp, is either modulated or pulsed and directed at a sample cell. The resultant pressure wave, which is created due to sample heating, is detected by a pressure transducer, in most cases a microphone, of the appropriate frequency response. The signal generated by the microphone is proportional to the amplitude

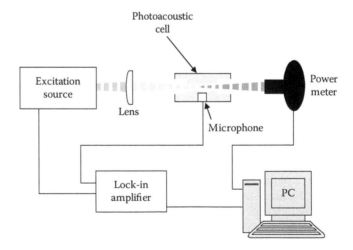

FIGURE 6.2 Simplified diagram of a typical photoacoustic sensor system with microphone detection. Electronic connections are indicated by a solid line.

of the pressure wave, but other information is contained in the phase and delay of the wave as well. This information is captured either by lock-in amplification or by use of a gated accumulation, such as a boxcar amplifier. Sample cells can take the form of a simple sealed tube or a complex resonant chamber or multipass cell (see Section "Acoustic Resonators"). Resonant PAS has gained favor due to the possible increase in signal using simple acoustic resonators with modulated excitation at the matched acoustic frequency. In the example shown in Figure 6.2, a personal computer is used to read and record the voltage outputs from the lock-in amplifier and a power meter is used to measure the transmitted laser power, which allows for normalization of the PA signal for any residual drift associated with the excitation source. Different adaptations of this basic scheme have been presented in the literature. The main features are summarized in the following subsections.

Radiation Sources

The first step in a PA experiment is the introduction of light. The main criterion for the starting point of PAS is that the molecules of interest need to be excited electronically, vibrationally, or rotationally. In the broadest sense, this light can be at many different wavelengths and even span large differences in imparted energy (e.g., ultraviolet through the infrared). Once excited, numerous pathways exist for the energy to dissipate, but a preponderance of the light energy is removed through a nonradiative pathway or through sample heating, which is the basis for all photothermal phenomena, including PAS. Light sources can take various forms depending on the application at hand. A selection of the common examples currently used for PA sensing is briefly outlined in the following paragraphs.

There are two main categories of light sources used for PAS: broadband sources, such as lamps, and narrow band laser sources. The general category of lamps includes the following: arc lamps, filament lamps, and glow bars. These sources were some of the first sources used to study the phenomenon of PAS [13] and have several advantages and disadvantages. The broadband output of these sources can be significant (e.g., ultraviolet to infrared) and in most cases can potentially cover all regions of interest for optical excitation and subsequent PA examinations. These sources are generally inexpensive and, depending on the overall wattage required, can be somewhat compact in size. Unfortunately, lamp sources also have low spectral brightness, require spectral selection through the use of filters or monochromators, and are usually restricted to low source modulation frequency and optical efficiencies. Despite the advent of the laser, lamp-based PAS is still a technique that is used with some regularity. The combination of PA measurements with Fourier transform infrared (FTIR) spectrometers is proof of this viability. In fact, PA attachments have become a standard part of FTIR instrumentation. It is also interesting to note that one of the few commercially available PA instruments is based on a lamp-filter architecture [23].

Although lamp–based PAS is still common, modern PAS research has been mainly carried out using laser sources. The remainder of this section focuses on these sources and their various formats in more recent PA examinations. As with lamps, lasers have numerous advantages, but also present some disadvantages. Some of the key advantages of these sources are their large spectral brightness, collimated output, ease of modulation, and narrow spectral linewidth. Disadvantages include expense and limited tunability; however, these are not a generic quality of every laser architecture. As mentioned previously, PAS can be performed as a pulsed or modulated measurement with respect to the light excitation. That allowance is seen vividly in the sources used for PA experimentation. Early work in the detection of gases used a variety of these sources, including pulsed and continuous wave (CW) dye laser sources [24,25] and CW laser sources, such as grating-tunable carbon monoxide (CO) and carbon dioxide (CO_2) [17] and helium–neon He–Ne [15] lasers. These sources usually provided reasonable or even high power levels, some limited tunability, and were based in the near-infrared or infrared wavelength regions. All these features allowed for some level of spectroscopic studies to be performed. This type of activity continued for years, but as laser technology matured, examinations using other types of lasers became more prevalent. Several PA studies involving liquid and solid samples, which require more laser irradiation, used solid-state sources such as neodymium-doped yttrium aluminum garnet (Nd:YAG) lasers. The harmonics or pumping of an optical parametric oscillator enables tuning of these sources. Semiconductor lasers based off of a direct bandgap transition with various feedback mechanisms (e.g., distributed Bragg reflectors, distributed feedback reflectors) were used for numerous studies, especially with regard to the study of atmospheric or small molecular gas targets that could be identified with tuning ranges from fractions to single-integer per centimeter (see Table 6.1). This type of tuning was easily accomplished by current and/or temperature tuning of the laser diode. Occasionally, efforts used other laser sources such as lead-salt diode lasers, which are centered in the infrared and theoretically can be tuned through mode-hopping over a larger spectral band; however, these sources were plagued by cryogenic cooling requirements and low output powers (e.g., 0.1 mW typically) [26].

In 1994, the introduction of the quantum cascade laser (QCL) changed the prospects of laser photoacoustics and, in general, infrared spectroscopy. In that year, Bell Labs first demonstrated the QCL as a new infrared laser source [27]. Since that time, continuing and aggressive evolution has been occurring. This type of laser was a complete departure from its semiconductor laser cousins. In this type of laser architecture, the sole use of material combinations to achieve changes in the bandgap, and thus emitting wavelengths, is abandoned for a more elegant approach of creating intersubband levels through the structuring of the active region with ultrathin layers known as quantum wells. Since these levels are created using methods for precise layering and solely involve movement of the electron in the active region, it can be recycled and injected into the next active region allowing the laser architecture the ability to have a "cascade" effect and produce numerous photons per electron [28,29]. Figure 6.3 provides a simplistic graphical picture of such a cascade structure.

There have been numerous government programs run by different agencies (e.g., Defense Advanced Research Projects Agency) that have addressed many aspects of QCL performance, including thermal management, overall optical efficiency, beam handling, and other advances in gain media production. This funding has provided investment to various researchers and companies to increase the QCL's capability and availability during this time period. In and around 2005, another breakthrough

TABLE 6.1

Examples of Photoacoustic Studies on Gases

Laser	Laser Power	Spectral Region	Analyte	Detector	Detection Limit	Reference
Nd:YAG	—	355 nm, 532 nm	H_2	Mic	200 ppm	[117]
CO_2	—	9–11 μm	CO_2 isotopes	Mic	ppb	[118]
EC-diode	2 mW	1125 nm	H_2O vapor	Mic	13 ppm	[119]
Diode	1 mW	1.67 μm	C_6H_6, $C_6H_5CH_3$, $C_6H_4(CH_3)_2$	Mic	70 ppb, 100 ppb, 160 ppb	[120]
Diode	1 mW	1.67 μm	$C_6H_5CH_3$, C_6H_6	Mic	1.1 ppm, 0.35 ppm	[121]
QCL	few mW	9.4 μm	C_2H_6O	Mic	1 ppm	[122]
CO_2	1.4 W	10.55 μm	SF_6	Mic	3.5 ppb	[123]
CO_2	6 W	10.58 μm	SF_6	Mic	65 ppt	[65]
CO_2	2 W	9.22 μm	NH_3	Mic	220 ppt	[124]
Argon	35 mW	Visible	NO_2	Mic	50–130 ppb	[125]
Q-switched Nd:YAG	≤40 mW	532 nm	NO_2	Mic	50 ppb	[126]
Diode	2 mW	1.65 μm	CH_4	Mic	0.3 ppm	[127]
DFB diode	38 mW	1.53 μm	NH_3	TF	0.65 ppm	[128]
DFB ICL	3.4 mW	3.53 μm	H_2CO	TF	0.6 ppm	[128]
DFB QCL	8 mW	5.3 μm	NO	Mic	500 ppb	[129]
QCL	19 mW	4.55 μm	N_2O	TF	4 ppb	[35]
Diode	20–30 mW	1531.7 nm	NH_3	Mic	120 ppb	[130]
Diode w/EDFA	750 mW	1532 nm	NH_3	Mic	3–6 ppb	[131]
Diode	18 mW, 22 mW, 16 mW	1651 nm, 1368.6 nm, 1737.9 nm	CH_4, H_2O Vapor, HCl	Piezo	80 ppb, 24 ppb, 30 ppb	[132]
DFB diode	mW	1370 nm, 1740 nm	H_2O, HCl	Mic	40 ppb, 60 ppb	[133]
DFB diode, VCSEL	30 mW, 0.5 mW	760 nm	O_2	CL	20 ppm, 5 ppt	[134]
Diode-pumped Nd:YAG	~70 mW	2.76–2.91 μm	N_2O	Mic	~313 ppb	[135]
QCL	25 mW	7.9 μm	CH_4	Mic	3 ppb	[136]
DFB QCL	2 mW, 5 mW	6.2 μm, 8 μm	NO_2, N_2O	Mic	80–100 ppb	[37]
DFB laser	3.4 mW	2–2.5 μm	NH_3	Mic	sub–ppm	[137]
Q-switched Nd:YAG	126 mW	2 μm	CO_2	Ultrasonic sensor	$\alpha_{min} = 3.3 \times 10^{-9}$	[135]

continued

TABLE 6.1 (continued)

Examples of Photoacoustic Studies on Gases

Laser	Laser Power	Spectral Region	Analyte	Detector	Detection Limit	Reference
DFB laser	8 mW	1648–1652 nm	CH_4	Mic	sub-ppm	[138]
Diode	40 mW	1574.5 nm	H_2S	Mic	0.5 ppm	[139]
QCL	5.3 mW	8.41 µm	Freon 134a	TF	$\alpha_{min} = 2.0 \times 10^{-8}$	[140]
CO_2	1 W	9–11 µm	O_3	Mic	5 ppb	[141]
DFB QCL	8 mW	8 µm	CH_4	Mic	34 ppb	[142]
DFB diode	30 mW	1572 nm	CO_2	CL	1.9 ppm	[92]
DFB QCL	1 mW	4.3 µm	CO_2	Mic	0.023% vol	[143]
DFB diode	6.2 mW	2.0 µm	CO_2, NH_3	TF	18 ppm, 3 ppm	[144]
Diode w/EDFA	1.17–1.89 W	1.53 µm	C_2H_2	Mic, CL	440 ppb, 14.5 ppb	[94]
DFB laser	16 mW	1738.9 nm	HCl	Mic	sub-ppm	[145]
DFB diode	8 mW	1.65 µm	CH_4	TF	$\alpha_{min} = 1.0 \times 10^{-8}$	[97]
DFB diode	1.5 mW	2.7 µm	CO_2	Mic	30 ppb	[146]
DFB diode	14 mW	1.53 µm	C_2H_2	CL	$\alpha_{min} = 1.2 \times 10^{-7}$	[147]
CO_2	10 W	9.22 µm	NH_3	Mic	—	[148]
Nd:YAG		266 nm	O_3	Mic	10 ppb	[149]
DFB QCL	4 mW	5.6 µm	CH_2O	Mic	150 ppb	[150]
DFB diode	8 mW	1.396 µm	H_2O vapor	TF	$\alpha_{min} = 1.68 \times 10^{-8}$	[151]
DFB diode	15 mW	1.62 µm	C_2H_4	TF	0.3–4 ppm	[152]
TEDFL w/EDFA	500 mW	1531.7 nm	NH_3	Mic	3 ppb	[153]
EC-QCL	~250 mW	5.26 µm	NO	Mic, TF	60 ppb, 15 ppb	[154]
DFB diode	46 mW	1.53 µm	C_2H_2	TF	$\alpha_{min} = 3.3 \times 10^{-9}$	[82]
Diode	3 mW	1450 nm	H_2O vapor	Mic	—	[155]
DFB diode	25 mW	1.53 µm	NH_3	TF	60 ppb	[156]
DFB diode	20 mW	1371 nm	H_2O vapor	Mic	80 ppm	[157]
EC-QCL	100 mW	5.26 µm	NO	TF	15 ppb	[158]

Note: CL, cantilever; EC, external cavity; EDFA, erbium-doped fiber amplifier; DFB, distributed feedback reflector; ICL, interband cascade laser; Mic, microphone; Nd:YAG, neodymium-doped yttrium aluminum garnet; QCL, quantum cascade laser; TF, tuning fork; TEDFL, tunable erbium-doped fiber laser; VCSEL, vertical-cavity surface-emitting.

α_{min} units are $cm^{-1} W\sqrt{Hz}$.

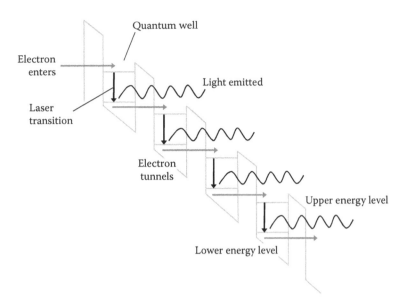

FIGURE 6.3 Simplified depiction of subband transitions within a quantum cascade struc-ture. Elections emit a cascade of photons as they undergo subband transitions while passing through a stack of quantum wells. (Adapted from Hecht, J., *Laser Focus World* 2010, 45–49.)

was realized by one of the originators of the QCL at Bell Labs, Federico Capasso [30]. His group, now at Harvard, demonstrated growth of the QCL commercial semi-conductor fabrication that is generally accomplished with a metal organic vapor pro-cess as opposed to the molecular beam epitaxy growth found previously. Finally, the research into these sources has experienced steady escalation, as can be seen with the increase in publications dedicated to QCL sources: 2006 (350 publications), 2007 (478), 2008 (551), 2009 (581). New developments are taking place at a ferocious pace with QCL arrays being one of the newer breakthroughs [31].

The QCL has matured to a level at which numerous companies can produce gain material for laser systems both in the United States (e.g., Adtech Optics, Maxion Technologies, Inc.) and abroad (e.g., Alpes Lasers, Hamamatsu). Along with this production, several companies have produced laser systems that are suitable for spectroscopic purposes (e.g., Daylight Solutions, Inc., Pranalytica, Inc., and Aples Lasers, Inc.). The current state of the art for external-cavity (EC) grating-tunable QCL systems has been shown by Daylight Solutions, Inc. (San Diego, California), which has demonstrated routine tuning of a single-gain element QCL system produc-ing 20% center bandwidth tuning. The current laser systems operating in our labora-tory produce emissions from 1015 to 1240 cm^{-1} (9.85–8.06 μm) (a 225 cm^{-1} span) and from 1351 to 1612 cm^{-1} (7.4–6.2 μm) (a 260 cm^{-1} span). The company has also demonstrated increased capability upward of 349 and 500 cm^{-1} tuning spans using a single element in the 8–12 μm and the 3–5 μm regions, respectively. Spectroscopically, these sources could have dramatic impact on PA research due to their wide tuning over pertinent regions in the infrared. The resolution of well-constructed systems can tune continuously and without mode-hoping over the whole tuning band with a nominal resolution of ~1 cm^{-1}. This resolution is a compromise associated with the output format, which is normally restricted to low-duty cycle pulsed or CW operation. This restriction is directly associated with the inefficiency of the laser source (<10%) and its thermal effect on the feedback mechanisms. The localized heating of the element will change output wavelength and "pull" the laser to hop unless it is pulsed short enough to avoid large heating or operated in a continuous fashion, which establishes thermal equilibrium. Power output of spectroscopic sources has generally been moderate; 10 s of milliwatt average power and of the order of 100 s of milliwatt peak pulsed power. This power has the potential to scale upward, but most scaling has been done with nonspectroscopic sources, and so the ultimate power potential is unclear.

PA sensing capability employing QCLs was identified early on and demonstrations by Paldus et al. [32] using these sources can be seen as early as in 1999. Although QCLs took years to evolve into their current state, work continued on PA studies using these sources throughout this development cycle [33–42]. Recently, we demonstrated that widely tunable QCL sources can achieve full spectroscopic discrimination of gas analytes, due to their large tuning ranges (see Section "Gaseous Samples") [43]. Furthermore, these sources, operating in low-duty cycles, have demonstrated that PAS based on lock-in amplification can still be performed and indeed shows great promise.

Acoustic Resonators

An essential element of a PA sensor is the cell, which serves as a container for the sample as well as the detector. Therefore, optimum design of a PA cell is necessary to facilitate signal generation and detection. To date, a variety of cell configurations have been reported for solid, liquid, and gas samples, including cells operated at acoustically resonant and nonresonant modes, single- and multipass cells, and cells for intracavity operation [1,11,44–51].

As is often the case for trace gas sensing, the detection sensitivity is limited by the signal-to-noise ratio (SNR). High sensitivity can be achieved in nonresonant gas cells; however, noise sources (e.g., amplifier noise, external acoustic noise) show a characteristic $1/f$ frequency dependence resulting in a small SNR. Furthermore, light absorption at the cell windows and walls results in a background signal, which is difficult to separate from the PA signal generated by the gas sample itself [45]. An improvement in the SNR of the cell can be achieved with cell design modifications. Stray reflections can be reduced using Brewster windows [52], and acoustic baffles [53] and "windowless" [54] or open cells [55] will minimize the effect of window heating noise. Meyer and Sigrist [56] modified the "windowless" PA cell by adding a buffer volume on either side of the central cylinder. In this design, both the laser beam and the gas sample flow enter and exit the cell at nodal positions of the cell's operating first radial mode. With this configuration, the authors minimized flow noise and allowed for gas flow rates of up to 1 L/min without a decrease in the SNR.

Applying higher modulation frequencies and acoustic amplification of the PA signal will also result in an improved SNR. When the modulation or pulse frequency is the same as an acoustic resonance frequency of the PA cell, the resonant eigenmodes (i.e., acoustic modes) of the cell can be excited, resulting in an amplification of the signal [57–60]. The resonance frequencies are dependent on the shape and size of the PA cell. The most commonly used resonator is the cylinder. Figure 6.4 illustrates the longitudinal, azimuthal, and radial eigenmodes that occur in a cylindrical cell. The theoretical determination of the corresponding eigenfrequencies of these modes has been discussed in detail elsewhere [61]. In order to achieve signal amplification, resonant PA cells operating on longitudinal, azimuthal, or radial resonances have been developed. Furthermore, Helmholtz resonators are widely used. A general description of these acoustic resonators has been discussed elsewhere [45]. Cells have also been designed for multipass or intracavity operation [44,62,63]. In the following paragraphs, some selected examples of PA cell designs are discussed in more detail.

The sensitivity of a PA sensor is strongly dependent on the geometry of the PA cell and the pressure distribution in the cell must be understood for optimization. Various PA cell modeling approaches have been investigated for the systematic optimization of PA cells. Bijnen et al. [1] examined a PA cell configuration similar to the design introduced by Meyer and Sigrist [56]. The authors applied acoustic transmission line theory [64] pertaining to experimental results from a cylindrical resonant PA cell excited in the first longitudinal mode. Numerous approaches to optimize a PA cell for trace gas detection were investigated, resulting in suggested parameters for the construction of a small and sensitive resonant cell with a fast response time. The criteria considered for the geometry of the cell were dependent on various background and noise sources (e.g., window absorption signal, AR window reflection, absorption of radiation at resonator wall, chopper noise, gas sample flow noise, laboratory noise), which lower the detection limit. The authors reported the relationship between buffer length (l_{buf}) and resonator length (l_{res}) for optimal window signal suppression; $l_{res} = 2\, l_{buf}$ ($l_{res} \gg r_{res}$), where r_{res} is the radius of the resonator. A larger gas absorption signal is achieved for a resonator having a small radius; however, the limit for r_{res} is mainly determined by the wings of the laser's Gaussian beam profile impinging on the cell wall. Therefore, r_{res} should be three times the beam waist. On the basis of theoretical results of acoustic response upon varying buffer parameters, optimal buffer length and radius (r_{buf}) were determined. The optimal buffer length

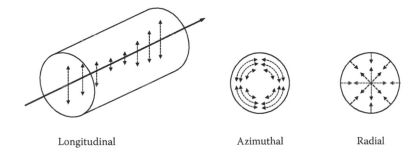

Longitudinal   Azimuthal   Radial

FIGURE 6.4 Longitudinal, azimuthal, and radial acoustic modes of a cylindrical resonator. (Adapted from Miklòs, A., Hess, P., and Bozoki, Z., *Rev. Sci. Instrum.* 2001, 72(4), 1937–1955.)

was reported to be $\lambda_a/4$, where λ_a is the wavelength of the acoustical wave. The radii of the buffer volumes should be deduced from $r_{buf} \approx 3r_{res}$. Further reduction of window signal was accomplished by positioning tunable air columns close to the windows, which provide constructive interference with noise signals originating from the windows. The length of these columns was best determined when the cell was operated in the experimental setup. The authors achieved a fast time response by placing the gas sample entry at the center of the resonator. Several notch filters were used in line to minimize gas flow noise.

In conjunction with research to examine performance and design issues associated with microelectromechanical systems (MEMS)-scale PA cells, we fabricated and tested a miniature non-MEMS (macro) resonant cell [65]. To date, limited research has been done to demonstrate the feasibility of a miniaturized PA sensor. The basic design for the macro PA cell is a modified version of the cell studied by Bijnen et al. [1] Our design is a 1/4 scale down from the "Bijnen" cell, with $r_{res} = 0.75$ mm and $l_{res} = 30$ mm. Experimental results suggested that miniaturization of a PA cell is viable without a significant loss in signal and no adverse effects of the size scaling were visualized in the optics or acoustics of the macro cell. We achieved a detection limit of 65 ppt for sulfur hexafluoride (SF_6) using this macro cell. We investigated finite element method (FEM) and transmission line theory to model the resonance structures for the macro PA cell [66]. Both models predicted a resonance frequency around 5900 Hz, which is in good agreement with the first longitudinal resonance of the acoustic cavity. Although we did observe the appearance of a large mode around 6000 Hz, we were not able to verify the predicted value experimentally, as this frequency was at the limit of our modulation ability.

Miklòs et al. [67] introduced a differential PA cell specifically designed for fast time response, low acoustic and electric noise characteristics, and high sensitivity. The authors developed a fully symmetrical design in order to reduce flow noise and electromagnetic disturbances. This configuration includes two acoustic resonators with $r_{res} = 2.75$ mm placed between two $\lambda_a/4$ band-rejecting acoustic filters. Figure 6.5 presents this optimized differential PA cell. The gas sample flows through both tubes, which produces about the same flow noise in both resonators; however, the laser light passes through only one of them, thus generating the PA signal in only one tube. The PA signal from each resonator is amplified with a differential amplifier, and thus all coherent noise components in the two tubes are effectively suppressed. In our PA sensor platforms for trace gas detection (see Section "Gaseous Samples"), we have used modified versions of the differential PA cell developed by Miklòs. In one configuration, we fabricated an MEMS-scale version of the "Miklòs" cell. Initial examination of the scaling principles associated with PAS with respect to MEMS dimensions indicated that PA signals would remain at similar sensitivities or even surpass those commonly found in macro-scale devices [68–73]. Figure 6.6 includes a photograph of the internal structure of the MEMS cell and the complete PA cell package. The resonators have square cross sections with $l_{res} = 8.5$ mm and $r_{res} = 0.465$ mm. The cell had two germanium windows, which were attached to the buffer volumes on either side of the resonator with epoxy. Tygon® tubing was connected to the buffer volumes to allow for gas sample inlet and outlet flow. The MEMS-scale PA cell was mounted on a printed circuit board, which allowed for

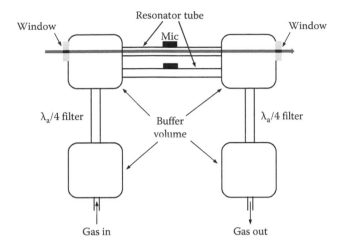

FIGURE 6.5 Optimized differential photoacoustic cell geometry with two resonator tubes and $\lambda_a/4$ filters. (Adapted from Front Miklòs, A. et al., *AIP Conf. Proc.* 1999 (463), 126–128.)

wiring the microphones to a power supply (AA battery) and a lock-in amplifier (via modified BNC cables). Another cell design consisted of two open resonators having square cross sections ($l_{res} = 10$ mm), each with $r_{res} = 0.432$ mm. The resonator is flanked on both sides by a buffer volume. To further suppress gas flow noise, the cell had a convoluted, split sample inlet/outlet design [42]. For each of these MEMS-scale PA cell designs, resonator and buffer volume dimensions were determined based on the criteria presented for the "Bijnen" cell.

Recently, Gorelik et al. [74,75] offered a miniaturized resonant PA cell with an internal volume of 0.6 cm³. The cell consisted of flat optical windows and a shell, which is a slanted section of a cylindrical tube. The length of the optical path inside

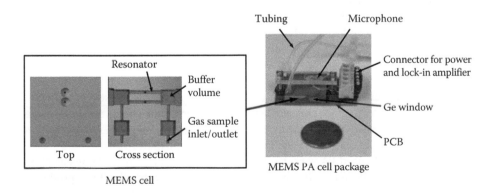

FIGURE 6.6 Photograph of the internal structure of the microelectromechanical system-scale cell and complete photoacoustic cell package.

the cell was ~12 mm. The face and end planes of the tube section were cut to the tube axis at an angle equal to the Brewster angle and the windows were attached to these planes. The inclined geometry of this cell did not allow for the acoustic eigenmodes to be classified as longitudinal or transverse. Instead, the cell eigenmodes were identified by the number of nodes given by intersection between the relevant standing wave and the longest cell diagonal. The PA cell parameters were selected, with the assistance of a numerical simulation, in order to enhance the cell performance of an individual acoustic eigenmode.

There are numerous examples in the literature of PA cell designs for pulsed PAS applications [24,76–80]. The authors suggest cell designs that differ from those utilized for modulated PAS. The MEMS-scale PA cell designs developed for our PA trace gas sensors have been used effectively with both modulated and pulsed sources. These results are discussed in more detail in Section "Gaseous Samples." Furthermore, for some applications such as quartz-enhanced photoacoustic spectroscopy (QEPAS), limitations imposed on the PA cell design are removed because the resonance frequency is determined by the transducer [81]. A commonly used QEPAS-based spectrophone consists of a quartz tuning fork (TF) and a microresonator, which is formed by one or two thin tubes. Initial studies demonstrated that the use of a microresonator increased the QEPAS sensitivity at least 10 times. The microresonator also isolates the TF from unintended acoustic resonances in the sensor enclosure, which can otherwise distort the QEPAS signal [82].

PA studies on liquid and solid samples utilizing modified cell designs have been reported. For example, Jalink and Bicanic [83] developed a PA heat pipe cell for use with liquid samples having low vapor pressures, and Schmid et al. [84] utilized a conventional 1 cm glass cuvette for construction of a PA sensor for opaque dyes. Sanchez et al. [85] attached an aluminum absorber to polymer thick films. An expanded helium–neon laser beam was directed to the aluminum surface to ensure that the incident light beam generated a surface sample heating, resulting in sample optical opaqueness. The sample-absorber system closed one of the 8-mm diameter openings of a 2-mm-long cylindrical cell cavity and a glass window closed the other cavity opening.

Finally, is it important to note the condition and quality of the PA cell surface. These characteristics influence the background signal due to scattering and molecule adsorption. Various cell materials and surface treatments and coatings have been investigated [86].

Detectors

The acoustic waves generated in a PA cell as a result of the absorption of radiation by a sample are detected by a pressure sensor. The appropriate choice depends on the application (e.g., sensitivity requirements, ease of operation, ruggedness). A selection of examples is briefly outlined in the following paragraphs.

The most widely used PA sensor scheme uses commercial microphones as pressure sensors. Typically, a lock-in amplifier is used to detect a small voltage produced by the microphone as a result of sample absorption of radiation. Microphone types include miniature electret microphones (e.g., Knowles, Sennheiser, Intricon Tibbetts) and condenser microphones (e.g., Brüel & Kjaer). These devices are easy to use, sensitive enough for PA studies of solids, liquids, and gases, and the responsivity only weakly depends on frequency. The electret microphones fabricated for hearing aids typically exhibit a flat frequency response between 20 Hz and 4 kHz. Certain models (e.g., Knowles FG23629) display increased responsivity at higher frequencies (~20 kHz). We have used this response to our advantage for our MEMS-scale PA cells, which exhibit higher resonance frequencies due to decreased resonator lengths. In most cases, the detection threshold of a PA system is determined neither by microphone responsivity nor by electrical noise, but instead by other noise sources (e.g., external noise, window heating, absorption of desorbing molecules from the cell walls, etc.).

In combination with our MEMS-scale PA sensor research, we examined the performance of MEMS microphones [65]. In comparison with conventional condenser and electret microphones, MEMS microphones provided a low-power alternative. Although the MEMS microphones exhibited good performance up to their resonance frequency, the sensitivity did not meet the levels obtainable by the electret and condenser varieties. More recently, Pedersen and McClelland [87] presented an optimized capacitive MEMS microphone for PA applications. Using analytical and numerical calculations, the authors demonstrated that an MEMS device could be designed with an SNR superior to that of state-of-the-art conventional microphones. Compared to the Brüel & Kjaer 4189 capacitive microphone, the vibration sensitivity of this MEMS microphone was reduced by more than 27 dB. The improvements in the vibration sensitivity and SNR were achieved by greatly limiting the acoustic bandwidth of the device.

Recently, alternative methods of transduction have been used. Kauppinen et al. [88] and Koskinen et al. [89,90] replaced a capacitive microphone with a cantilever-type pressure sensor made out of silicon. In this configuration, the sensor in the cantilever microphone is a flexible bar. The typical dimensions for width and length are a few millimeters, with a thickness of 5–10 μm. The cantilever is separated on three sides from a thicker frame with a narrow gap (3–5 μm) and moves like a flexible door due to the pressure variations in the surrounding gas (Figure 6.7). The fabrication and characterization of the cantilever sensor is described in detail elsewhere [91]. As the pressure changes, the cantilever bends but does not stretch. A laser interferometer was used to measure the displacement of the cantilever. The authors have achieved a limit of detection of 0.3 ppm and a normalized noise equivalent sensitivity (NNEA) of 1.7×10^{-10} $cm^{-1}W/\sqrt{Hz}$ for CO_2 [92]. More recently, this group used a dual-cantilever PA detection scheme and achieved a 0.5 ppm detection limit for CH_4 gas [93]. Lindley et al. [94] compared three PA cells containing a single electret microphone, a differential dual microphone, or a cantilever pressure sensor. The authors reported normalized sensitivities of 3.1×10^{-7}, 1.7×10^{-7}, and 2.2×10^{-9} $cm^{-1}W/\sqrt{Hz}$, respectively, for acetylene. The cantilever pressure transducer demonstrated increased sensitivity close to two orders of magnitude better than the electret single and differential microphone arrangements. Although cantilevers exhibit superior sensitivity, PA cells containing these devices and the associated interferometric detection system are much more expensive and fragile compared to a cell equipped with a conventional capacitive microphone.

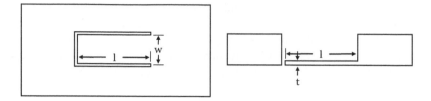

FIGURE 6.7 Dimensions of a cantilever-type pressure sensor. (Adapted from Kauppinen, J. et al., *Microchem J.* 2004, *76*(1–2), 151–159.)

QEPAS is another PA detection approach in which the microphone is replaced with a quartz crystal TF. Kosterev et al. [81,95] suggested inverting the typical resonant PA approach in which the absorbed energy is accumulated in the gas, and instead accumulate the energy in a sharply resonant acoustic transducer. This approach removes limitations imposed on the PA cell by acoustic resonance conditions because the resonant frequency is determined by the TF. Therefore, the cell is optional in QEPAS and is utilized only to separate the gas sample from the environment and control its pressure. This approach allows for gas samples that are 1 mm³ in volume. Crystal quartz was chosen as a suitable material because it is a common low-loss piezoelectric material and is mass produced and inexpensive. Furthermore, quartz TFs have become common devices for atomic force and optical near-field microscopy and therefore are well characterized [96]. Only the symmetric vibration of a TF (i.e., the two prongs bend in opposite directions) is piezoelectrically active. Therefore, efficient excitation of this vibration is achieved when the excitation beam passes through the gap between the TF prongs. The pressure wave generated when the optical radiation interacts with a gas excites a resonant vibration of the TF. This event is converted into an electric signal due to the piezoelectric effect. This electric signal is proportional to the concentration of the gas and is measured by a transimpedance amplifier. The initial feasibility experiments performed by Kosterev et al. [81] utilized a quartz-watch TF and demonstrated an NNEA of 1.2×10^{-7} cm⁻¹W/√Hz for CH_4 gas. More recently, sensors based on QEPAS were used to achieve NNEAs of 1.0×10^{-8} and 3.3×10^{-9} cm⁻¹W/√Hz for CH_4 in humid gas and acetylene, respectively [82,97]. Petra et al. [98] have completed a theoretical analysis of QEPAS to calculate the optimal position of the laser beam with respect to the TF and the phase of the piezoelectric current.

An interesting technique employing five stacked electromechanical films (EMFi) as a microphone in a PA cell was introduced by Saarela et al [99]. The EMFi is a flexible, cellular polypropylene film, approximately 70 μm thick, with an internal charge [100]. The film has two thin metal electrodes, which generate electric charges when a dynamic force is applied to the film. The authors used a five-layer EMFi as a transducer in a multipass PA cell and achieved a detection limit of 22 ppb and NNEA of 3.2×10^{-9} cm⁻¹W/√Hz for flowing nitrogen dioxide (NO_2). The sensitivity can be further enhanced by optimizing the number of layered films [101].

For PA applications such as studies on liquid and solid samples, the use of conventional microphones was reported to be inefficient by Hordvik and Schlossberg [102] and Farrow et al [103]. Both groups were concerned with acoustic impedance mismatching between the solid–gas or liquid–gas interface, resulting in most of the acoustic energy being reflected or absorbed back into the sample rather than transferred across the boundary. The authors demonstrated improvements in sensitivity with the use of piezoelectric transducers in contact with solid and liquid samples, respectively. Piezoelectric elements used in this manner offer the advantage of good impedance matching. Similar to a conventional microphone detection scheme, a lock-in amplifier is used to detect the voltage change produced by the piezoelectric sensor. This direct coupling method is simple and there have been numerous studies using piezoelectric elements in contact with liquids or solids for PA detection [5,102,104–106]. More recent reports describing PA detection of solid samples utilize conventional microphones [107–110] and the indirect coupling method described by Rosencwaig [9,10]; however, piezoelectric transducers are still widely used for liquid studies [84,111–114].

PA SENSING APPLICATIONS

There are numerous publications with thorough discussions on the use of PAS for various applications [11,21,63,115,116]. We focus on conventional PA sensing, including detection in gas, liquid, and solid media. We aim to summarize the state of the art in detection by laser PA techniques in the laboratory and in the field. Because PA sensing has evolved tremendously since the initial investigations, we have not attempted to include every important reference.

GASEOUS SAMPLES

Investigations of gaseous species continue to be the most common application for PAS. In Table 6.1, numerous examples of PA studies on gases are given. The table lists the laser source with the corresponding wavelength or wavelength range, the laser power, the analyte studied, the type of detector, and the minimum detectable concentration. In the past, many investigations were performed with CO_2 lasers [63]; however, more recently, as evidenced by this truncated list, distributed feedback reflector lasers, including diodes and QCLs, have become a popular choice for PA studies due to their smaller size, room temperature operation, and the availability of a variety of wavelengths. For example, Grossel et al. [159] used a CW distributed feedback reflector QCL for the PA detection of nitric oxide (NO). Wavelength tuning of the laser was done with temperature scans, which afforded tuning of several cm^{-1}. Specifically, the laser was operated between 100 and 150 K, resulting in tuning from 1856 to 1862 cm^{-1} (5.4 μm region), which includes two doublets of the P branch of the NO band. The power of this source was ≤3 mW. The PA cell was a Helmholtz resonator with a path length of 10 cm; however, the authors placed a mirror behind the output window to allow for a second pass of the light, resulting in an increased signal. The cell was filled with certified mixtures of nitrogen (N_2) and low concentrations of NO (102, 10.7, and 1.2 ppm) at atmospheric pressure. A laser PA spectrum was recorded for 100 ppm NO and was in excellent agreement with the calculated spectrum using the parameters of the HITRAN database. The NO absorbance feature at ~1857.3 cm^{-1} was used to determine sensitivity. A detection limit of 20 ppb was achieved for this double-pass PA sensor configuration. More recently, the same authors used a room temperature QCL for the PA detection of CH_4 [160]. A Helmholtz resonator was again used for the PA cell in these experiments. The thermoelectrically cooled laser was operated between 243 and 303 K, and emitted between 1276 and 1283 cm^{-1} (7.8 μm region). The laser power was 8 mW for a working temperature of 283 K. Two CH_4 absorption features appear in this wavelength region at 1276.84 and 1277.47 cm^{-1}. Data were collected using certified mixtures of CH_4 in air (2, 10, 20, and 100 ppm). The authors achieved a CH_4 detection limit of 17 ppb. The ability of the sensor to detect CH_4 and N_2O in ambient air was also demonstrated and the laser PA spectrum recorded for these species (315 ppb N_2O and 1.85 ppm CH_4) was in excellent agreement with the calculated spectrum using the parameters of the HITRAN database.

Recent advances in QCL technology (see Section "Radiation Sources") have allowed for continuous wavelength tuning ranges of ≥200 cm^{-1}. These broad tuning ranges permit laser PA absorbance spectra of various analytes to be collected. Figure 6.8 shows the laser PA spectrum for the standard nerve agent stimulant, dimethyl methyl phosphonate (DMMP). The spectrum in Figure 6.8a was collected as a pulsed EC-QCL was continuously tuned from 990 to 1075 cm^{-1} (10.10–9.30 μm), in 1 cm^{-1} increments. DMMP has known absorption features in this wavelength range, assigned to phosphorus–oxygen–carbon stretching vibrations. The source operated at room temperature (thermoelectrically cooled) and the average optical power was 1.35 mW. The spectrum in Figure 6.8b was collected as a CW-modulated EC-QCL was continuously tuned from 1032 to 1070 cm^{-1} (9.69–9.34 μm). Due to the mode hopping [161] nature of this source, we reduced the number of measured wavelengths, which was possible due to the broad absorption features of DMMP in this wavelength region. The source required a compact chiller for cooling and had an average optical power of 11.97 mW. The absorbance spectrum for DMMP, which was recorded using an FTIR, is also provided in Figure 6.8a and b for comparison with the PA data. There is good agreement between the laser PA data and the FTIR spectroscopy absorbance spectrum. PA data were collected using a certified permeation tube and N_2 as the carrier gas. The DMMP permeation device was placed in a generator oven held at a constant temperature of 100°C. Varying calibrated flow

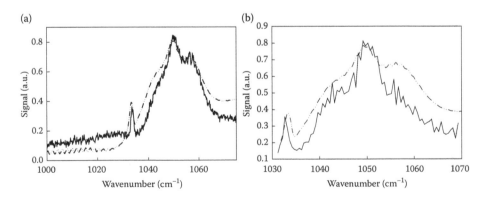

FIGURE 6.8 Measured (a) pulsed and (b) continuous wave modulated laser photoacoustic (PA) spectra (—) of DDMP. Data derived from our own PA measurements are compared to FTIR reference spectra (– ·· –).

rates of the N_2 carrier gas from 200 to 600 mL/min governed the analyte concentration. A MEMS-scale differential PA cell (described in Section "Acoustic Resonators") was used for spectroscopic and sensitivity data collection. The cell had two resonator tubes, both housing a commercial microphone. The DMMP absorbance maximum in the QCL wavelength tuning range is 1053 cm^{-1} (9.50 µm). Therefore, this wavelength is best suited for continuous DMMP sensitivity monitoring. Minimal detectable DMMP concentrations of 54 and 20 ppb were achieved with the pulsed QCL and CW-modulated QCL-based PA sensor systems, respectively. Although the increased output power of the modulated CW laser afforded a better detection limit than that obtained using the pulsed source, consideration of power consumption and heat management favors the use of a pulsed QCL for future sensor development.

We used a similar PA sensing system to study Freon 116, a propellant analog and a component commonly found in refrigerants. In Figure 6.9, the spectrum of Freon 116 was collected as a pulsed EC-QCL was continuously tuned from 1050 to 1240 cm^{-1} (9.52–8.06 µm), in 1 cm^{-1} increments. Freon 116 has known absorption features in this spectral region, assigned to carbon–fluorine stretching vibrations in the molecules. The source operated at room temperature (thermoelectrically cooled) and the average optical power was 1.32 mW. The absorbance spectrum for this species, recorded using an FTIR, is also provided in Figure 6.9 for comparison with the PA data. There is excellent agreement between the PA and FTIR data. Both the FTIR and PA spectra were recorded using a certified permeation tube and N_2 as the carrier gas. The Freon 116 vapor was contained in a stainless-steel cylinder maintained at a constant temperature of 100°C. Varying calibrated flow rates of the N_2 carrier gas from 50 to 1000 mL/min governed the analyte concentration. An MEMS-scale differential PA cell was used for spectroscopic and sensitivity data collection. The Freon 116

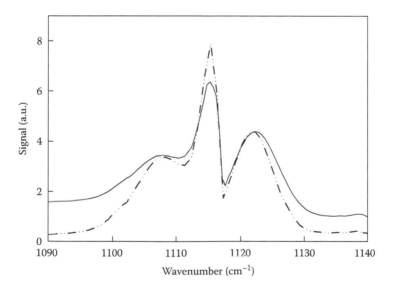

FIGURE 6.9   Measured laser photoacoustic (PA) spectrum (—) of Freon 116. Data derived from our own PA measurements are compared to FTIR reference spectrum (– · · –).

absorbance maximum in the QCL wavelength tuning range is 1115 cm^{-1} (8.97 µm). We achieved a minimal detectable Freon 116 concentration of 19 ppb.

Although the microphone continues to be the most commonly used detector for gaseous sensors, quartz crystal TFs and microcantilevers have also emerged as desired transducers for trace vapor detection. Specifically, the combination of tunable QCLs used with QEPAS has become an attractive sensing scheme. Lewicki et al. [36] reported a QEPAS-based detection of Freon 125 using a widely tunable CW EC-QCL. The source was tunable between 1122 and 1257 cm^{-1} (8.91–7.96 µm) and had an average optical power of 6.6 mW. Freon 125 has two broad unresolved absorption bands within the tuning range of the laser. A calibrated reference mixture of Freon 125 in N$_2$ (5 ppm) was used for the QEPAS experiments. The authors measured the laser PA absorption spectrum for the laser wavelength tuning range and reported excellent agreement with a database reference. A minimal detectable concentration limit of 3 ppb was demonstrated for Freon 125. More recently, the same authors developed a mid-infrared QEPAS sensor device for triacetone triperoxide (TATP) detection [162]. The authors used two different EC-QCLs (pulsed and CW), having different center wavelengths and tuning ranges. The average laser output power was 5 mW. The two C–O stretch bands of the TATP molecule were detected using an 8.4 µm pulsed QCL with 75 cm^{-1} tuning range. The O–O stretch band of the molecule was detected using a 10.5 µm CW QCL that could be tuned over 60 cm^{-1}. These tuning ranges allowed for these absorption features to be clearly seen in the laser PA spectra. The authors report a TATP detection limit of 1 ppm using this QEPAS sensor.

As discussed above, the wide tuning ranges of diode and QC lasers allow for laser PA absorbance spectra of various analytes to be collected. Furthermore, this tunability permits the simultaneous detection of several components in a gas mixture and, if a proper wavelength region is chosen in which the absorption spectral features are clearly identified, increased molecular discrimination. Boschetti et al. [163] presented the simultaneous PA detection of CH$_4$ and ethylene (C$_2$H$_4$) using a pulsed diode laser. The source was mounted in an EC configuration allowing for a tuning range of 5900–6250 cm^{-1} (1.7–1.6 µm) and a maximum output power of 3 mW. The PA cell was a brass tube flanked on either side by a $\lambda_a/4$ buffer volume and the acoustic signal was detected by a microphone. The authors presented experimental spectra of CH$_4$, C$_2$H$_4$, and a mixture of the two in the same wavelength interval. Two clearly separated features, the C$_2$H$_4$ triplet at 6154.5 cm^{-1} and the CH$_4$ triplet at 6157.2 cm^{-1}, were observed, along with a partially convoluted spectral feature between 6156.2 and 6156.7 cm^{-1}. Sensitivities of 8 and 40 ppm were reported for the simultaneous detection of C$_2$H$_4$ and CH$_4$, respectively. Mukherjee et al. [38] reported multigas detection using QCL-based PAS. The authors demonstrated a nondispersive beam-combining scheme by multiplexing five tunable CW EC-QCLs in the 6–11 µm spectral range. The power output of each laser was ~100 mW at room temperature. A gas mixture containing ~3 ppb ammonia (NH$_3$), ~8 ppb nitrogen dioxide (NO$_2$), ~20 ppb DMMP, ~30 ppb acetone (C$_3$H$_6$O), and ~40 ppb ethylene glycol (C$_2$H$_6$O$_2$) was contained in a PA cell and the PA signal of each species was detected in the mixture using a piezoelectric transducer. Hanyecz et al. [164] also demonstrated a multicomponent PA gas sensor in which the analysis of water vapor (H$_2$O vapor), CO, CO$_2$, and CH_4 was carried out using two CW distributed feedback reflector diode lasers [164]. One of the diode lasers had a 20 mW output power and emission around 1371 nm. This source was used for measuring H_2O vapor and CH_4. The second laser had a 40 mW output power and emission around 1581 nm, and was used for measuring CO and CO_2. The laser wavelengths were tuned by tuning the temperature. The authors used a stainless-steel resonant PA cell and a microphone for PA signal detection. The geometry of the resonator was designed specifically for data collection in hydrogen carrier gas. Calibration gases were produced by mixing high-purity gases obtained from cylinders and regulating the flow rates with mass flow controllers. The authors collected PA absorption spectra for the gas mixture and reported minimum detectable concentrations of 0.003% for H_2O vapor, 0.13% for CO, 0.16% for CO_2, and 0.06% for CH_4.

Multivariate analysis permits simultaneous analysis of multiple components and is therefore an attractive tool for distinguishing several analytes based on spectral differences. Recently, we have utilized the partial least-squares 2 regression method to develop a model for the simultaneous differentiation of acetic acid ($C_2H_4O_2$), C_3H_6O, 1,4-dioxane ($C_4H_8O_2$), and vinyl acetate ($C_4H_6O_2$) based on their absorbance features from 1050 to 1240 cm^{-1} (9.52–8.06 μm) acquired experimentally using a pulsed EC-QCL and a MEMS-scale differential PA cell [41]. Figure 6.10 presents the laser PA spectra collected for these species in this wavelength region. The absorption features for these molecules in this wavelength region are assigned to carbon–carbon and carbon–oxygen stretching vibrations. We used training samples (PA spectra) having a wide range of analyte concentrations as the basis to develop this model. The model simultaneously uses four algorithms, each devoted to the identification of a particular species. It is the concurrent application of the algorithms that facilitates

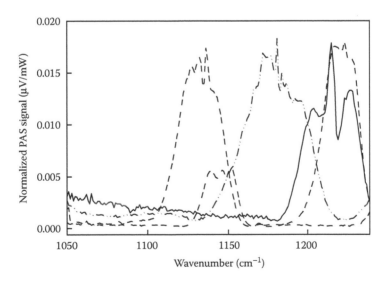

FIGURE 6.10 Laser photoacoustic spectral absorption features of acetic acid (– · · –), acetone (——), 1,4-dioxane (– – – –), and vinyl acetate (— —) in the 1050–1240 cm^{-1} (9.52–8.06 μm) region.

the classification (i.e., identification) of unknown spectra. A model using 1–10 factors or principal components was constructed for the prediction of the analytes of interest. Scores for the first three principal components are plotted in Figure 6.11.

This plot illustrates four distinct groups, defined for C_2H_4O, C_3H_6O, $C_4H_8O_2$, and $C_4H_6O_2$. These results confirm that the spectral features of the four analytes, and therefore their molecular compositions, are different. We evaluated the performance of the model and the ability of our QCL-based miniaturized PA sensing platform for the simultaneous detection and discrimination of multiple species. We introduced mixtures composed of known concentrations of the four analytes of interest, which were used to challenge the model. The model accurately discriminated between species having similar molecular components and identified specific analytes in mixtures and their concentrations.

LIQUID SAMPLES

Investigating liquids using the PA technique is attractive as it allows optical absorption measurements to be made for optically opaque samples. This capability of PAS, along with its insensitivity to scattered light, makes this technique a very attractive spectroscopic tool for the investigation of liquids. For instance, in industrial applications where ultraviolet–visible online process monitoring is hampered by light scattering and opacity of samples, PAS allows for the measurements of both high and low absorptions and no need for sample preparations. Schmid et al. [84] described the construction and characterization of a PA sensor for optical absorption measurements in transparent and opaque liquids. A Q-switched frequency-doubled Nd:YAG

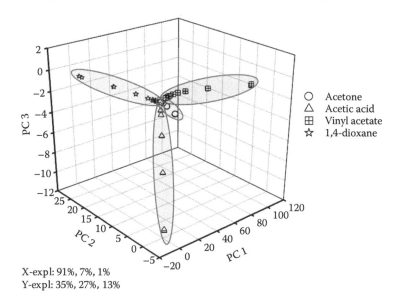

FIGURE 6.11 Scores plot for the partial least-squares 2 model, using the first three calculated principal components (PC). Data points for each analyte are encircled in lobes as a guide to the eye and convey no direct information.

laser was used for excitation. The pulsed source had an emission wavelength of 532 nm. A glass cuvette with piezoelectric transducers placed on two sides was used to contain the samples for PA data collection. The samples were aqueous solutions of a black textile dye with concentrations in the range of 20 mg/L to 1.34 g/L, aqueous solutions of red textile dye in the range of 100 mg/L to 50 g/L, and suspensions of TiO_2 in water. The PA sensor allowed for the determination of absorption coefficients ranging from 0.1 to 1000 cm^{-1}.

Furthermore, PAS is attractive for industrial or environmental applications for the identification of contaminants in solution. For example, lubrication and hydraulic oils used to maintain heavy equipment can degrade by contamination. Water is the most common contaminant and therefore knowledge of the condition of oil in equipment is necessary. Foster et al. [165] used PAS to detect trace levels of water in oil. Raman shifting deuterium in a 1 m Raman cell was used to generate 2.93 μm excitation light and 1064 nm light was produced using a pulsed Nd:YAG laser. PA data were collected using a flow through a layered-prism cell. The authors used an ultrasonic couplant fluid to facilitate acoustic contact between the cell and transducer. PA data were recorded for clean samples of unused transmission, hydraulic, and engine oil samples. Numerous samples of the clean oil with known relative water content were prepared. Detection limits for water were reported to be 60 ppm in hydraulic oil, 45 ppm in transmission oil, and 515 ppm in engine oil. The results of these experiments demonstrated that water in petroleum oils can be detected at levels at least two times lower than those obtained with FTIR. Hodgson et al. [166] developed a laser PA measurement technique for the detection of oil contamination in water for the continuous monitoring of hydrocarbons in return process water from oil production installations. Two pulsed diode lasers were used. One had an emission wavelength of 0.904 μm, which coincides well with a small absorption feature present in both methanol (CH_3OH) and pentane (C_5H_{12}). The second laser emitted at 1.55 μm, which is in the wings of the strong water absorption feature centered at 1.44 μm. A quartz cuvette equipped with a piezoelectric transducer (PZT-5A) ceramic disk was used for PA data collection. The authors studied crude oil emulsions and hydrocarbons in solution in the range of 0–900 mg/L and demonstrated that dispersed and dissolved hydrocarbon components give an additive PA response whereas with most optical instrumentation the components must be measured separately. Hernández-Valle et al. [167] presented a PA technique for the trace analysis of pesticides in water. The light from a 460 nm pulsed dye laser was directed into a double crystal in which 230 nm light was generated. Different concentrations of the pesticides atrazine and methyl parathion were prepared. Samples were poured into quartz cells and two piezoelectric sensors were used to monitor PA signals. The authors reported the ability to monitor concentrations as low as tenths of nanograms per liter.

SOLID SAMPLES

Similar to liquid samples, the initial PA studies on solid samples demonstrated that optical absorption measurements could be obtained for optically opaque materials, therefore making this technique an attractive spectroscopic tool for the investigation of solid materials. Although this topic is discussed extensively in the literature, specifically with regard to applications of PA FTIR spectroscopy, the use of lasers in PA experiments on solids has not been widely reported. Petzold and Niessner [168] presented a PA soot sensor for monitoring black carbon *in situ*. The sensor included a 450 mW laser diode emitting at 802 nm, a novel cylindrical azimuthal resonator, and two electret microphones to achieve a portable sensor system for black carbon measurements. The authors reported a detection limit of 0.5 μg black carbon per cubic meter. Wen and Michaelian [109] used an EC-QCL for the measurement of the PA spectrum of acetyl polystyrene beads. There have been several reports discussing the use of these tunable infrared lasers, such as QC and CO_2, to obtain PA spectra of gases; however, the application of these sources in PA experiments on solids is minimal. In this work, the pulsed EC-QCL was tunable from 990 to 1075 cm^{-1} (10.10–9.30 μm) and the results demonstrated better peak definition in the PA spectrum than in the FTIR spectrum, allowing identification of bands at ~1005 and 1030 cm^{-1} for the solid material.

Numerous PA studies on explosive materials have been reported due to an increased interest in the identification and quantification of these substances. The detection of these compounds in the solid form is attractive because the majority of explosives have extremely low vapor pressures, making vapor detection difficult. Chaudhary et al. [169] presented the low-limit PA detection of solid RDX and TNT using a CO_2 laser with an output power of 12 W. The source was grating-tunable and could be precisely tuned in the 9.25–10.74 μm spectral region. The laser beam was chopped at 22 Hz using an electrical chopper. The chopped beam was intensity-modulated and incident on the explosive sample housed in an aluminum nonresonant PA cell. The detector was a condenser microphone. The authors successfully demonstrated for the first time the use of a CO_2 laser-based PA technique to record the high-resolution PA absorption spectra of RDX and TNT in solid form at room temperature. Detection limits of 16.5 and 10.0 ppb were reported for TNT and RDX, respectively. Giubileo and Puiu [108] characterized standard explosives in the solid phase using the same PA technique offered by Chaudhary et al. [169]. The authors studied classical explosives, including 2,4-DNT, 2,6-DNT, HMX, TATP, PETN, TNT, and RDX, using a CO_2 laser operating in the 9–11 μm spectral region. The output power of the laser was 10 W. Trace detection and molecular composition analysis were successfully performed with high sensitivity (less than 100 μg) and reproducible PA signals in the entire laser wavelength range. All the investigated materials exhibited different spectral behavior, allowing for simple principal component analysis and identification of each explosive. Van Neste et al. [170] demonstrated the detection of trace amounts of RDX adsorbed on a quartz crystal TF using illumination from a QCL. The authors used three pulsed EC-QCLs having an overall optical tuning range from 7.83 to 10.93 μm when used in combination. The QCLs had peak powers ranging from 100 to 400 mW with a duty cycle of 5%. The RDX samples were derived from Non-Hazardous Explosives for Security Training and Testing stimulants suspended in acetonitrile. The RDX signatures recorded using the tunable sources had excellent agreement with published infrared absorption spectra.

Furthermore, PAS of solid samples is often applied to depth profiling of layered samples and two- and three-dimensional tomographic imaging for biomedical applications [171–173]. Viator et al. [174] developed an *in vivo* PA probe that used an

Nd:YAG laser operating at 532 nm to generate acoustic pulses in skin. A piezoelectric element was used to detect the acoustic waves arising from thermoelastic expansion, which were analyzed for epidermal melanin content using a photoacoustic melanin index (PAMI). Melanin content was compared to results obtained using visible reflectance spectroscopy (VRS), which is currently used to estimate epidermal melanin content. Although VRS provides reliable measurements of melanin content, it gives no depth information and utilizes an integrating sphere that averages skin reflectance over a large area, which makes local estimates of melanin concentration impossible. The authors reported a good correlation between PAMI and VRS measurements and the 200 μm active area of the PA method allowed for pinpoint measurements. Schmid et al. [175] used PAS to observe the interception of iron (III) oxide particles by a biofilm, which can influence the stability of the biofilm and lead to a partial detachment. The PA sensors for biofilm monitoring consisted of a piezoelectric poly(vinylidene fluoride) film coupled to a transparent prism by conductive epoxy resin. The biofilm grew directly on the surface of the prism. Short pulses from an Nd:YAG laser operating at 420 nm irradiated the biofilm through the prism and the resulting pressure waves were detected by the poly(vinylidene fluoride) film. By investigating the differences in signal shape before and after the addition of iron particles, the particle distribution inside the biofilm could be estimated.

STANDOFF DETECTION

Standoff detection of chemical residues has rapidly gained attention due to its relevance for security and commercial applications. Standoff refers to instances in which the excitation source, detector, and system operator are located at a safe distance from the target sample. Reports of *in situ* and short-range PA experiments are more prevalent in the literature than standoff PA techniques. This is mainly due to the higher sensitivities achieved with *in situ* PA methods as well as the numerous challenges associated with PAS of samples in open air. In the absence of a sealed PA cell, acoustic waves spread, broadening their already minimal energy below the detection limits of acoustic detectors. Additionally, direct microphone detection is met with difficulty due to the influence of wind effects, such as air turbulence or constant winds [176]. Perrett et al. [177] attempted to overcome these challenges using a pulsed indirect photoacoustic spectroscopy (PIPAS) technique, which employed a stronger light source and a parabolic focusing mirror. The authors suggested that a more powerful light source would increase the amplitude of the acoustic waves as the parabolic mirror captured and refocused the sound back onto a microphone. Furthermore, enhanced sensitivity was also brought about by signal processing that allowed improved discrimination of the signal from noise levels through prior knowledge of the laser pulse shape and repetition frequency. To test this technique, ethanol (C_2H_6O) and diffusion pump oil samples were selected. Both species have absorption features in the 10 μm wavelength region of the CO_2 laser used for these experiments. The laser pulse energy was 300 μJ, which was achieved using an amplifier. Samples were placed at a range of up to 8 m from the output of the laser. The laser output passed through a telescope to ensure an appropriate illuminating spot size. A microphone was placed at the focus of a parabolic reflector, which was used to improve the acoustic detection sensitivity of the system. The detection equipment was collocated with the laser on an independent tripod, which allowed for the microphone to be aimed at the sample. In comparison with earlier PIPAS studies, the authors achieved a gain factor of 28 using the parabolic reflector; however, significant losses in the system and field environment effects ultimately limited the range and sensitivity of this technique.

Although the acoustic waves generated at standoff distances are affected by atmospheric conditions, the reflected or scattered light is not. Van Neste et al. [176] demonstrated a standoff PAS technique used for analyzing surface-adsorbed chemicals. In this experiment, pulsed light scattered or reflected off of a target excited an acoustic resonator, and the variation of the resonance amplitude as a function of wavelength resulted in absorption data of the target. The authors used an EC-QCL to illuminate a target located 0.5–20 m away. The QCL had a tuning range from 1020.41 to 1080.08 cm^{-1} (9.8–9.25 µm) and the laser power was 10 mW with a 5% duty cycle. The laser beam spot size was ~6 mm and was expanded to ~25 mm at 20 m. The light reflected or scattered off the target was collected using a spherical mirror and focused onto a quartz crystal TF, producing an acoustic wave on the TF surface. The pulse rate of the laser light was adjusted to match the resonance frequency of the TF. The authors chose tributyl phosphate (TBP), RDX, TNT, and PETN as chemical targets. The sample solutions were prepared in acetonitrile (1 mg/mL) and deposited on a target surface and allowed to dry. Experiments were conducted at standoff distances of 0.5, 4, 10, and 20 m. Signal amplification was needed at distances of 10 and 20 m. A detection limit of the order of 100 ng/cm^2 was achieved with this standoff PA detection system. More recently, these authors used the same standoff PAS technique but added an additional EC-QCL and a TF array to demonstrate the detection of trace quantities of TBP [178]. A schematic diagram of the experimental setup is shown in Figure 6.12. Each QCL was pulsed at a different frequency that matched the resonant frequency of a single TF in the TF array. The additional laser was tunable between 1250 and 1355.01 cm^{-1} (8.0–7.38 µm). There was a ~0.5 mm gap between each TF and the authors experimentally determined that the resonant frequency difference between each TF must be at least 10 Hz. This difference minimized the development of cross talk between the sensors. In this demonstration, the resonance frequency difference between two TFs was 143 Hz. The spectral peaks observed with the standoff PAS method agreed very well with the spectrum collected with an FTIR. The authors reported a similar detection limit of 100 ng/cm^2 for this dual-QCL detection scheme.

CONCLUSIONS AND OUTLOOK

The versatility of laser-based PAS for sensing applications has been demonstrated by numerous state-of-the-art examples. Detection limits for gaseous and condensed media are often in the ppb or sub-ppb range, with some reports of ppt level sensing capabilities. Furthermore, PA detection schemes are not limited to laboratory measurements and can be used at standoff distances, which has been demonstrated recently for the detection of explosive materials.

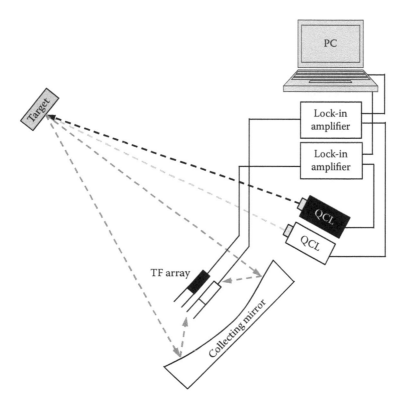

FIGURE 6.12 Schematic diagram of an experimental setup used for standoff photoacoustic spectroscopy. (Adapted from Van Neste, C. W., Senesac, L. R., and Thundat, T., *Anal. Chem.*, 2009, *81*, 1952–1956.) Electronic connections are indicated by a solid line.

The development of continuously tunable sources, such as QCLs, allows for PA absorption spectra to be collected. Data collected with a simple hearing aid microphone have been presented for numerous chemical species and are in excellent agreement with absorption spectra collected for the same analytes using an FTIR spectrometer. Additionally, these broadly tunable sources allow for the simultaneous detection of several molecules of interest as well as increased molecular discrimination, which assists in overcoming the absorption interference problem often associated with PAS.

Although numerous reports have verified the sensitivity of PA sensors at trace levels, the total system size represents a large logistics burden in terms of bulk, cost, and power consumption. The future of PAS for sensing applications includes the continued development of laser sources with respect to broad continuous tunability and decreased system sizes and power requirements. In addition, the successful demonstration of PA sensing platforms using miniaturized PA cells is an important step toward the development of man-portable sensor systems. We expect continued success from PA-based sensing applications for the detection of a diverse range of chemical and biological agents and for use in environmental monitoring.

REFERENCES

1. Bijnen, F. G. C., Reuss, J., and Harren, F. J. M., Geometrical optimization of a longitudinal resonant photoacoustic cell for sensitive and fast trace gas detection. *Rev. Sci. Instrum.* 1996, *67*(8), 2914–2923.
2. Nägele, M. and Sigrist, M. W., Mobile laser spectrometer with novel resonant multipass photoacoustic cell for trace-gas sensing. *Appl. Phys. B* 2001, *70*, 895–901.
3. Bialkowski, S. E., *Photothermal Spectroscopy Methods for Chemical Analysis*, Vol. 134. New York: John Wiley & Sons, 1996.
4. Kinney, J. B. and Staley, R. H., Applications of photoacoustic spectroscopy. *Ann. Rev. Mater. Sci.* 1982, *12*, 295–321.
5. Patel, C. K. N. and Tam, A. C., Pulsed optoacoustic spectroscopy of condensed matter. *Rev. Mod. Phys.* 1981, *53*(3), 517–550.
6. Tam, A. C., Applications of photoacoustic sensing techniques. *Rev. Mod. Phys.* 1986, *58*(2), 381–431.
7. Miklòs, A. and Hess, P., Modulated and pulsed photoacoustics in trace gas analysis. *Anal. Chem.* 2000, *72*, 30A–37A.
8. Kreuzer, L. B., The physics of signal generation and detection. In Y.-H. Pao (ed.), *Optoacoustic Spectroscopy and Detection*. New York: Academic Press, 1977.
9. Rosencwaig, A., Theoretical aspects of photoacoustic spectroscopy. *J. Appl. Phys.* 1978, *49*(5), 2905–2910.
10. Rosencwaig, A. and Gersho, A., Theory of photoacoustic effect with solids. *J. Appl. Phys.* 1976, *47*(1), 64–69.
11. Tam, A. C., Photoacoustics: Spectroscopy and other applications. In D. S. Klinger (ed.), *Ultrasensitive Laser Spectroscopy*. New York: Academic Press, 1983.
12. Bell, A. G., The photophone. *Science* 1880, *1*(11), 130–134.
13. Viengerov, M. L., Eine methode der gasanalyze beruhend auf dem optischakustischen Tyndall-Röntgeneffekt. *Dolk. Akad. Nauk. SSSR* 1938, *19*, 687.
14. Kerr, E. L. and Atwood, J. G., Laser illuminated absorptivity spectrophone—A method for measurement of weak absorptivity in gases at laser wavelengths. *Appl. Optics* 1968, *7*(5), 915–922.
15. Kreuzer, L. B., Ultralow gas concentration infrared absorption spectroscopy. *J. Appl. Phys.* 1971, *42*(7), 2934–2943.
16. Kreuzer, L. B., Laser optoacoustic spectroscopy—New technique of gas-analysis. *Anal. Chem.* 1974, *46*(2), A237–A244.
17. Kreuzer, L. B., Kenyon, N. D., and Patel, C. K. N., Air-pollution—Sensitive detection of 10 pollutant gases by carbon-monoxide and carbon-dioxide lasers. *Science* 1972, *177*(4046), 347–349.
18. Kreuzer, L. B. and Patel, C. K. N., Nitric oxide air pollution—Detection by optoacoustic spectroscopy. *Science* 1971, *173*(3991), 45–47.
19. Parker, J. G., Optical-absorption in glass—Investigation using an acoustic technique. *Appl. Optics* 1973, *12*(12), 2974–2977.
20. West, G. A., Barrett, J. J., Siebert, D. R., and Reddy, K. V., Photo-acoustic spectroscopy. *Rev. Sci. Instrum.* 1983, *54*(7), 797–817.
21. Rosencwaig, A., *Photoacoustics and Photoacoustic Spectroscopy*. New York: John Wiley & Sons, 1980.
22. Sigrist, M. W., Laser generation of acoustic-waves in liquids and gases. *J. Appl. Phys.* 1986, *60*(7), R83–R121.
23. LumaSense Technologies Innova 1412. Available at: www.innova.dk.
24. Claspy, P. C., Ha, C., and Pao, Y. H., Optoacoustic detection of NO_2 using a pulsed dye-laser. *Appl. Optics* 1977, *16*(11), 2972–2973.

25. Angus, A. M., Marinero, E. E., and Colles, M. J., Opto-acoustic spectroscopy with a visible CW dye laser. *Opt. Commun.* 1975, *14*(2), 223–225.

26. Vansteenkiste, T. H., Faxvog, F. R., and Roessler, D. M., Photoacoustic measurement of carbon-monoxide using a semiconductor diode-laser. *Appl. Spectrosc.* 1981, *35*(2), 194–196.

27. Faist, J., Capasso, F., Sivco, D. L., Sirtori, C., Hutchinson, A. L., and Cho, A. Y., Quantum cascade laser. *Science* 1994, *264*(5158), 553–556.

28. Curl, R. F., Capasso, F., Gmachl, C., Kosterev, A. A., McManus, B., Lewicki, R., Pushkarsky, M., Wysocki, G., and Tittel, F. K., Quantum cascade lasers in chemical physics. *Chem. Phys. Lett.* 2010, *487*(1–3), 1–18.

29. Hecht, J., Quantum cascade lasers prepare to compete for terahertz applications. *Laser Focus World* 2010, *6*(10), 45–49.

30. Troccoli, M., Corzine, S., Bour, D., Zhu, J., Assayag, O., Diehl, L., Lee, B. G., Holfer, G., and Capasso, F., Room temperature continuous-wave operation of quantum-cascade lasers grown by metal organic vapour phase epitaxy. *Electron. Lett.* 2005, *41*, 1059–1061.

31. Soibel, A., Mansour, K., Sivco, D. L., and Cho, A. Y., Evaluation of thermal crosstalk in quantum cascade laser arrays. *IEEE Photonics Technol. Lett.* 2007, *19*(5–8), 375–377.

32. Paldus, B. A., Spence, T. G., Zare, R. N., Oomens, J., Harren, F. J. M., Parker, D. H.; Gmachl, C. et al., Photoacoustic spectroscopy using quantum-cascade lasers. *Opt. Lett.* 1999, *24*(3), 178–180.

33. da Silva, M. G., Vargas, H., Miklòs, A., and Hess, P., Photoacoustic detection of ozone using a quantum cascade laser. *Appl. Phys. B: Lasers Opt.* 2004, *78*(6), 677–680.

34. Hofstetter, D., Beck, M., Faist, J., Nägele, M., and Sigrist, M. W., Photoacoustic spectroscopy with quantum cascade distributed-feedback lasers. *Opt. Lett.* 2001, *26*(12), 887–889.

35. Kosterev, A. A., Bakhirkin, Y. A., and Tittel, F. K., Ultrasensitive gas detection by quartz-enhanced photoacoustic spectroscopy in the fundamental molecular absorption bands region. *Appl. Phys. B: Lasers Opt.* 2005, *80*(1), 133–138.

36. Lewicki, R., Wysocki, G., Kosterev, A. A., and Tittel, F. K., QEPAS based detection of broadband absorbing molecules using a widely tunable, CW quantum cascade laser at 8.4 μm. *Opt. Express* 2007, *15*(12), 7357–7366.

37. Lima, J. P., Vargas, H., Miklòs, A., Angelmahr, M., and Hess, P., Photoacoustic detection of NO_2 and N_2O using quantum cascade lasers. *Appl. Phys. B: Lasers Opt.* 2006, *85*(2–3), 279–284.

38. Mukherjee, A., Prasanna, M., Lane, M., Go, R., Dunayevskiy, I., Tsekoun, A., and Patel, C. K. N., Optically multiplexed multi-gas detection using quantum cascade laser photoacoustic spectroscopy. *Appl. Optics* 2008, *47*(27), 4884–4887.

39. Taslakov, M., Simeonov, M., Froidevaux, M., and van den Bergh, H., Open-path ozone detection by quantum-cascade laser. *Appl. Phys. B-Lasers Opt.* 2006, *82*(3), 501–506.

40. Weidmann, D., Kosterev, A., Roller, C., Curl, R. F., Fraser, M. P., and Tittel, F. K., Monitoring of ethylene by a pulsed quantum cascade laser. *Appl. Optics* 2004, *43*(6), 3329–3334.

41. Holthoff, E., Bender, J., Pellegrino, P., Fisher, A., and Stoffel, N., Photoacoustic spectroscopy for trace vapor detection and molecular discrimination. *Proc. SPIE: Int. Soc. Opt. Eng.* 2010, *7665*, 766510-1–766510-7.

42. Holthoff, E. L., Heaps, D. A., and Pellegrino, P. M., Development of a MEMS-scale photoacoustic chemical sensor using a quantum cascade laser. *IEEE Sens. J.* 2010, *10*(3), 572–577.

43. Holthoff, E., Bender, J., Pellegrino, P., and Fisher, A., Quantum cascade laser-based photoacoustic spectroscopy for trace vapor detection and molecular discrimination. *Sensors* 2010, *10*, 1986–2002.

44. Harren, F. J. M., Bijnen, F. G. C., Reuss, J., Voesenek, L., and Blom, C., Sensitive intra-cavity photoacoustic measurements with a CO2 wave-guide laser. *Appl. Phys. B: Photophys. Laser Chem.* 1990, *50*(2), 137–144.

45. Miklòs, A., Hess, P., and Bozoki, Z., Application of acoustic resonators in photoacoustic trace gas analysis and metrology. *Rev. Sci. Instrum.* 2001, *72*(4), 1937–1955.

46. Monchalin, J. P., Bertrand, L., Rousset, G., and Lepoutre, F., Photoacoustic-spectroscopy of thick powdered or porous samples at low-frequency. *J. Appl. Phys.* 1984, *56*(1), 190–210.

47. Oda, S. and Sawada, T., Laser-induced photoacoustic detector for high-performance liquid-chromatography. *Anal. Chem.* 1981, *53*(3), 471–474.

48. Rosencwaig, A., Photoacoustic spectroscopy of solids. *Opt. Commun.* 1973, *7*(4), 305–308.

49. Rosengren, L. G., Optimal optoacoustic detector design. *Appl. Optics* 1975, *14*(8), 1960–1976.

50. Veeken, K., Dam, N., and Reuss, J., A multipass transverse photoacoustic cell. *Infrared Physics* 1985, *25*(5), 683–696.

51. Hodgson, R. J. W., Regularization techniques applied to depth profiling with photoacoustic spectroscopy. *J. Appl. Phys.* 1994, *76*(11), 7524–7529.

52. Patel, C. K. N. and Kerl, R. J., New optoacoustic cell with improved performance. *Appl. Phys. Lett.* 1977, *30*(11), 578–579.

53. Dewey, C. F., Jr., Design of optoacoustic system. In Y.-H. Pao (ed.), *Optoacoustic Spectroscopy and Detection.* New York: Academic Press, 1977.

54. Gerlach, R. and Amer, N. M., Brewster window and windowless resonant spectrophones for intra-cavity operation. *Appl. Phys.* 1980, *23*(3), 319–326.

55. Miklòs, A. and Lorincz, A., Windowless resonant acoustic chamber for laser-photoacoustic applications. *Appl. Phys. B: Photophys. Laser Chem.* 1989, *48*(3), 213–218.

56. Meyer, P. L. and Sigrist, M. W., Atmospheric-pollution monitoring using CO_2-laser photoacoustic-spectroscopy and other techniques. *Rev. Sci. Instrum.* 1990, *61*(7), 1779–1807.

57. Dewey, C. F., Kamm, R. D., and Hackett, C. E., Acoustic amplifier for detection of atmospheric pollutants. *Appl. Phys. Lett.* 1973, *23*, 633–635.

58. Kamm, R. D., Detection of weakly absorbing gases using a resonant optoacoustic method. *J. Appl. Phys.* 1976, *47*, 3550–3558.

59. Karbach, A. and Hess, P., High precision acoustic spectroscopy by laser excitation of resonator modes. *J. Chem. Phys.* 1985, *83*, 1075–1084.

60. Miklòs, A., Hess, P., and Bozoki, Z., Application of acoustic resonators in photoacoustic trace gas analysis and metrology. *Rev. Sci. Instrum.* 2001, *72*, 1937–1955.

61. Hess, P., Resonant photoacoustic spectroscopy. In K. N. Houk et al. (eds), *Topics in Current Chemistry*, Vol. 111. Berlin: Springer-Verlag, 1983, pp. 1–32.

62. Rey, J. M., Marinov, D., Vogler, D. E., and Sigrist, M. W., Investigation and optimization of a multipass resonant photoacoustic cell at high absorption levels. *Appl. Phys. B: Lasers Opt.* 2005, *80*(2), 261–266.

63. Sigrist, M. W., Air monitoring by laser photoacoustic spectroscopy. In M. W. Sigrist (ed.), *Air Monitoring by Spectroscopic Techniques*, Vol. 127. New York: Wiley Interscience, 1994.

64. Bernegger, S. and Sigrist, M. W., Co-laser photoacoustic-spectroscopy of gases and vapors for trace gas-analysis. *Infrared Phys.* 1990, *30*(5), 375–429.

65. Pellegrino, P. M. and Polcawich, R. G., Advancement of a MEMS photoacoustic chemical sensor. *Proc. SPIE: Int. Soc. Opt. Eng.* 2003, *5085*, 52–63.

66. Pellegrino, P. M., Polcawich, R. G., and Firebaugh, S. L., Miniature photoacoustic chemical sensor using microelectromechanical structures. *Proc. SPIE: Int. Soc. Opt. Eng.* 2004, *5416*(1), 42–53.

67. Miklòs, A., Hess, P., Mohacsi, A., Sneide, J., Kamm, S., and Schafer, S., Improved photoacoustic detector for monitoring polar molecules such as ammonia with a 1.53 μm DFB diode laser. *AIP Conf. Proc.* 1999 (463), 126–128.
68. Firebaugh, S. L., Jensen, K. F., and Schmidt, M. A., Miniaturization and integration of photoacoustic detection with a microfabricated chemical reactor system. *JMEMS* 2001, *10*, 232–237.
69. Firebaugh, S. L., Jensen, K. F., and Schmidt, M. A., Miniaturization and integration of photoacoustic detection. *J. Appl. Phys.* 2002, *92*, 1555–1563.
70. Heaps, D. A. and Pellegrino, P., Examination of a quantum cascade laser source for a MEMS-scale photoacoustic chemical sensor. *Proc. SPIE* 2006, *6218*, 621805-1–621805-9.
71. Heaps, D. A. and Pellegrino, P., Investigations of intraband quantum cascade laser source for a MEMS-scale photoacoustic sensor. *Proc. SPIE* 2007, *6554*, 65540F-1–65540F-9.
72. Pellegrino, P. and Polcawich, R., Advancement of a MEMS photoacoustic chemical sensor. *Proc. SPIE* 2003, *5085*, 52–63.
73. Pellegrino, P., Polcawich, R., and Firebaugh, S. L., Miniature photoacoustic chemical sensor using microelectromechanical structures. *Proc. SPIE* 2004, *5416*, 42–53.
74. Gorelik, A. V. and Starovoitov, V. S., Small-size resonant photoacoustic cell with reduced window background for laser detection of gases. *Opt. Spectrosc.* 2009, *107*(5), 830–835.
75. Gorelik, A. V., Ulasevich, A. L., Nikonovich, F. N., Zakharich, M. P., Firago, V. A., Kazak, N. S., and Starovoitov, V. S., Miniaturized resonant photoacoustic cell of inclined geometry for trace-gas detection. *Appl. Phys. B: Lasers Opt.* 2010, *100*(2), 283–289.
76. Schafer, S., Miklòs, A., and Hess, P., Quantitative signal analysis in pulsed resonant photoacoustics. *Appl. Optics* 1997, *36*(15), 3202–3211.
77. Schafer, S., Miklòs, A., Pusel, A., and Hess, P., Absolute measurement of gas concentrations and saturation behavior in pulsed photoacoustics. *Chem. Phys. Lett.* 1998, *285*(3–4), 235–239.
78. Brand, C., Winkler, A., Hess, P., Miklòs, A., Bozoki, Z., and Sneider, J., Pulsed-laser excitation of acoustic modes in open high-Q photoacoustic resonators for trace gas monitoring—Results for C_2H_4. *Appl. Optics* 1995, *34*(18), 3257–3266.
79. Miklòs, A., Brand, C., Winkler, A., and Hess, P., Effective noise-reduction on pulsed-laser excitation of modes in a high-Q photoacoustic resonator. *J. Phys. IV* 1994, *4*(C7), 781–784.
80. Leugers, M. A. and Atkinson, G. H., Quantitative-determination of acetaldehyde by pulsed laser photoacoustic-spectroscopy. *Anal. Chem.* 1984, *56*(6), 925–929.
81. Kosterev, A. A., Bakhirkin, Y. A., Curl, R. F., and Tittel, F. K., Quartz-enhanced photoacoustic spectroscopy. *Opt. Lett.* 2002, *27*(21), 1902–1904.
82. Dong, L., Kosterev, A. A., Thomazy, D., and Tittel, F. K., QEPAS spectrophones: Design, optimization, and performance. *Appl. Phys. B: Lasers Opt.* 2010, *100*(3), 627–635.
83. Jalink, H. and Bicanic, D., Concept, design, and use of the photoacoustic heat pipe cell. *Appl. Phys. Lett.* 1989, *55*(15), 1507–1509.
84. Schmid, T., Panne, U., Niessner, R., and Haisch, C., Optical absorbance measurements of opaque liquids by pulsed laser photoacoustic spectroscopy. *Anal. Chem.* 2009, *81*(6), 2403–2409.
85. Sanchez, R. R., Rieumont, J. B., Cardoso, S. L., da Silva, M. G., Sthel, M. S., Massunaga, M. S. O., Gatts, C. N., and Vargas, H., Photoacoustic monitoring of internal plastification in poly(3-hydroxybutyrate-co-3-hydroxyvalerate) copolymers: Measurements of thermal parameters. *J. Braz. Chem. Soc.* 1999, *10*(2), 97–103.
86. Beck, S. M., Cell coatings to minimize sample (NH_3 and N_2H_4) adsorption for low-level photoacoustic detection. *Appl. Optics* 1985, *24*(12), 1761–1763.

87. Pedersen, M. and McClelland, J., Optimized capacitive MEMS microphone for photo-acoustic spectroscopy (PAS) applications. *Proc. SPIE: Int. Soc. Opt. Eng.* 2005, *5732*(1), 108–121.

88. Kauppinen, J., Wilcken, K., Kauppinen, I., and Koskinen, V., High sensitivity in gas analysis with photoacoustic detection. *Microchem J.* 2004, *76*(1–2), 151–159.

89. Koskinen, V., Fonsen, J., Kauppinen, J., and Kauppinen, I., Extremely sensitive trace gas analysis with modern photoacoustic spectroscopy. *Vib. Spectrosc.* 2006, *42*(2), 239–242.

90. Koskinen, V., Fonsen, J., Roth, K., and Kauppinen, J., Progress in cantilever enhanced photoacoustic spectroscopy. *Vib. Spectrosc.* 2008, *48*(1), 16–21.

91. Sievila, P., Rytkonen, V. P., Hahtela, O., Chekurov, N., Kauppinen, J., and Tittonen, I., Fabrication and characterization of an ultrasensitive acousto-optical cantilever. *J. Micromech. Microeng.* 2007, *17*(5), 852–859.

92. Koskinen, V., Fonsen, J., Roth, K., and Kauppinen, J., Cantilever enhanced photoacoustic detection of carbon dioxide using a tunable diode laser source. *Appl. Phys. B: Lasers Opt.* 2007, *86*(3), 451–454.

93. Fonsen, J., Koskinen, V., Roth, K., and Kauppinen, J., Dual cantilever enhanced photoacoustic detector with pulsed broadband IR-source. *Vib. Spectrosc.* 2009, *50*(2), 214–217.

94. Lindley, R. E., Parkes, A. M., Keen, K. A., McNaghten, E. D., and Orr-Ewing, A. J., A sensitivity comparison of three photoacoustic cells containing a single microphone, a differential dual microphone or a cantilever pressure sensor. *Appl. Phys. B: Lasers Opt.* 2007, *86*(4), 707–713.

95. Kosterev, A. A., Tittel, F. K., Serebryakov, D. V., Malinovsky, A. L., and Morozov, I. V., Applications of quartz tuning forks in spectroscopic gas sensing. *Rev. Sci. Instrum.* 2005, *76*(4), 9.

96. Serebryakov, D. V., Cherkun, A. P., Loginov, B. A., and Letokhov, V. S., Tuning-fork-based fast highly sensitive surface-contact sensor for atomic force microscopy/near-field scanning optical microscopy. *Rev. Sci. Instrum.* 2002, *73*(4), 1795–1802.

97. Kosterev, A. A., Bakhirkin, Y. A., Tittel, F. K., McWhorter, S., and Ashcraft, B., QEPAS methane sensor performance for humidified gases. *Appl. Phys. B: Lasers Opt.* 2008, *92*(1), 103–109.

98. Petra, N., Zweck, J., Kosterev, A. A., Minkoff, S. E., and Thomazy, D., Theoretical analysis of a quartz-enhanced photoacoustic spectroscopy sensor. *Appl. Phys. B-Lasers Opt.* 2009, *94*(4), 673–680.

99. Saarela, J., Sand, J., Sorvajarvi, T., Manninen, A., and Toivonen, J., Transversely excited multipass photoacoustic cell using electromechanical film as microphone. *Sensors* 2010, *10*, 5294–5307.

100. Paajanen, M., Lekkala, J., and Kirjavainen, K., ElectroMechanical Film (EMFi)—A new multipurpose electret material. *Sens. Actuator A: Phys.* 2000, *84*(1–2), 95–102.

101. Hillenbrand, J. and Sessler, G. M., High-sensitivity piezoelectric microphones based on stacked cellular polymer films. *J. Acoust. Soc. Am.* 2004, *116*(6), 3267–3270.

102. Hordvik, A. and Schlossberg, H., Photoacoustic technique for determining optical-absorption coefficients in solids. *Appl. Optics* 1977, *16*(1), 101–107.

103. Farrow, M. M., Burnham, R. K., Auzanneau, M., Olsen, S. L., Purdie, N., and Eyring, E. M., Piezoelectric detection of photoacoustic signals. *Appl. Optics* 1978, *17*(7), 1093–1098.

104. Burt, J. A., Response of a fluid-filled piezoceramic cylinder to pressure generated by an axial laser-pulse. *J. Acoust. Soc. Am.* 1979, *65*(5), 1164–1170.

105. Emmony, D. C., Sigrist, M., and Kneubuhl, F. K., Laser-induced shock-waves in liquids. *Appl. Phys. Lett.* 1976, *29*(9), 547–549.

106. White, R. M., Generation of elastic waves by transient surface heating. *J. Appl. Phys.* 1963, *34*(12), 3559–3567.

107. Goncalves, S. S., Da Silva, M. G., Sthel, M. S., Cardoso, S. L., Sanchez, R. R., Rieumont, J. B., and Vargas, H., Determination of thermal and sorption properties of poly-3-hydroxy octanoate using photothermal methods. *Phys. Status Solidi A: Appl. Res.* 2001, *187*(1), 289–295.

108. Giubileo, G. and Puiu, A., Photoacoustic spectroscopy of standard explosives in the MIR region. *Nucl. Instrum. Methods Phys. Res. A* 2010, *623*(2), 771–777.

109. Wen, Q. and Michaelian, K. H., Mid-infrared photoacoustic spectroscopy of solids using an external-cavity quantum-cascade laser. *Opt. Lett.* 2008, *33*(16), 1875–1877.

110. Rabasovic, M. D., Nikolic, M. G., Dramicanin, M. D., Franko, M., and Markushev, D. D., Low-cost, portable photoacoustic setup for solid samples. *Meas. Sci. Technol.* 2009, *20*(9), 6.

111. Park, H. K., Kim, D., Grigoropoulos, C. P., Tam, A. C., Pressure generation and measurement in the rapid vaporization of water on a pulsed-laser-heated surface. *J. Appl. Phys.* 1996, *80*(7), 4072–4081.

112. Kaneko, S., Yotoriyama, S., Koda, H., and Tobita, S., Excited-state proton transfer to solvent from phenol and cyanophenols in water. *J. Phys. Chem. A* 2009, *113*(13), 3021–3028.

113. Samokhin, A. A., Vovchenko, V. I., Il'ichev, N. N., and Shapkin, P. V., Explosive boiling in water exposed to q-switched erbium laser pulses. *Laser Phys.* 2009, *19*(5), 1187–1191.

114. Kim, H., Yu, S. J., and Cho, S. H., Detection of pressure waves in water by using optical techniques. *J. Korean Phys. Soc.* 2008, *53*(4), 1906–1909.

115. Michaelian, K. H., *Photoacoustic Infrared Spectroscopy*. Hoboken, NJ: Wiley-Interscience, 2003.

116. Mandelis, A. and Hess, P., *Progress in Photothermal and Photoacoustic Science and Technology: Life and Earth Sciences*, Vol. 3. Bellinghan, WA: SPIE—The International Society for Optical Engineering, 1997.

117. Wan, J. K. S., Ioffe, M. S., and Depew, M. C., A novel acoustic sensing system for on-line hydrogen measurements. *Sens. Actuator B: Chem.* 1996, *32*(3), 233–237.

118. Calasso, I. G., Funtov, V., and Sigrist, M. W., Analysis of isotopic CO_2 mixtures by laser photoacoustic spectroscopy. *Appl. Optics* 1997, *36*(15), 3212–3216.

119. Bozoki, Z., Sneider, J., Gingl, Z., Mohacsi, A., Szakall, M., Bor, Z., and Szabo, G., A high-sensitivity, near-infrared tunable-diode-laser-based photoacoustic water-vapour-detection system for automated operation. *Meas. Sci. Technol.* 1999, *10*(11), 999–1003.

120. Beenen, A. and Niessner, R., Development of a photoacoustic trace gas sensor based on fiber-optically coupled NIR laser diodes. *Appl. Spectrosc.* 1999, *53*(9), 1040–1044.

121. Mohacsi, A., Bozoki, Z., and Niessner, R., Direct diffusion sampling-based photoacoustic cell for *in situ* and on-line monitoring of benzene and toluene concentrations in water. *Sens. Actuator B: Chem.* 2001, *79* (2–3), 127–131.

122. Barbieri, S., Pellaux, J. P., Studemann, E., and Rosset, D., Gas detection with quantum cascade lasers: An adapted photoacoustic sensor based on Helmholtz resonance. *Rev. Sci. Instrum.* 2002, *73*(6), 2458–2461.

123. Gondal, M. A., Baig, M. A., and Shwehdi, M. H., Laser sensor for detection of SF6 leaks in high power insulated switchgear systems. *IEEE Trans. Dielectr. Electr. Insul.* 2002, *9*(3), 421–427.

124. Pushkarsky, M. B., Webber, M. E., Baghdassarian, O., Narasimhan, L. R., and Patel, C. K. N., Laser-based photoacoustic ammonia sensors for industrial applications. *Appl. Phys. B: Lasers Opt.* 2002, *75*(2–3), 391–396.

125. Santiago, G., Slezak, V., and Peuriot, A. L., Resonant photoacoustic gas sensing by PC-based audio detection. *Appl. Phys. B: Lasers Opt.* 2003, *77*(4), 463–465.

126. Slezak, V., Codnia, J., Peuriot, A. L., and Santiago, G., Resonant photoacoustic detection of NO_2 traces with a Q-switched green laser. *Rev. Sci. Instrum.* 2003, *74*(1), 516–518.

127. Zeninari, V., Parvitte, B., Courtois, D., Kapitanov, V. A., and Ponomarev, Y. N., Methane detection on the sub-ppm level with a near-infrared diode laser photoacoustic sensor. *Infrared Phys. Technol.* 2003, *44*(4), 253–261.

128. Horstjann, M., Bakhirkin, Y. A., Kosterev, A. A., Curl, R. F., Tittel, F. K., Wong, C. M., Hill, C. J., and Yang, R. Q., Formaldehyde sensor using interband cascade laser based quartz-enhanced photoacoustic spectroscopy. *Appl. Phys. B: Lasers Opt.* 2004, *79*(7), 799–803.

129. Elia, A., Lugara, P. M., and Giancaspro, C., Photoacoustic detection of nitric oxide by use of a quantum-cascade laser. *Opt. Lett.* 2005, *30*(9), 988–990.

130. Webber, M. E., MacDonald, T. S., Pushkarsky, M. B., Patel, C. K. N., Zhao, Y. J., Marcillac, N., and Mitloehner, F. M., Agricultural ammonia sensor using diode lasers and photoacoustic spectroscopy. *Meas. Sci. Technol.* 2005, *16*(8), 1547–1553.

131. Besson, J. P., Schilt, S., Rochat, E., and Thevenaz, L., Ammonia trace measurements at ppb level based on near-IR photoacoustic spectroscopy. *Appl. Phys. B: Lasers Opt.* 2006, *85*(2–3), 323–328.

132. Besson, J. P., Schilt, S., and Thevenaz, L., Sub-ppm multi-gas photoacoustic sensor. *Spectrosc. Acta A: Molec. Biomolec. Spectr.* 2006, *63*(5), 899–904.

133. Besson, J. P., Schilt, S., Sauser, F., Rochat, E., Hamel, P., Sandoz, F., Nikles, M., and Thevenaz, L., Multi-hydrogenated compounds monitoring in optical fibre manufacturing process by photoacoustic spectroscopy. *Appl. Phys. B: Lasers Opt.* 2006, *85*(2–3), 343–348.

134. Cattaneo, H., Laurila, T., and Hernberg, R., Photoacoustic detection of oxygen using cantilever enhanced technique. *Appl. Phys. B: Lasers Opt.* 2006, *85*(2–3), 337–341.

135. da Silva, M. G., Miklòs, A., Falkenroth, A., and Hess, P., Photoacoustic measurement of N_2O concentrations in ambient air with a pulsed optical parametric oscillator. *Appl. Phys. B: Lasers Opt.* 2006, *82*(2), 329–336.

136. Grossel, A., Zeninari, V., Joly, L., Parvitte, B., Courtois, D., and Durry, G., New improvements in methane detection using a Helmholtz resonant photoacoustic laser sensor: A comparison between near-IR diode lasers and mid-IR quantum cascade lasers. *Spectroc. Acta A: Molec. Biomolec. Spectr.* 2006, *63*(5), 1021–1028.

137. Mattiello, M., Nikles, M., Schilt, S., Thevenaz, L., Salhi, A., Barat, D., Vicet, A., Rouillard, Y., Werner, R., and Koeth, J., Novel Helmholtz-based photoacoustic sensor for trace gas detection at ppm level using GaInAsSb/GaAlAsSb DFB lasers. *Spectroc. Acta A: Molec. Biomolec. Spectr.* 2006, *63*(5), 952–958.

138. Schilt, S., Besson, J. P., and Thevenaz, L., Near-infrared laser photoacoustic detection of methane: The impact of molecular relaxation. *Appl. Phys. B: Lasers Opt.* 2006, *82*(2), 319–328.

139. Varga, A., Bozoki, Z., Szakall, M., and Szabo, G., Photoacoustic system for on-line process monitoring of hydrogen sulfide (H_2S) concentration in natural gas streams. *Appl. Phys. B: Lasers Opt.* 2006, *85*(2–3), 315–321.

140. Wojcik, M. D., Phillips, M. C., Cannon, B. D., and Taubman, M. S., Gas-phase photoacoustic sensor at 8.41 μm using quartz tuning forks and amplitude-modulated quantum cascade lasers. *Appl. Phys. B: Lasers Opt.* 2006, *85*(2–3), 307–313.

141. Gondal, M. A. and Yamani, Z. H., Highly sensitive electronically modulated photoacoustic spectrometer for ozone detection. *Appl. Optics* 2007, *46*(29), 7083–7090.

142. Grossel, A., Zeninari, V., Parvitte, B., Joly, L., and Courtois, D., Optimization of a compact photoacoustic quantum cascade laser spectrometer for atmospheric flux measurements: Application to the detection of methane and nitrous oxide. *Appl. Phys. B: Lasers Opt.* 2007, *88*(3), 483–492.

143. Lendl, B., Ritter, W., Harasek, M., Niessner, R., and Haisch, C. Photoacoustic monitoring of CO/sub 2/ in biogas matrix using a quantum cascade laser, In *Proceedings of the IEEE Sensors 2006*, Daegu, South Korea, October 22–25. Daegu, South Korea: IEEE, 2007, pp. 338–341.
144. Lewicki, R., Wysocki, G., Kosterev, A. A., and Tittel, F. K., Carbon dioxide and ammonia detection using 2 mu m diode laser based quartz-enhanced photoacoustic spectroscopy. *Appl. Phys. B: Lasers Opt.* 2007, *87*(1), 157–162.
145. Besson, J. P., Schilt, S., and Thevenaz, L., Molecular relaxation effects in hydrogen chloride photoacoustic detection. *Appl. Phys. B: Lasers Opt.* 2008, *90*(2), 191–196.
146. Wolff, M., Germer, M., Groninga, H. G., and Harde, H., Photoacoustic CO_2 sensor based on a DFB diode laser at 2.7 µm. *Eur. Phys. J.: Spec. Top.* 2008, *153*, 409–413.
147. Adamson, B. D., Sader, J. E., and Bieske, E. J., Photoacoustic detection of gases using microcantilevers. *J. Appl. Phys.* 2009, *106*(11), 4.
148. Giubileo, G., Puiu, A., Dell'Unto, F., Tomasi, M., and Fagnani, A., High resolution laser-based detection of ammonia. *Laser Phys.* 2009, *19*(2), 245–251.
149. Gondal, M. A., Dastageer, A., and Yamani, Z. H., Laser-induced photoacoustic detection of ozone at 266 nm using resonant cells of different configuration. *J. Environ. Sci. Health A: Toxic/Hazard. Subst. Environ. Eng.* 2009, *44*(13), 1457–1464.
150. Elia, A., Di Franco, C., Spagnolo, V., Lugara, P. M., and Scamarcio, G., Quantum cascade laser-based photoacoustic sensor for trace detection of formaldehyde gas. *Sensors* 2009, *9*(4), 2697–2705.
151. Liu, K., Li, J., Wang, L., Tan, T., Zhang, W., Gao, X., Chen, W., and Tittel, F. K., Trace gas sensor based on quartz tuning fork enhanced laser photoacoustic spectroscopy. *Appl. Phys. B: Lasers Opt.* 2009, *94*(3), 527–533.
152. Schilt, S., Kosterev, A. A., and Tittel, F. K., Performance evaluation of a near infrared QEPAS based ethylene sensor. *Appl. Phys. B: Lasers Opt.* 2009, *95*(4), 813–824.
153. Yong, P., Wang, Z., Liang, L., and Qingxu, Y., Tunable fiber laser and fiber amplifier based photoacoustic spectrometer for trace gas detection. *Spectrochim. Acta A: Mol. Biomol. Spectrosc.* 2009, *74*(4), 924–927.
154. Di Franco, C., Elia, A., Spagnolo, V., Lugara, P. M., and Scamarcio, G., Advanced optoacoustic sensor designs for environmental applications. *Proc. SPIE: Int. Soc. Opt. Eng.* 2010, *7808*, 78081A (8pp.).
155. Rey, J. M., Romer, C., Gianella, M., and Sigrist, M. W., Near-infrared resonant photoacoustic gas measurement using simultaneous dual-frequency excitation. *Appl. Phys. B: Lasers Opt.* 2010, *100*(1), 189–194.
156. Serebryakov, D. V., Morozov, I. V., Kosterev, A. A., and Letokhov, V. S., Laser microphotoacoustic sensor of ammonia traces in the atmosphere. *Quantum Electron.* 2010, *40*(2), 167–172.
157. Bozoki, Z., Szabo, A., Mohacsi, A., and Szabo, G., A fully opened photoacoustic resonator based system for fast response gas concentration measurements. *Sens. Actuator B: Chem.* 2010, *147*(1), 206–212.
158. Spagnolo, V., Kosterev, A. A., Dong, L., Lewicki, R., and Tittel, F. K., NO trace gas sensor based on quartz-enhanced photoacoustic spectroscopy and external cavity quantum cascade laser. *Appl. Phys. B: Lasers Opt.* 2010, *100*(1), 125–130.
159. Grossel, A., Zeninari, V., Joly, L., Parvitte, B., Durry, G., and Courtois, D., Photoacoustic detection of nitric oxide with a Helmholtz resonant quantum cascade laser sensor. *Infrared Phys. Technol.* 2007, *51*(2), 95–101.
160. Zeninari, V., Grossel, A., Joly, L., Decarpenterie, T., Grouiez, B., Bonno, B., and Parvitte, B., Photoacoustic spectroscopy for trace gas detection with cryogenic and room-temperature continuous-wave quantum cascade lasers. *Cent. Eur. J. Phys.* 2010, *8*(2), 194–201.
161. Heumier, T. A. and Carlsten, J. L. Mode hopping in semiconductor lasers. Application Note No. 8. Bozeman, MT: ILX Lightwave Corporation, 1992. Available at: http://www.ilxlightwave.com/appnotes/mode_hopping_semiconductor_lasers.pdf

162. Bauer, C., Willer, U., Lewicki, R., Pohlkotter, A., Kosterev, A., Kosynkin, D., Tittel, F. K., and Schade, W., A mid-infrared QEPAS sensor device for TATP detection. *J. Phys. Conf. Ser.* 2009, *157*, 012002 (6pp.).

163. Boschetti, A., Bassi, D., Iacob, E., Iannotta, S., Ricci, L., and Scotoni, M., Resonant photoacoustic simultaneous detection of methane and ethylene by means of a 1.63-μm diode laser. *Appl. Phys. B: Lasers Opt.* 2002, *74*(3), 273–278.

164. Hanyecz, V., Mohacsi, A., Pogany, A., Varga, A., Bozoki, Z., Kovacs, I., and Szabo, G., Multi-component photoacoustic gas analyzer for industrial applications. *Vib. Spectrosc.* 2010, *52*(1), 63–68.

165. Foster, N. S., Amonette, J. E., Autrey, T., and Ho, J. T., Detection of trace levels of water in oil by photoacoustic spectroscopy. *Sens. Actuator B: Chem.* 2001, *77*(3), 620–624.

166. Hodgson, P., Quan, K. M., Mackenzie, H. A., Freeborn, S. S., Hannigan, J., Johnston, E. M., Greig, F., and Binnie, T. D., Application of pulsed-laser photoacoustic sensors in monitoring oil contamination in water. *Sens. Actuator B: Chem.* 1995, *29*(1–3), 339–344.

167. Hernández-Valle, F., Navarrete, M., Mejia, E. V., and Villagran-Muniz, M., Trace analysis of pesticides in water using pulsed photoacoustic technique. *Eur. Phys. J.: Spec. Top.* 2008, *153*, 507–510.

168. Petzold, A. and Niessner, R., Photoacoustic soot sensor for in-situ black carbon monitoring. *Appl. Phys. B: Lasers Opt.* 1996, *63*(2), 191–197.

169. Chaudhary, A. K., Bhar, G. C., and Das, S., Low-limit photo-acoustic detection of solid RDX and TNT explosives with carbon dioxide laser. *J. Appl. Spectrosc.* 2006, *73*(1), 123–129.

170. Van Neste, C. W., Morales-Rodriguez, M. E., Senesac, L. R., Mahajan, S. M., and Thundat, T., Quartz crystal tuning fork photoacoustic point sensing. *Sens. Actuator B: Chem.* 2010, *150*(1), 402–405.

171. De Albuquerque, J. E., Balogh, D. T., and Faria, R. M., Quantitative depth profile study of polyaniline films by photothermal spectroscopies. *Appl. Phys. A: Mater. Sci. Process.* 2007, *86*(3), 395–401.

172. Xu, M. H. and Wang, L. H. V., Photoacoustic imaging in biomedicine. *Rev. Sci. Instrum.* 2006, *77*(4), 22.

173. Uchiyama, K., Yoshida, K., Wu, X. Z., and Hobo, T., Open-ended photoacoustic cells: Application to two-layer samples using pulse laser-induced photoacoustics. *Anal. Chem.* 1998, *70*(3), 651–657.

174. Viator, J. A., Komadina, J., Svaasand, L. O., Aguilar, G., Choi, B., and Nelson, J. S., A comparative study of photoacoustic and reflectance methods for determination of epidermal melanin content. *J. Invest. Dermatol.* 2004, *122*(6), 1432–1439.

175. Schmid, T., Helmbrecht, C., Panne, U., Haisch, C., and Niessner, R., Process analysis of biofilms by photoacoustic spectroscopy. *Anal. Bioanal. Chem.* 2003, *375*(8), 1124–1129.

176. Van Neste, C. W., Senesac, L. R., and Thundat, T., Standoff photoacoustic spectroscopy. *Appl. Phys. Lett.* 2008, *92*, 234102-1–234102-3.

177. Perrett, B., Harris, M., Pearson, G. N., Willetts, D. V., and Pitter, M. C., Remote photoacoustic detection of liquid contamination of a surface. *Appl. Opt.* 2003, *42*(24), 4901–4908.

178. Van Neste, C. W., Senesac, L. R., and Thundat, T., Standoff spectroscopy of surface adsorbed chemicals. *Anal. Chem.* 2009, *81*, 1952–1956.

7 Design of a Low-Cost Underwater Acoustic Modem

Bridget Benson and Ryan Kastner

CONTENTS

INTRODUCTION

Small, dense, wireless sensor networks are beginning to revolutionize our understanding of the physical world by providing fine resolution sampling of the surrounding environment. The ability to have many small devices streaming real-time data physically distributed near the objects being sensed brings new opportunities to observe and act on the world, which could provide significant benefits to mankind. For example, dense wireless sensor networks have been used in agriculture to improve the quality, yield, and value of crops, by tracking soil temperatures and informing farmers of fruit maturity and potential damages from freezing temperatures [1]. They have been deployed in sensitive habitats to monitor the causes for mortality in endangered species [2]. Dense wireless sensor networks have also been used to detect structural damages in bridges and other civil structures to inform authorities of needed repair [3] and have been used to monitor the vibration signatures of industrial equipment in fabrication plants to predict mechanical failures [4].

While wireless sensor-net systems are beginning to be fielded in applications on the ground, underwater sensor nets remain quite limited by comparison [5]. Still, a large portion of ocean research is conducted by placing sensors (that measure current speeds, temperature, salinity, pressure, bioluminescence, chemicals, etc.) into the ocean and later physically retrieving them to download and analyze their collected data. This method does not provide for real-time analysis of data which is critical for event prediction. Real-time underwater wireless sensor networks that do exist are often sparsely deployed over wide areas. For example, the Deep-ocean Assessment and Reporting of Tsunami (DART) project consists of 39 stations worldwide acquiring critical data for early detection of tsunamis [6]. The FRONT network consists of about 10 subsurface wirelessly networked sensors spaced about 9 km apart in the inner continental shelf outside Block Island Sound to increase scientific understanding of the coastal ocean [7]. The SeaWeb network consists of tens of nodes spaced 2–5 km apart for oceanographic telemetry, underwater vehicle control, and other uses of underwater wireless digital communications [8,9]. Other real-time networks that exist are wired and extremely expensive [10–13].

The existence of small, dense wireless sensor networks on land was made possible by the advent of low-cost radio platforms such as PicoRadio and Mica2 [14,15]. These radio platforms cost a few hundred U.S. dollars enabling researchers to purchase many nodes with a fixed budget allowing for dense, short-range deployment. The aquatic counterpart to the terrestrial radio is the underwater acoustic modem. A number of acoustic modems are currently available including commercial offerings from companies like Teledyne Benthos, DSPComm, LinkQuest, and Tritech, as well as academic projects, most notably the WHOI MicroModem. Unfortunately, these existing modems' power consumption, ranges, and price points are all designed for sparse, long-range, expensive systems rather than small, dense, and inexpensive sensor nets [5,16,17]. It is widely recognized that an aquatic counterpart to inexpensive terrestrial radio is required to enable deployment of small, dense, underwater wireless sensor networks for advanced underwater ecological analyses.

This chapter describes the design of a short-range underwater acoustic modem starting with the most critical component from a cost perspective—the transducer. The transducer is a device that converts electrical energy to/from acoustic energy, which is equivalent to the antenna in radios. The design substitutes a commercial underwater transducer with a homemade underwater transducer using cheap piezoceramic material and builds the rest of the modem's components around the properties of the transducer to extract as much performance as possible. We describe the modem's transducer design, followed by its analog transceiver design and digital transceiver design. We end the chapter describing real-world tests performed on the complete low-cost modem design which illustrate the modem provides bit rates of up to 200 bps for ranges up to 400 m at a components cost of U.S. $350.

This chapter consists of excerpts from the author's PhD thesis [18].

TRANSDUCER DESIGN

This section describes the design of a low-cost transducer (an electromagnetic device responsible for converting electrical energy to mechanical energy—sound pressure—and vice versa) used as the basis for the design of our low-cost underwater acoustic modem. We first describe the selection of the transducer's piezoceramic material based on its type and geometry. We then describe our transducer construction techniques including the selection of wiring and potting compound. We finally describe the calibration procedure used to measure the electromechanical properties of our homemade transducer and present the experimentally determined electromechanical properties that are used to govern the rest of the low-cost modem design.

Piezo Ceramics

In 1880, Jacques and Pierre Curie discovered that certain naturally occurring crystalline substances (such as quartz) exhibit an unusual characteristic: when subjected to a mechanical force, the crystals became polarized and when exposed to an electric field the crystals lengthened or shortened according to the polarity and in proportion to the strength of the field. These behaviors were labeled the piezoelectric effect and the inverse piezoelectric effect, respectively [19–21].

In the twentieth century, researchers began to manufacture synthetic materials that exhibit the piezoelectric effect using polycrystalline ceramics or certain synthetic polymers. These materials are relatively inexpensive to manufacture, physically strong, and chemically inert. Common compositions include lead zirconate-titanate (PZT) and barium-titanate [19]. The type of the ceramic and its geometry affect the ceramic's piezoelectric properties and are described in more detail in the following subsections.

Type

Although several standards exist, in the United States, piezoceramics are popularly classified into six types created by DOD-STD-1376A [22] which was replaced by MIL-STD-1376B [23] in 1995 and discontinued in 1999. Although the standard is no longer officially used, most ceramics manufacturers still use it as a guideline and have options of their product that comply with the standard. The six types of ceramics can be further lumped into two very general groups: hard and soft ceramics. Hard ceramics have low dielectric and mechanical loss and are generally better at producing a signal whereas soft ceramics have large dielectric losses, low mechanic quality factors, and poor linearity, but are generally better at receiving a signal [19,24]. Soft ceramics produce large displacements and wider signal bandwidth, but they exhibit greater hysteresis and are more susceptible to depolarization [19]. Either of these ceramics is still capable of producing and receiving signals regardless of which group it is in because of piezoelectric reciprocity. For underwater network communication where one transducer is used for transmitting and receiving for cost effectiveness, a piezoelectric element that is good at doing both is desired.

We selected to use a "hard" modified Navy Type I ceramic due to its low dielectric and mechanical loss and high electromechanical coupling efficiency making it suitable as both a transmitter and receiver. Also, ceramics manufactured from formulations of PZT are the most widely used because they exhibit greater sensitivity and higher operating temperatures, relative to ceramics of other compositions [20]. Typical Type I PZTs can experience up to 12 V_{pp} AC per 0.001 in. wall thickness without much effect to its electromechanical properties [25]. A thickness of 0.1 in. (2.54 mm) gives a maximum voltage of 1200 V_{pp} or 425 V_{rms}.

Geometry

After selecting the type of ceramic necessary for the application, the geometry selection is the next important step. The cost of the PZT element can vary greatly and is significantly affected by geometry. Not only are some shapes harder to make, more intricate shapes make poling the ceramic more difficult as well. Element geometry and polarization direction determine the radiated direction of acoustic signals as well as the electromechanical properties of the ceramic element itself, such as resonance frequency, capacitance, generated voltage under load, and displacement. Exactly how they are affected depends on the geometry selected. Simplified equations for commonly used shapes can be found on many ceramic manufacturers websites [20,24]. Geometry-independent properties include the electromechanical coupling coefficient, piezoelectric constants, dielectric constant, Curie temperature, and dissipation factor, which depend on the material.

For underwater communication, ceramics are usually omnidirectional in the horizontal plane to reduce reflection off the surface and bottom [21]. A radial expanding ring provides two-dimensional omnidirectionality in the plane perpendicular to the axis and near omnidirectionality in planes through the axis only if the height of the ring is small compared to the wavelength of sound being sent through the medium. Note, if the ring is too tall, extension along the axis and bending modes of the ring may become a problem [21].

A radial expanding ring ceramic element can be made of several ceramics cemented together, providing greater electromechanical coupling, power output, and electrical efficiency than one-piece ceramic rings. The piezoelectric constant and coupling coefficient are approximately double that of a one-piece ceramic ring [21]. They work better because the polarization can be done in the direction of primary stresses and strains along the circumference. However, these are much more difficult to manufacture and are therefore much more expensive than a one-piece radial expanding piezoelectric ceramic.

Thus, a one-piece modified Type I ring transducer with radial resonance mode, 26 mm outer diameter, 22 mm inner diameter, and 2.54 mm wall thickness was selected. Specifically, part SMC26D22H13SMQA from Steiner and Martins, Inc.* was purchased for approximately $10 per element with no minimum purchase. This is much less than many other piezoelectric manufacturers charge for ceramics with very similar geometry and electromechanical properties. Much of the cost difference stems from the difficulty of manufacturing PZTs with consistent properties under tight tolerances. Many piezoceramics for underwater communication are being used for oil and military use where they must comply with tight specifications and operate under extreme conditions. These extreme conditions include deep-ocean and long-range operation. Thus, many companies cater toward this market selling high-quality parts that are not necessarily needed for short-range underwater communication research. In this context lower quality does not mean lower performance, but a looser control on tolerances and increase in availability to drive down costs.

For a single radial expanding ceramic ring, the resonance frequency occurs when the circumference approximately equals the operating wavelength [21]. The resonance frequency and antiresonance frequency occur at minimum and maximum impedances, respectively [21,26]. In air, the resonance frequency is about 41 kHz for every inch in diameter; for the ring made of several ceramics cemented together, in the case that there is no inactive material (such as electrodes or cement), the resonance frequency is approx 37 kHz for every inch [21]. If the ring is too tall, problems with length extensional and bending modes of vibration may be experienced.

---

* Steminc, Steiner & Martins, Inc., http://www.steminc.com/

The bandwidth of the ceramic is broader if the thickness is much smaller than the radius [21]. The SMC26D22H13SMQA has an outer diameter of 26 mm (approx 1 in.) and has a nominal resonance frequency of about 43 kHz. Steiner and Martins, Inc. specifies that the ceramic ring has a nominal resonance frequency of 43 ± 1.5 kHz.

TRANSDUCER CONSTRUCTION

Although the piezoelectric element is a key component of the transducer, there are other aspects to manufacturing a transducer that are important to its performance. Wiring electrical leads, potting the piezoceramics, and reducing unwanted acoustic radiation should be paid special attention. Figure 7.1 depicts our raw piezoceramic, transducer before potting, and fully potted transducer.

Wiring

Using shielded cables to attach to the ceramic will greatly enhance the performance of the transducer. Unshielded wires can act as antennas and pick up much unwanted electromagnetic noise that can bury small signals received by the transducer. When soldering to the ceramic's electrodes, care should be taken to prevent contact of the soldering tip to the ceramic for more than a few seconds. Heating the ceramic above its Curie temperature will damage its electromechanical properties. If possible, the soldering iron's temperature should also be adjusted accordingly. Solder with 3% silver is recommended by Steiner and Martins, Inc.

Potting

The piezoelectric ceramic needs to be encapsulated in a potting compound to prevent contact with any conductive fluids. Urethanes are the most common material used for potting because of their versatility. The most important design consideration is to find a urethane that is acoustically transparent in the medium that the transducer will be used; this is more important for higher frequency or more sensitive applications where wavelength and amplitude are smaller than the thickness of the potting material. Many urethane manufacturers do not know the acoustical properties of their

FIGURE 7.1 (a) Raw lead zirconate-titanate; (b) Prepotted transducer; and (c) Potted transducer.

urethanes. However, information on similarities to the acoustical properties of water is becoming easier to find. Generally, similar density provides similar acoustical properties. Mineral oil is another good way to pot the ceramics because it is inert and has acoustical properties similar to those of water. Some prefer using mineral oil to urethane because it is not permanent. However, the oil still needs to be contained by something, which is often a urethane tube. A two-part urethane potting compound, EN12, manufactured by Cytec Industries* was selected as it has a density identical to that of water, providing for efficient mechanical to acoustical energy coupling.

Creating a transducer by potting, the ceramic shifts its resonant frequency due to the additional mass moving immediately around the transducer. The extent of the shift depends on the type, age, and amount of potting as well as the temperature and mixing method of the compound. Having tight control over these variables to ensure exact reproducibility requires expensive equipment. To keep costs low, we used the following simplistic potting method.

As urethanes are toxic, when potting the ceramic, touching or inhaling the fumes of the urethane compound should be avoided as specified by the Material Safety Data Sheet provided by the manufacturer. Urethanes must have the correct ratio of both parts. Adding more of one will not help it cure faster, will change its physical properties, and many times will prevent it from fully curing at all. A scale was used to measure the exact proportions of the two-part compound and the parts were hand-poured into a plastic container for mixing. Larger batches help keep mixing ratios more consistent between separate batches.

Urethanes are very susceptible to moisture absorption. The moisture creates bubbles within the urethane which are undesired because they absorb acoustic waves propagating from the piezoceramic material. Much of the moisture is introduced during mixing. Using plastic or metal spoons and containers over paper or wooden ones during mixing reduces the amount of bubbles that appear in the cured urethane. Unmixed urethane in opened containers are often stored in metal cans and covered with a blanket of nitrogen to reduce moisture absorption [27].

Although this was not done on our own transducer, placing a vacuum on the urethane mixture while curing will cause many of the bubbles to enlarge and rise out of the curing liquid. In addition, the curing urethane can subsequently be placed under pressure to reduce the size of the remaining bubbles. Higher vacuums and putting unmixed urethane components into a vacuum separately may cause boiling of certain chemicals that need to remain in the liquid components in order to cure [27]. Again, processing parameters in the urethane's technical specifications should be followed for best results. Tips on making molds and other potting tips can be found in Reference [28].

A tennis ball with a hole cut in it and an ABS pipe cap with holes drilled in it were glued together using silicone and used as our transducer's mold. The ABS cap was only used for mounting. Mixed and uncured urethane was poured into the mold through the holes in the ABS cap. After the urethane was fully cured, the tennis ball and silicone were removed.

The next section describes the procedures used to determine the electromechanical properties of the homemade transducer.

---

* Cytec Industries, http://www.cytec.com/

CALIBRATION PROCEDURE

The calibration of a transducer consists of the determination of its electromechanical response as a function of frequency, namely its transmitting voltage response (TVR) and its receiving voltage response (RVR). The TVR is defined as the sound pressure level experienced at 1 m range, generated by the transducer per 1 V of input voltage as a function of frequency. The RVR is a measure of the voltage generated by a plane wave of unit acoustic pressure at the receiver and is a function of frequency. The units of the transmitting response are typically expressed in dB/1 µPa/m and the units of the receiving response are typically expressed in dB/1 V/µPa. Although numerous calibration procedures exist [29], we used the "comparison method" as it is the simplest transducer calibration procedure wherein the output of the unknown transducer is compared with that of a previously calibrated reference transducer.

In the comparison method, the unknown and reference transducers are placed in a tank of water at a known separation (typically 1 m). To obtain the RVR, the reference transducer sends sinusoidal signals of a known duration across the desired frequency range and the unknown transducer collects the sinusoidal signals over the same duration at each frequency. The collected data represent the combination of the transmitting response of the reference transducer plus the receiving response of the unknown transducer and the effects of attenuation at the separation distance. The RVR may be calculated using the following equations:

$$D = 20\log_{10}\left(\frac{V_{receivier}}{V_{transmitter}}\right) \tag{7.1}$$

$$RVR = D_{from\,ref} + TVR_{ref} + A \tag{7.2}$$

where $V_{receiver}$ is the average amplitude voltage of the receiver over the known duration, $V_{transmitter}$ is the average amplitude voltage of the transmitter over the known duration, and A is the attenuation of the signal due to the separation distance.

To obtain the TVR, the unknown transducer sends sinusoidal signals of reference duration across the desired frequency range and the reference transducer collects the sinusoidal signals over the reference duration at each frequency. The collected data represent the combination of the transmitting response of the unknown transducer plus the receiving response of the reference transducer and the effects of attenuation at the separation distance. The TVR may similarly be calculated as

$$TVR = D_{to\,ref} - RVR_{ref} + A \tag{7.3}$$

EXPERIMENTAL MEASUREMENTS

To execute the comparison method, we suspended both the homemade transducer and the reference transducer, a spherical ITC 1042,* 0.18 m apart in the middle of a

---

* ITC-1042, Deep-water omnidirectional transducer, http://www.itc-transducers.com

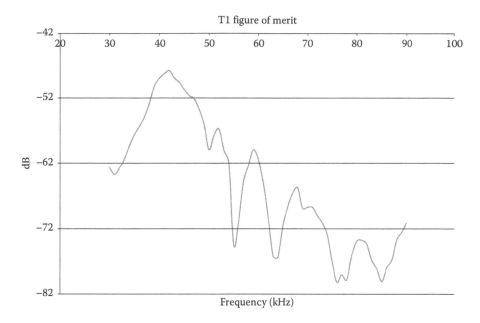

FIGURE 7.2   Transducer figure of merit.

3 m-deep, 2 m-wide cylindrical test tank filled with salt water. Burst signals of duration 2 ms, across frequencies in 1 kHz increments were sent from the reference transducer to the homemade transducer and vice versa. Signals were sent and collected via LabView. Adding the TVR and RVR gives a quantity known as the "Figure of Merit" which gives an indication of the transducer's best operating frequencies when acting as both a transmitter and a receiver. Figure 7.2 shows the Figure of Merit calculated from the LabView data indicating an operating frequency range around 40 kHz for transducer T1. The peaks and valleys of the Figure of Merit can be attributed to constructive and destructive interference caused by reflections off the sides of the small calibration tank.

Summary

This section described the design of our low-cost transducer, its ceramic type and geometry, its potting compound and procedure, and its electromechanical properties. The low-cost transducer costs ~$40, is omnidirectional in the horizontal plane, operates in a narrow frequency band around 40 kHz, and has a source level of about 140 dB re 1 μPa at 1 m. The next section describes the modem's analog transceiver that was designed to operate in the transducer's operating frequency range.[*]

---

[*] The text of this section is currently being prepared for submission for publication of the material. The section author was a coprimary researcher and author (with Kenneth Domond). Ryan Kastner and Don Kimball directed and supervised the research which forms the basis for this section.

ANALOG TRANSCEIVER DESIGN

The modem's analog transceiver consists of a power amplifier, a power management circuit, an impedance matching circuit, and a preamplifier (Figure 7.3). The power amplifier is responsible for amplifying the modulated signal from the digital hardware platform. It sends the signal to the power management circuit which further amplifies the signal to a power level that matches the actual distance between the transmitter and receiver. The power management circuit then sends the amplified signal to the impedance matching circuit, which is matched to the transducer's resonance frequency, so that the signal may be transmitted efficiently to the water. The preamplifier amplifies the signal that is detected by the transducer so that the digital hardware platform can effectively demodulate the signal and analyze the received data. This section describes the design of the power amplifier, power management circuit, impedance matching circuit, and preamplifier of the analog transceiver.

POWER AMPLIFIER DESIGN

When designing the power amplifier we considered the following requirements:

- The amplifier should provide a linear, undistorted output over a relatively wide bandwidth (10–100 kHz) to allow for use with a variety of underwater transducers.
- The amplifier must be power efficient (especially for large output power) as a deployed modem must be powered from batteries.

An amplifier is said to be linear if it preserves the details of the signal waveform, that is

$$V_0(t) = AV_i(t) \tag{7.4}$$

FIGURE 7.3   Analog transceiver.

The amplifier is said to be efficient if it can convert the majority of the DC power of the supply into the signal power delivered to the load. Efficiency is defined as

$$\text{Efficiency} = \frac{\text{Signal power delivered to load}}{\text{DC power supplied to output circuit}} \tag{7.5}$$

We designed a unique architecture that consists of a Class AB (known for being linear) and a Class D (known for being efficient) amplifier working in parallel to meet our design requirements. The Class AB amplifier provides a highly linear voltage gain of 27 across input voltages and frequencies. The output of the Class AB amplifier is connected to current sense circuitry that in turn controls the secondary amplifier, which is a Class D switching amplifier. The Class D amplifier is inherently nonlinear, but, when working in tandem with the Class AB amplifier, it produces a linear output for input voltages greater than 500 mV_{pp} across frequencies. The Class D amplifier provides high-power efficiency to the complete amplifier for large power outputs (where the load resistance is below 15 Ω) but must be turned off for lower-power outputs where its efficiency drops below that of the Class AB amplifier alone.

POWER MANAGEMENT CIRCUIT

The power management circuit is provided to adjust the output power of the transceiver in real time to match the actual distance between the transmitter and the receiver. The power management circuit takes the output of the power amplifier and further amplifies it to one of five power levels depending on how far the modem must transmit the signal. This is enabled through five different outputs at the taps on the secondary coil of the transformer. The number of windings on the secondary coil (N) divides the effective resistance of the load by N^2 thus increasing the power. The different transformer taps are connected to the impedance matching network (explained in the next section) by a series of relays. The relays are single pole, single throw, and are controlled by a 5 V, 40 mA signal. Each of the relay outputs is connected to one another but none are open unless the relay is energized.

The number of windings (N), voltage output, and transceiver power consumption of each power setting when connected to the homemade transducer and given a 1 V_{pp} input is given in Table 7.1. Note that because the output load of the transducer is high (750 Ω divided by the appropriate N^2), only the class AB amplifier is used in this current configuration of the transceiver.

IMPEDANCE MATCHING

Impedance matching is the practice of setting the input impedance of an electrical load equal to the output impedance of the signal source to which it is connected in order to maximize power transfer. Thus, the output impedance of the power amplifier must match the input impedance of the transducer. Matching is obtained when $Z_S = Z_L^*$, where Z_S is the impedance of the source (or power amplifier) and Z_L^* is the complex conjugate of the impedance of the load (or transducer).

TABLE 7.1
Power Management Characteristics

Power Level	No. Windings (N)	V_0 (V_{pp})	$P_{consumed}$ (W)
0	0	23	1.2
1	2.5	62	1.8
2	5	122	2.7
3	7.5	180	4.5
4	10	230	6.9

The transducer's impedance varies across frequencies as it is an active element that can be modeled with the RLC circuit shown in Figure 7.4 [21]. The static capacitance is the only physical component and is a direct consequence of the type of piezoelectric material and transducer geometry. It can be measured by an RLC meter and for the homemade transducer equals approximately 6.0 nF across frequencies.

To experimentally determine the transducer's electrical impedance, we measured the voltage, current, and phase difference between the voltage and current across frequencies. We then modeled the circuit in Figure 7.4 in PSpice [30] with values of the RLC circuit selected to match the characteristics of the measured values. The RLC values of 750 Ω, 2.3 mH, and 700 pF, respectively, provided good agreement between the measured and simulated values.

As the transducer is mostly capacitive, the impedance matching circuit consists of a single 2.5 mH inductor. This inductor value was experimentally chosen to make the circuit look mostly resistive (0 phase) around the operating frequency (40 kHz).

Although the power coupled into the transducer cannot directly be measured, we can use the equivalent circuit in PSpice to estimate the power delivered to the load. Figure 7.5 shows that the majority of the amplifier's output power is coupled into the transducer between 35 and 45 kHz.

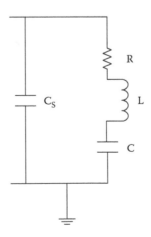

FIGURE 7.4 Electrical equivalent circuit model for a transducer.

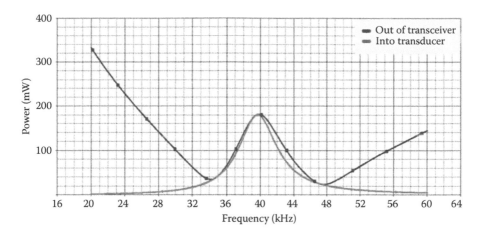

FIGURE 7.5   Estimated power coupled into the transducer.

PREAMPLIFIER DESIGN

When designing the preamplifier for the receiver, we considered the following requirements:

- The preamplifier must amplify signals around the transducer's resonance frequency (40 kHz) and filter out all other frequencies.
- The preamplifier must provide high gain to pick up signals as small as a couple of hundred microvolts.
- The design must be easily modifiable to accommodate different transducers with different resonance frequencies and bandwidths.

To meet the above design requirements of a highly sensitive, high-gain, narrow-band receiver, the architecture consists of two main components: (1) a 40-dB-per-decade roll-off high-pass filter and (2) an 80-dB-per-decade roll-off band-pass filter.

As underwater noise is concentrated in low frequencies, the first stage (a high-pass filter) cancels out a majority of unwanted noise. The high-pass filter consists of two cascaded filters, each with a 20-dB-per-decade roll-off. Each filter has a gain of 10 and a cutoff frequency of 16 kHz, thus giving a total gain of 100 (40 dB). The second stage is a band-pass filter used to further amplify signals in the transducer's operating band. It consists of four cascaded biquad filters, each with a 20-dB-per-decade roll-off. The current configuration has the center frequency of the first and third filters set to 40 kHz and the center frequency of the second and fourth filters set to 41 kHz to obtain a flatter frequency response in the pass band. Thus, the combined preamplifier provides a gain of ~80 dB around 40 kHz while attenuating low frequencies at a rate of 120 dB per decade and high frequencies at a rate of 80 dB per decade (Figure 7.6). The preamplifier's gain, cutoff, and center frequencies can be easily modified by replacing a few standard resistor and capacitor components.

The current receiver configuration consumes about 240 mW when in standby mode and less about 275 mW when fully engaged.

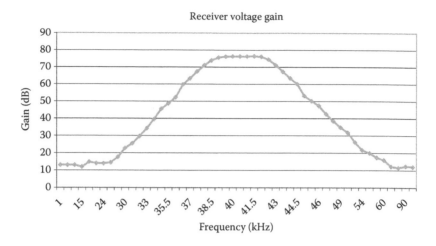

FIGURE 7.6   Overall receiver gain.

SUMMARY

This section described the full design of the analog transceiver including the power amplifier, power management circuit, impedance matching circuit, and preamplifier. The power amplifier is linear in the 10–100 kHz band for inputs greater than 500 mV$_{pp}$ and up to 95% efficient for high power outputs. The impedance matching circuit makes the transducer look resistive between 35 and 45 kHz coupling the majority of the amplifier's output power to the transducer in that frequency band. The preamplifier provides a flat, high gain for frequencies between 38 and 42 kHz matching the operating frequency of the transducer and allowing for reception of a signal as low as 200 µV. The power management circuit provides five different output power levels that consume 1.2, 1.8, 2.7, 4.5, and 6.9 W, respectively, allowing the modem to adjust the output power in real time to match it to the actual distance between the transmitter and the receiver. All components can be easily modified by replacing a few standard components.

The next section describes the design of the digital transceiver.

DIGITAL DESIGN

This section describes the digital design of the low-cost acoustic modem. We begin with a discussion on different modulation schemes, describing the reason for the selection of the use of frequency shift keying (FSK). We then discuss various digital hardware platforms and the selection of a field programmable gate array (FPGA) for our design. These discussions are followed by a brief description of the FPGA implementation of the FSK digital transceiver. The section concludes by presenting the resource requirements of the complete digital design.

MODULATION SCHEMES

Various modulation schemes have been implemented in existing commercial and research modems. These schemes all attempt to combat the performance limitations induced by the underwater acoustic channel while at the same time improving the bandwidth efficiency and bit rate as much as possible. This section briefly describes the characteristics of some of these schemes and reasons for our selection of the use of FSK for our modem design.

Frequency Shift Keying

FSK is a simple modulation scheme that has been widely used in underwater communications over the past two decades [31,32]. In M-ary FSK, the data are transmitted by shifting the frequency of a continuous carrier to one of M discrete frequencies. The simplistic receiver typically compares the energy at different frequencies to infer what data have been sent. Using only noncoherent energy detection at the receiver, this scheme bypasses the need for phase tracking which is a very difficult task because of Doppler spread in the underwater channel [33]. To combat intersymbol interference caused by multipath and frequency spreading, guard intervals are typically inserted between successive symbol transmissions for channel clearing [34] or symbols with durations longer than the multipath spread are used. As a result, the data rate of FSK is very low. Frequency-hopped FSK improves the data rate as it does not need to wait for channel clearing but requires a larger bandwidth. The minimum theoretical bandwidth for a binary FSK scheme is twice the bit rate (in Hertz).

Phase Shift Keying

In M-ary phase shift keying (PSK), the data are transmitted by shifting the phase of a continuous carrier to one of M discrete phases. PSK signals require less transmitted power for a given probability of error than FSK systems and require less bandwidth (the minimum theoretical bandwidth for a PSK scheme is bit rate, in Hertz), but require more coherent detection, either by regenerating a local carrier in the receiver or by using differential detection in which the previous bit is used as the phase reference for the current bit [35]. Although coherent detection could be used with FSK, the complexity required in the receiver for carrier regeneration justifies the better bit error rate (BER) performance of PSK. To combat intersymbol interference caused by multipath and frequency spreading, channel equalization techniques are exploited. Decision feedback equalizers are used to track slowly varying and faster varying channels when combined with a phase-locked loop [36]. Parameters in the equalizer may have to be fine-tuned to meet channel conditions. As coherent modulation does not have to wait for channel clearing, higher bit rates may be achieved at the expense of a more complex receiver.

Direct Sequence Spread Spectrum

In direct sequence spread spectrum (DSSS) modulation, the transmitted signal takes up more bandwidth than the information signal that is being modulated by multiplying each symbol with a spreading code and transmitting the resulting sequence at a rate allowed by the wider bandwidth. Due to the autocorrelation properties of the spreading sequence, intersymbol interference caused by the underwater channel is suppressed by the despreading operation at the receiver. Channel estimation and tracking are needed if phase-coherent modulation such as PSK is used to map information bits to symbols before spreading [37]. For noncoherent DSSS, channel estimation and tracking may be avoided by using the information bits to select different spreading codes and comparing the amplitudes of the outputs from different matched filters (each matched to one spreading code) at the receiver [34]. DSSS techniques such as those seen in the study by Iltis et al. [38] require less transmitted power for a given probability of error in comparison with FSK.

Orthogonal Frequency Division Multiplexing

Orthogonal frequency division multiplexing (OFDM) is often referred to as multi-carrier modulation because it transmits signals over overlapping subcarriers simultaneously. Subcarriers (specific frequency bands) that experience higher signal-to-noise ratio (SNR) are allotted with a higher number of bits, whereas less bits are allotted to subcarriers experiencing lower SNR. Underwater OFDM has been shown to provide data rates of 12, 25, and 50 kbps with bandwidths of 12, 25, and 50 kHz, respectively [39]. The subcarriers are divided in a way so as to ensure that each carrier is long compared to the multipath spread in the channel [40,41] so that intersymbol interference may be ignored, greatly simplifying the receiver complexity. Although intersymbol interference may be ignored, the large Doppler spread in the underwater channel introduces significant interference between subcarriers; thus, receivers must be designed to overcome intercarrier interference increasing their complexity.

Selection of FSK

The proven robustness of FSK and its simplicity make it an attractive modulation scheme for our low-cost modem design for short-range, low-data rate applications. The scheme requires a simple receiver that can fit onto a small, low-power device as evidenced by the FSK mode of the WHOI modem that only uses 0.18 W for the receive processing as compared to its PSK receiver which uses 2 W for processing [42], or the UCSB DSSS AquaModem that uses 1.6 W for processing (D. Doonan, AquaModem Electronics Engineer, pers. comm., May 2006). Binary FSK also does not require a large bandwidth which is suitable for our narrow-band low-cost transducer and corresponding analog transceiver design. Although data rates are low, they are suitable for the low-data rate applications of interest.

HARDWARE PLATFORMS

This section briefly discusses four possible digital hardware platforms for our digital transceiver design including Microcontrollers (MCUs), Digital Signal Processors (DSPs), Application-Specific Processors (ASICs), and Field Programmable Gate Arrays (FPGAs). We highlight the advantages and disadvantages of each platform and describe the selection of an FPGA for the low-cost modem.

Microcontrollers

MCUs are general-purpose devices consisting of a relatively small CPU, clocks, timers, I/O ports, and memory. They are used for information processing and control and can be adapted to a wide variety of applications by software. MCUs offer design flexibility and nonrecurring engineering costs as they can be easily reprogrammed in software and are widely available. The disadvantage is that an MCU has limited computation abilities and is only a viable solution for relatively simple applications at low sample rates. A compact code that makes the most efficient use of the MCU architecture is essential.

A few research underwater acoustic modem designs make use of MCUs, including the Atmega128L [43,44] and the Blackfin 533 [45]. The design in the study by Wills et al. [43] is claimed to have a power consumption of only 25 mW, but it relies on an RF ASIC to do most of the processing. The Blackfin 533 used in the study by Vasilescu et al. [45] is claimed to have a power consumption of 280 mW [46].

Digital Signal Processors

DSPs are specialized microprocessors with an optimized architecture for the fast operational needs of digital signal processing applications [47]. DSPs offer many architectural features that reduce the number of instructions necessary for efficient signal processing. Integrated specialized compute engines increase performance by executing complex functions in hardware. Like MCUs, DSPs offer design flexibility and nonrecurring engineering costs as they can be easily reprogrammed in software and are widely available. Until recent years, DSPs were notoriously power hungry, but with the advent of new low-power, fixed-point DSPs, power consumption has come down.

Most existing research on underwater modem designs (both commercial and research) make use of DSPs as they are relatively easy to program and can meet the computational requirements of more complex digital signal processing algorithms. The DSSS design in the study by Iltis et al. [38] and the PSK design in the study by Freitag et al. [31] both make use of a floating-point TI TMS320C6713 processor and consume 2 and 1.6 W, respectively. The FSK mode used by Freitag et al. [31] makes use of a low-power fixed-point DSP consuming only 180 mW of power.

Application-Specific Processors

An ASIC is custom-designed for a particular application, with as much system functionality implemented on a single die. ASICs offer exceptional performance, small size, and low power as they optimize transistor use and clock cycles at the expense of flexibility. They have been shown to offer power consumption 20 times [48] lower than any competing platform [49]. ASICs have a long time to market and high nonrecurring engineering costs making them practical only for high-volume production or for designs that demand extremely tight size and power requirements. To the best of our knowledge, no ASIC underwater acoustic modem exists.

Field Programmable Gate Arrays

An FPGA is an integrated circuit designed to be configured by the customer or designer after being manufactured. FPGAs strike a balance between solely hardware

(ASIC) and solely software (DSP or MCU) solutions as they have the programmability of software with performance capacity approaching that of a custom hardware implementation. They also present designers with substantially more parallelism allowing for a more efficient application implementation [50–55]. Studies have shown that FPGAs have evolved into highly valued digital signal processing solution platforms that reduce overall system costs and power consumption for high-throughput applications [56,57]. (Further comparison studies are required to determine whether an FPGA provides power and cost benefits for simpler, lower-throughput applications [57].) However, they require specialized knowledge and increased design time over DSPs and MCUs.

No current underwater acoustic modem designs solely make use of an FPGA, although two designs use an FPGA for preprocessing [58,59]. Power consumption estimates are not provided.

Selection of FPGA

As FSK-based underwater acoustic modems have already been implemented on a DSP [31] and an MCU [46], we chose to implement our FSK modem on an FPGA to serve as a comparison (particularly in terms of cost and power consumption) to these designs. Also, FPGAs provide a relatively easy transition to ASIC should a large volume of modems need to be produced.

DIGITAL TRANSCEIVER

Now that we have described the selection of the modulation scheme and the hardware platform, this section provides a brief description of our FPGA implementation of the FSK digital transceiver including the digital down converter, the modulator/demodulator, the symbol synchronizer, and the hardware/software (HW/SW) codesign controller. The complete transceiver design is shown in Figure 7.7. Each component was designed in Verilog and initially tested individually in ModelSim* to verify its operation prior to system integration.

The digital transceiver design makes use of the parameters given in Table 7.2. The carrier frequency and frequency separation were selected to match the transducer's resonance frequency and narrow bandwidth. Nyquist sampling necessitates that the signal be sampled at twice the frequency of the highest-frequency component of the signal, but in practical applications four to six times sampling is desired. Thus, the given sampling frequency and processing frequency were selected to provide sufficient oversampling of the desired frequency component while being integer multiples of one another. The symbol duration was selected to provide a suitable raw bit rate for low-data rate sensor networking applications.

Digital Down Converter

The digital down converter is responsible for converting high-resolution signals to lower-resolution signals to simplify subsequent processing. It takes the incoming

---

* ModelSim SE 6.4a. http://model.com/content/modelsim-downloads

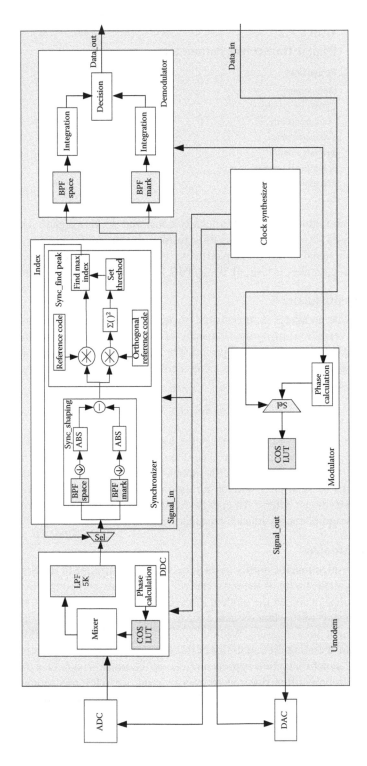

FIGURE 7.7 Block diagram of complete digital receiver.

TABLE 7.2
Digital Transceiver Parameters

Properties	Assignment
Modulation	FSK
Carrier frequency	40 kHz
Mark frequency	2 kHz
Space frequency	1 kHz
Symbol duration	5 ms
Sampling frequency	192 kHz
Processing frequency	16 kHz

signal *adc_in* and multiplies it with a locally generated 40 kHz signal. The mixed signal then passes through a low-pass filter to filter out the high-frequency components. Then the signal is down-sampled from 192 to 16 kHz to reduce processing power. The low-pass filter is a small 20-tap FIR filter designed using Spiral tool.*

Modulator/Demodulator

The modulator/demodulator is responsible for translating a bit stream into a waveform and vice versa by shifting the frequency of a continuous carrier to the "mark" or "space" frequency of each symbol period. The modulator takes a binary input and chooses to generate a sinusoidal wave using a cosine look-up table. The phase angle offset is calculated using the following formula:

$$\text{Offset} = \text{round}\left(\text{size} * F/F_s\right) \tag{7.6}$$

where "size" is the number of elements in the look-up table, F is the mark or space frequency, and F_s is the sampling rate. The demodulator uses the classic "matched" filter structure, which is optimal for FSK detection with white Gaussian noise interference. It works by sending a symbol duration of the received signal through two add-and-shift band-pass filters. An energy detection block is applied to determine the relative amount of energy in each frequency band.

Symbol Synchronizer

Symbol synchronization, the ability of the receiver to synchronize to the first symbol of an incoming data stream, is the most critical and complex component in our digital transceiver design. When the modem receiver obtains an input stream, it must be able to find the start of the data sequence to set accurate sampling and decision timing for subsequent demodulation. Without accurate symbol synchronization, higher BERs incur thus reducing the reliability of the wireless network.

Our symbol synchronization approach relies on the transmission of a predefined sequence of symbols, often referred to as a training, or reference sequence. The transmitter sends a packet that begins with the reference sequence and the receiver correlates the received sequence and the known reference sequence in order to locate

---

* Spiral. http://spiral.net/hardware/filter.html

the start of the packet (and start of the first symbol). When the reference and receiving sequence exactly align with each other, the correlation result reaches a maximum value and the synchronization point can be located as the maximum point above a predetermined threshold. We use a 15-bit Gold code as our reference sequence and perform a correlation with a 15-bit orthogonal Gold code to set a dynamic threshold. Details of our symbol synchronization implementation and design considerations can be found in the study by Li et al. [60].

HW/SW Codesign Controller

Xilinx Platform Studio 10.1 is applied to build an HW/SW codesign for accurate control and I/O of the digital transceiver. The codesign consists of the digital transceiver, a Universal Asynchronous Receiver Transmitter (UART), to connect to serial sensors or to a computer serial port for debugging, an interrupt controller to process interrupts received by the UART or the transceiver, logic to configure the onboard ADC, DAC, and clock generator, and MicroBlaze, an embedded microprocessor to control the system.

The MicroBlaze processor interfaces to the digital transceiver through two fast simplex links (FSLs), point-to-point, unidirectional asynchronous first-in, first-out (FIFOs) that can perform fast communication between any two design elements on the FPGA that implement the FSL interface. The MicroBlaze interfaces to the interrupt controller and UART core over a peripheral local bus (PLB), based on the IBM standard 64-bit PLB architecture specification.

Upon start-up, the MicroBlaze initializes communication with the digital transceiver by sending a command signal through the FSL bus signaling the transceiver to turn on. When the transceiver is ready to begin receiving signals, it sends an interrupt back to MicroBlaze to indicate initialization is complete. The transceiver then begins the down conversion and synchronization process, processing the signal received from the ADC and looking for a peak above the threshold to indicate a packet has been received. If the transceiver finds a peak above the threshold, it finds the synchronization point, and demodulates the packet. The demodulated bits are stored in the FSL FIFO. When the full packet has been demodulated, the transceiver sends an interrupt indicating a packet has been received and the MicroBlaze may retrieve the packet from the FSL. The transceiver then returns to synchronization, searching for the next incoming packet.

After initialization, the MicroBlaze remains idle, waiting for interrupts either from the transceiver or the UART. If it receives an interrupt from the transceiver indicating that a packet has been demodulated, the MicroBlaze reads the bits from the FSL FIFO and sends the bits over the UART to be printed on a computer's HyperTerminal for verification. If the MicroBlaze receives an interrupt from the UART, indicating that the user would like to send data, the MicroBlaze sends a command to the transceiver to send the bitstream the MicroBlaze places in the FSL. The transceiver then modulates the data from the FSL and sends the modulated waveform to the DAC for transmission. The MicroBlaze then returns to waiting for interrupts from the transceiver or the UART and the transceiver returns to synchronization, searching for the next incoming packet. This control flow is depicted in Figure 7.8.

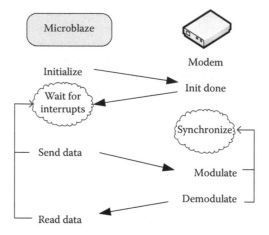

FIGURE 7.8   Digital transceiver control flow. Interrupts are shown in gray.

RESOURCE REQUIREMENTS

Table 7.3 shows the FPGA hardware resources occupied for each component of the digital transceiver. The resources reported for "Total" include the complete digital transceiver and HW/SW codesign controller. These resources were mapped for a Spartan 3 XC3S4000, the smallest device in the established Spartan 3 family that can fit the design.

We obtained a power estimate of the complete FSK modem design on various FPGA devices by entering the resource values of the total modem into the Xilinx XPower Estimator 9.1.03 and the Altera Cyclone IV PowerPlay Early Power Estimator. We acknowledge that the resource values determined for one device are not an exact estimate of the resource values needed for another device, but they do provide us with a reasonable first-order resource (and power) estimate. Also note that the estimates reported for the Xilinx devices are more accurate than those reported for the Altera device as the current design makes use of a Xilinx MicroBlaze core which would have to be replaced with the Altera NIOS processor should the design be implemented on an Altera chip. The devices reported are in device families known for their low power consumption (the Xilinx Spartan 6 and the Altera Cyclone

TABLE 7.3
Digital Design Resource Usage

	Occupied Slices	LUTs	BRAMs
Modulator	95	184	9
Down converter	284	541	9
Demodulator	1025	1980	1
Synchronizer	12,000	22,101	2
Total (% Spartan3)	16,706 (60%)	29,076 (51%)	55 (57%)

TABLE 7.4
Field Programmable Gate Array Power Consumption

Family	Device	Q Power (W)	D Power (W)	T Power (W)	Technology Size (nm)
Spartan 3	XC3S4000	0.274	0.105	0.379	90
Spartan 6	XC6SLX150T	0.212	0.021	0.255	45
Cyclone IV	EP4CE30	0.087	0.06	0.147	60

IV being some of the newest FPGA device families on the market). The particular devices reported are the smallest devices in their family that fit the total modem design. The letters "Q," "D," and "T" in Table 7.4 stand for "quiescent," "dynamic," and "total" power, respectively. The last column of Table 7.4 reports the size of the device family's CMOS technology.

From Table 7.4 we observe that the quiescent power contributes significantly to the total power of the design. Thus, it is important to select a device that offers low quiescent power. Altera has invested considerable resources in reducing static power in their products as evidenced by our design's low quiescent power on the Cyclone IV. Furthermore, the newer technologies (Spartan 6 and Cyclone IV) consume less power than the older technology, suggesting that, with continued technological advancements, FPGA power consumption will continue to decrease.

Table 7.5 compares the total digital hardware platform processing power and cost of our modem design with existing research underwater modem digital designs that report their total processing power. As previously described, these research modem designs use various modulation schemes (FSK, PSK, or DSSS) and are implemented on various hardware platforms.

TABLE 7.5
Digital Transceiver Design Comparison

Modem	Modulation	Category	Platform	Total Power (W)	Cost ($)
[42]	FSK	Fixed DSP	TMS320C5416	0.180	45
[42]	PSK	FP DSP	TMS320C6713	2.0	25
D. Doonan[a]	DSSS	FP DSP	TMS320C6713	1.6	25
[45]	FSK	MCU	Blackfin 533	0.280	25
Ours	FSK	FPGA	XC6SLX150T	0.233	14
			EP4CE30	0.147	40

Note: DSP, Digital Signal Processor; DSSS, direct sequence spread spectrum; FPGA, Field Programmable Gate Array; FSK, frequency shift keying; MCU, microcontroller; PSK, phase shift keying.

[a] AquaModem Electronics Engineer, pers. comm., May 2006.

From Table 7.5 we notice that the designs implemented on floating point DSPs consume considerably larger power than any of the other designs. However, these designs use more complex modulation schemes and thus require more resources. The FSK design in the "Micro-Modem Overview" [42] on the fixed-point DSP and the FSK design in the study by Vasilescu et al. [45] on an MCU provide comparable power consumption to our FSK design on an FPGA.

The costs reported for the DSP, MCU, and FPGA devices were all found on electronic distributer websites. These estimates suggest that the FPGA design also provides a comparable cost to other digital hardware platforms.

SUMMARY

This section described the design of the low-cost modem's FSK FPGA-based digital transceiver including the modulator, digital down converter, symbol synchronizer, demodulator, clock generator, and HW/SW codesign. The design operates at 200 bps in a narrow band around a 40 kHz carrier (to match the operating frequency of the transducer and the design of the analog transceiver) and provides comparable cost and power consumption to other FSK-based modem designs. The next section describes the system tests used to evaluate the functionality and performance of the complete modem design.[*]

SYSTEM TESTS

In order to verify the operation of the low-cost modem, we first tested the analog components and digital components separately and then tested the full, integrated system in three different underwater environments at various ranges. This section describes the results of each test and summarizes the performance of our modem.

ANALOG TESTING

To test the operation of our analog components (the transducer and analog transceiver), we took our analog hardware to Mission Bay, a salt water bay near San Diego, California. One analog transceiver and transducer was placed on the dock to act as the transmitter and another analog transceiver and transducer was placed on a boat to act as the receiver. The transmitter was powered by power supplies on the dock and the receiver was powered by a power supply connected to an inexpensive RadioShack AC/DC converter connected to the boat's power that unfortunately produced a substantial amount of noise (200 mV$_{pp}$).

We sent a 40 kHz sinusoid from the transmitter to the receiver placed at three different locations 1. 75, 2. 235, and 3. 350 m away. We were able to successfully detect the signal at 350 m at the fourth output power level; however, the received

---

[*] The text of Section "Digital Transceiver" is in part a reprint of the material as it appears in the proceedings [61]. The dissertation author is a co-primary researcher and author (with Ying Li). Ryan Kastner and Lan Chen directed and supervised the research which forms the basis for Section "Digital Transceiver."

signal was just above 200 mV$_{pp}$ at this distance and hence could just be detected above the converter's noise. This test proved that our analog hardware could transmit to a considerable distance and would likely be able to transmit to a farther distance given a low-noise power supply at the receiver and further improvements to the analog transceiver.

Digital Testing

For digital testing, we purchased a prototype test platform, the DINI DMEG-AD/ DA* that includes analog-to-digital and digital-to-analog converters, a Xilinx Virtex-4 FPGA, an onboard oscillator, and a serial port and downloaded the HW/SW codesign to the board. We also purchased two M-Audio ProFire 610 devices† to collect and store received waveforms for postanalysis. We conducted hard-wired and bucket tests of the digital transceiver as described in the subsections below.

Hard-Wired Tests

To verify the functionality of our complete FPGA digital transceiver, we downloaded the HW/SW codesign to the DINI board, connected the output of the DAC directly to the input of the ADC, and sent packets consisting of the 15-bit Gold code of "011001010111101" followed by 8 bits (for easy visual verification of the data) or 100 bits (a data length comparable to that sent by sensors) of randomized ones and zeros. We used HyperTerminal to give the command to send a packet at random times and to display the decoded results of the modem. We used ChipScope Pro‡ to view the internal waveforms.

Figure 7.9 shows some of the internal waveforms and signals from a 100-bit data length test. The blue signal shows the output of the band-pass filter centered on the space frequency in the demodulator and the green signal shows the output of the band-pass filter centered on the mark frequency in the demodulator. Note that the *data_in* signal exactly aligns with the waveforms. The decoded result (*decode_result*) is a delayed version of the input signal due to the 80, 16 kHz clock cycles required for synchronization initialization.

Fifty 100-bit data length packets were sent, all achieving perfect synchronization and 0% BER, thus verifying the correct operation of the hardware.

Bucket Tests

For our initial in-water tests of the digital hardware, we sent a packet consisting of the 15-bit Gold code of "011001010111101" followed by a 100-bit packet of randomized ones and zeros with the M-Audio device through a 12-in. bucket of fresh tap water and used the DINI board to synchronize and demodulate the data. Figure 7.10 shows a snapshot of the test result from postanalysis with ModelSim.

The four signals in the figure are: the output signal of the down converter ($DDC out$), the output of the reference cross-correlation block ($correlation$) used for

---

* DINI Group, DNMEG_ADDA, http://www.dinigroup.com/index.php?product=DNMEG_ADDA
† M-Audio Ltd, http://www.m-audio.com
‡ Xilinx Chipscope Pro, http://www.xilinx.com/tools/cspro.htm

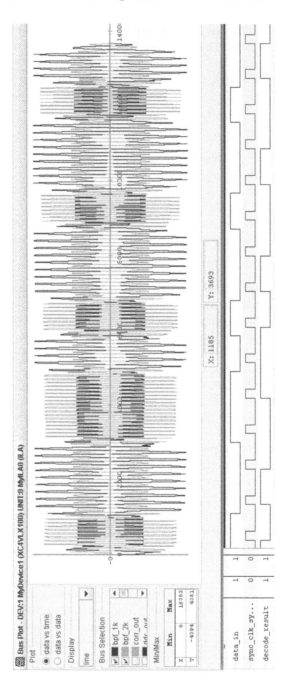

FIGURE 7.9 ChipScope internal waveforms and signals from a 100-bit data length test.

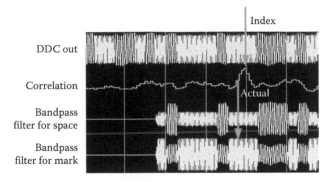

FIGURE 7.10   Snapshot of hardware simulation result for a 12-in. bucket test.

synchronization, and the output of the two band-pass filters in the demodulator. In the *DDC out* signal one can observe the FSK realization of the Gold code followed by the first 8 bits of data (the digital "0" being represented by the sparse waveform and the digital "1" being represented by the dense waveform). The vertical arrow labeled "Index" illustrates the synchronized peak found by the hardware which is a known clock delay from the start of the data (vertical arrow labeled "Actual"). The bits written to the HyperTerminal revealed 0% BER for the 100-bit packet from the 12-in. plastic bucket. Calculating the SNR as

$$SNR = 10 * \log_{10}\left(\frac{\mathrm{var(signal)}}{\mathrm{var(noise)}}\right) \tag{7.7}$$

at the input to the digital down converter, the *SNR* in the bucket was 33 dB. The test was repeated with different data bits 10 times, all producing 0% error.

INTEGRATED TESTS

After verifying the correct operation of the analog and digital hardware separately, we conducted integrated system tests of the complete modem design in a tank, pool, and lake. To protect the digital electronics, we added a voltage limiter to the output of the preamplifier to clip all signals above 1.3 V_{pp}. To characterize the multipath in the different environments, we sent a 200-ms 35 k–45 kHz chirp signal from the transmitter to the receiver to measure the multipath delay spread. The multipath measurements and test results are described in the following subsections.

Multipath Measurements

Underwater, there exist multiple paths from the transmitter to the receiver, or multipath. Two fundamental mechanisms of multipath formation are reflection at the boundaries (bottom, surface, and any objects in the water) and ray bending (where rays of sound bend toward regions of lower propagation speed). The amount of multipath seen at the receiver depends on the locations of the transmitter and the receiver and the geometric and physical properties of the environment.

The extent of the multipath at a receiver can be characterized by the multipath delay spread. Delay spread can be interpreted as the difference between the time of arrival of the first significant path (typically the line of site component) and the time of arrival of the last multipath component. Given the amplitude delay profile, $A_c(\tau)$, with effective signal length, M, the mean delay, $\bar{\tau}$, and the root mean square (rms) delay spread, τ_{rms}, are given as [62]:

$$\bar{\tau} = \frac{\sum_{n=1}^{m} \tau A_o(\tau)^2}{\sum_{n=1}^{m} \tau A_o(\tau)^2} \tag{7.8}$$

$$\tau_{rms} = \sqrt{\frac{\sum_{n=1}^{M} (\tau - \bar{\tau})^2 A_o(\tau)^2}{\sum_{n=1}^{M} A_o(\tau)^2}} \tag{7.9}$$

For FSK, multipath will cause intersymbol interference when the multipath delay spread is larger than the symbol duration. Intersymbol interference is a form of distortion of a signal in which one symbol interferes with subsequent symbols. This is an unwanted phenomenon as the previous symbols have similar effect as noise, thus making the communication less reliable [63]. Therefore, because the modem has a symbol duration of 5 ms, the delay spread of the channel must be less than 5 ms to ensure reliable communication.

To measure the multipath delay spread of the different test environments, we sent a 200-ms 35 k–40 kHz chirp signal from the transmitter to the receiver and used the DAQ device to collect 5 s of the received signal containing the chirp. We then post-processed the received signal, correlating the transmitted waveform (the 2 ms chirp) with the received waveform to form the amplitude delay profile. We then used Equations 7.8 and 7.9 to compute the rms delay spread. The amplitude delay profiles and rms delay spread of each test environment are given in the following subsections along with performance results.

Tank Tests

We conducted an initial full integrated system test in a $0.5 \times 1 \times 0.5$-m tank filled with fresh water with the transducers spaced 50 cm apart. Five packets consisting of the reference code followed by 1000 randomized bits were sent from the transmitter to the receiver using all power levels. Calculating the *SNR* as

$$SNR = 10 * \log_{10}\left(\frac{\text{var(signal)}}{\text{var(noise)}}\right) \tag{7.10}$$

at the input to the digital transceiver was 28 dB for all power levels as the signal was clipped to 1.3 V_{pp} at each level. All tests revealed 0% BER.

Canyon View Pool Tests

We conducted another full integrated system test at the UCSD Canyon View Pool, a 50 m × 25 m concrete pool with a depth of 1 m at the shallow end and 5 m at the deep end filled with chlorinated fresh water. At a distance of 50 m, we sent a packet of 400 symbols followed by a 400-symbol clearing period followed by another packet of 400 symbols using all power levels. The transducers were submerged to a depth of 10 cm and placed along the 50-m side of the pool to avoid swimmers. The digital hardware was able to successfully detect the start of each packet, but failed to accurately demodulate the data, achieving only 25% BER with an SNR of 14 dB (for all power levels). The rms delay spread in the pool was computed to be 21.3 ms, thus causing severe intersymbol interference. Concrete pools are one of the most difficult underwater channels due to extremely strong multipath. As expected, a longer symbol duration or a channel equalizer would have to be added to the modem to improve the performance of the modem in high-multipath environments.

Westlake Tests

After completing tests in a tank and a pool, we conducted full integrated system tests in Westlake, a freshwater lake in Westlake Village, California, at distances of 5, 50, 95, and 380 m. The transmitter was located on a dock, and the receiver was located on a boat. Five packets consisting of the reference code followed by 1000 randomized bits were sent from the transmitter to the receiver using all power levels. Tests at distances of 95 and 380 m were incomplete due to the inability to remain near a private dock for a length of time. The calculated delay spreads for the tests at distances of 5, 50, and 95 m were 2, 1.5, and 2.66 ms, respectively. Multipath at the distance of 380 m was not measured.

The SNR at the input to the digital transceiver was 34 dB for all power levels at a distance of 5 m, 21 dB for all power levels at a distance of 50 m, and 9 dB for power level 4 at a distance of 380 m (power levels 0–3 could not reach 380 m). BERs averaged 2.95%, 0.20%, and 4.13% at distances of 5, 50, and 380 m, respectively. BER and SNR measurements were not measured at 95 m.

Integrated System Test Summary

The integrated system test results in terms of distance, multipath delay spread, BER, and SNR are summarized in Figure 7.11. As anticipated, the modem performed well (having a BER of <5%) in environments with a multipath spread less than 5 ms. The results also suggest that higher SNR will only improve performance for environments with low multipath.

SUMMARY

This section described the analog, digital, and integrated system tests used to evaluate the functionality and performance of the complete modem design. These tests prove that a short-range underwater acoustic modem can be designed from low-cost components. The tests indicate that the modem can support data rates of 200 bps for ranges up to ~400 m with the power characteristics given in Table 7.1. Tables 7.6 and

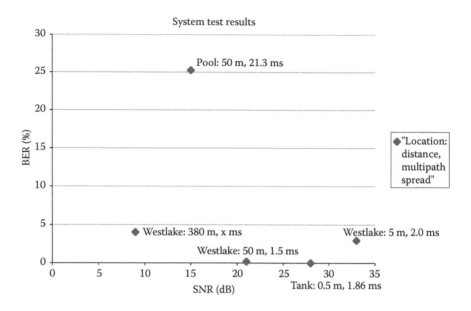

FIGURE 7.11   System test results.

TABLE 7.6

Commercial Underwater Acoustic Modem Comparison

Company	Modem	Frequency (kHz)	Tx Power (W)	Range (km)	Rx Power (W)	Modulation	Bit Rate (bps)	Cost ($)
Aquatec	AQUAModem	8–16	20	10	0.6	DSSS	300–2000	>7600
DSPComm	AquaComm	16–30	Varied	3	Varied	DSSS/OFDM	480	6600
TriTech	MicronModem	20–24	7.92	0.5	0.72	DSSS	40	3500
WHOI	MicroModem	25	<50	1–10	0.23/2	FSK/PSK	80/5400	8100/9400
Benthos	ATM885	16–21	28–84	2–6	0.7	FSK/PSK	140–15,360	7200–11,000
EvoLogics	S2CM48/78	48–78	2.5–80	1	0.5	S2C	15,000	12,500
LinkQuest	UWM2000H	NS	1.5	0.8	NS	Proprietary	9600	7000
Ours	Low cost	40	1.3–7.0	400	0.42	FSK	200	350[a]

Note: DSSS, direct sequence spread spectrum; FSK, frequency shift keying; PSK, phase shift keying; OFDM, orthogonal frequency division multiplexing.

[a]   Component cost only.

TABLE 7.7
Research Underwater Acoustic Modem Comparison

Modem	Platform	Modulation	Bit Rate (bps)	Range (km)	Bit Error Rate	Reference
USC	MCU	FSK	NS	NS	10^{-5} (coded)	[43]
UCI	Tmote	FSK	12	5	10%	[65]
uConn	DSP	Varied	Varied	Varied	NS	[66]
AquaModem	DSP	DSSS	133	440	1%	[38]
Kookmin	MCU	NS	5000	30	NS	[67]
AquaNode	MCU	FSK	300	400	NS	[45]
Ours	FPGA	FSK	200	400	4%	[18]

Note: DSP, Digital Signal Processor; DSSS, direct sequence spread spectrum; FPGA, Field Programmable Gate Array; FSK, frequency shift keying; MCU, microcontroller.

7.7 show how the modem compares with existing commercial and research modems, respectively.*

CONCLUSION

This section described the design and initial testing of a functional low-cost underwater acoustic modem prototype for short-range underwater sensor networks. The modem can support data rates of 200 bps for ranges up to ~400 m in environments with multipath delay spread of less than 5 ms. We conclude by discussing possible future improvements to the modem to make it more versatile for a wider variety of sensor network applications and underwater environments.

Power Reduction: Reducing power consumption (particularly idle power consumption) is essential to ensure a longer deployment lifetime on a limited battery supply. Further improvements could be made to the analog and digital transceivers to make them more power efficient and a low power wake-up circuit could be added to greatly reduce listening/idle power.

Wider Bandwidth, Higher Bit Rate: The existing FSK modulation scheme uses only a 1 kHz bandwidth that achieves a theoretical maximum bit rate of 500 bps and a practical bit rate of 200 bps. As the transducer is power efficient over a 5 kHz band, the bandwidth of the receiver could be widened to amplify signals over the 5 kHz range allowing for a modulation scheme that uses more bandwidth allowing higher data rates. However, the widening of the receiver bandwidth comes at a cost of reducing gain, thus reducing transmission range.

Low-Power DSP: The existing digital transceiver design was implemented on an FPGA that provides comparable power consumption to other low-power research

---

* Selections of Section "System Tests" are in part a reprint of the material as it appears in the Proceedings of the IEEE Oceans Conference. The dissertation author was a coprimary researcher and author along with Ying Li, Brian Faunce, and Kenneth Domond. The other coauthors listed in Reference [64] directed and supervised the research that forms the basis of Section "System Tests."

modem designs. Because the FPGA does not offer considerable power savings over the other designs (see Table 7.5) and because FPGA design time is so high, it may make sense to move the design to a low-power DSP. DSPs can be much more easily configured and reprogrammed allowing the user to easily modify the modulation scheme and its parameters.

Channel Adaptive Modem: As shown in the system test results, the current modem design can only perform well in environments with a low-multipath delay spread. An adaptive algorithm could be programmed into the modem to measure the channel characteristics and apply channel equalization and/or lengthen the symbol period for channels with high multipath.

REFERENCES

1. R. Beckwith, D. Teibel, and P. Bowen, Unwired wine: Sensor networks in vineyards, In *Proceedings of IEEE Sensors*, 2004, 2: 561–564, http://www.scopus.com/inward/record. url?eid=2-s2.0-27944461498&partnerID=40&md5=8feb4efc61403ac391fe2699fccf674e
2. A. Mainwaring, J. Polastre, R. Szewczyk, and D. Culler, Wireless sensor networks for habitat monitoring, In *Proceedings of the First ACM International Workshop on Wireless Sensor Networks and Applications*, Atlanta, Georgia, September 2002, pp. 88–97; doi: 10.1145/570738.570751.
3. J. Lynch, K. Law, E. Straser, A. Kiremidjian, and T. Kenny, The development of a wireless modular health monitoring system for civil structures, *Proceedings of the MCEER Mitigation of Earthquake Disaster by Advanced Technologies (MEDAT-2) Workshop*, Las Vegas, Nevada, November 30–31, 2000.
4. N. Ramanathan, M. Yarvis, J. Chhabra, N. Kushalnagar, L. Krishnamurthy, and D. Estrin, A stream-oriented power management protocol for low duty cycle sensor network applications, In *Proceedings of the IEEE Workshop on Embedded Networked Sensors*, May 2005, pp. 53–61, http://portal.acm.org/citation.cfm?id=1253400
5. J. Heidemann, W. Ye, J. Wills, A. Syed, and Y. Li, Research challenges and applications for underwater sensor networking, *Proceedings of IEEE Wireless Communications and Networking Conference (WCNC)*, Las Vegas, Nevada, pp. 228–235, April 2006, http://ieeexplore.ieee.org/xpls/abs_all.jsp?arnumber=1683469&tag=1
6. H. Milburn, A. Nakamura, and F. I. Gonzalez, Deep Ocean Assessment and Reporting of Tsunamis (DART): Real-time tsunami reporting from the deep ocean, *NOAA Online Report*, 1996, http://www.ndbc.noaa.gov/Dart/milburn_1996.shtml
7. D. Codiga, J. A. Rice, and P. S. Bogden, Real-time wireless delivery of subsurface coastal circulation measurements from distributed instruments using networked acoustic modems, *Proceedings of MTS/IEEE Oceans*, 2000, 1: 575–582.
8. J. G. Proakis, E. M. Sozer, J. A. Rice, and M. Stojanovic, Shallow water acoustic networks, *IEEE Communications Magazine*, 2001, 39(11): 114–119.
9. E. M. Sozer, M. Stojanovic, and J. G. Proakis, Underwater acoustic networks, *IEEE Journal of Oceanic Engineering*, 2000, 25: 72–83.
10. Ocean Research Interactive Observatory Networks (ORION) website, http://www. orionprogram.org
11. Laboratory for the Ocean Observatory Knowledge Integration Grid (LOOKING) website, http://lookingtosea.ucsd.edu.
12. Monterey Accelerated Research System (MARS) website, http://www.mbari.org/mars/
13. NEPTURE website, http://www.neptune.washington.edu

14. D. Estrin, L. Girod, G. Pottie, and M. Srivastava, Instrumenting the world with wireless sensor networks, *IEEE International Conference on Acoustics, Speech, and Signal Processing*, Salt Lake City, Utah, May 2001.
15. J. M. Rabaey, M. J. Ammer, J. L. da Silva, D. Patel, and S. Roundy, PicoRadio supports ad-hoc ultra-low power wireless networking, *IEEE Computer*, 2000, 33(7): 42–48.
16. I. F. Akyildiz, D. Pompili, and T. Melodia, Challenges for efficient communication in underwater acoustic sensor networks, *ACM Sigbed Review*, July 2004, 1(2).
17. R. Jurdak, C. V. Lopes, and P. Baldi, Battery lifetime estimation and optimization for underwater sensor networks, In S. Phoha, T. F. LaPorta, and C. Griffin (eds), *Sensor Network Operations*. Piscataway, NJ: IEEE Press, 2004, pp. 397–416.
18. Benson, Bridget. *Design of a Low-Cost Underwater Acoustic Modem for Short-Range Sensor Networks*. PhD thesis, Department of Computer Science and Engineering, University of California San Diego, 2010.
19. *Sensor Design Fundamentals: Piezoelectric Transducer Design for Marine Use*, Airmar Technology Corporation, 2000.
20. APC International Ltd. Piezo theory, http://www.americanpiezo.com/piezo_theory/index.html
21. C. H. Sherman and J. L. Butler, *Transducers and Arrays for Underwater Sound*. New York: Springer, 2007.
22. Allan C. Tims (ed.). Ad Hoc Subcommittee Report on Piezoceramics—Revision of DOD-STD-1376A. NRL Memorandum Report No. 5687. Washington DC: Naval Research Laboratory, April 1, 1986.
23. Navy SH. MIL-STD-1376B (Notice 1), *Military Standard Piezoelectric Ceramic Material and Measurements Guidelines for Sonar Transducers*, July 13, 1999.
24. Channel Industries Inc. *Piezoelectric Ceramic Catalog*, June 2009, www.channelindustries.com
25. Morgan Technical Ceramics (MTC ElectroCeramics), http://www.morganelectroceramics.com
26. O.B. Wilson, *An Introduction to Theory and Design of Sonar Transducers*. Los Altos, CA: Peninsula Publishing, 1985.
27. DeepSea Power & Light Facilities. *Urethanes, Silicones, and Epoxies Seminar*. San Diego, CA: DeepSea Power & Light, May 2009.
28. Smooth-On, Inc. *How-To Guide for Mold Making and Casting*, www.smooth-on.com.
29. R. Urick, *Principles of Underwater Sound for Engineers*. New York: McGraw-Hill Book Company, 1967.
30. Cadence PSpice A/D and Advanced Analysis, http://www.cadence.com/products/orcad/pspice_simulation/Pages/default.aspx
31. L. Freitag, M. Grund, S. Singh, J. Partan, P. Koski, and K. Ball, The WHOI micro-modem: An acoustic communications and navigation system for multiple platforms, *Proceedings of MTS/IEEE Oceans*, 2005, 2: 1086–1092.
32. D. B. Kilfoyle and A. B. Baggeroer, The state of the art in underwater acoustic telemetry, *IEEE Journal of Oceanic Engineering*, 2000, 25(1): 4–27.
33. I. F. Akyildiz, D. Pompili, and T. Melodia, Underwater acoustic sensor networks: Research challenges, *Ad Hoc Networks (Elsevier)*, 2005, 3(3): 257–279.
34. L. Liu, S. Zhou, and J.-H. Cui, Prospects and problems of wireless communication for underwater sensor networks, *Wireless Communications and Mobile Computing*, 2008, 8: 977–994.
35. A. Phadke, *Handbook of Electrical Engineering Calculations*. Technology and Engineering Series. New York: Marcel Dekker, 1999.
36. M. Stojanovic, J. Catipovic, J. G. Proakis, Phase coherent digital communications for underwater acoustic channels, *IEEE Journal of Oceanic Engineering*, 1994, 19(1): 100–111.

37. L. Freitag, M. Stojanovic, S. Singh, and M. Johnson, Analysis of channel effects on direct-sequence and frequency-hopped spread spectrum acoustic communications, *IEEE Journal of Oceanic Engineering*, 2001, 26(4): 586–593.
38. R. A. Iltis, H. Lee, R. Kastner, D. Doonan, T. Fu, R. Moore, and M. Chin. An underwater acoustic telemetry modem for eco-sensing, *Proceedings of MTS/IEEE Oceans*, 2005, 2: 1844–1850.
39. B. Li, S. Zhou, J. Huang, and P. Willett, Scalable OFDM design for underwater acoustic communications, In *Proceedings of the International Conference on ASSP*, Las Vegas, Nevada, March 3–April 4, 2008.
40. J. A. C. Bingham, Multicarrier modulation for data transmission: An idea whose time has come, *IEEE Communications Magazine*, 1990, 28(5): 5–14.
41. Z. Wang and G. B. Giannakis, Wireless multicarrier communications: Where Fourier meets Shannon, *IEEE Signal Processing Magazine*, 2000, 17(3): 29–48.
42. Woods Hole Oceanographic Institution, Micro-Modem Overview, http://acomms.whoi.edu/umodem/
43. J. Wills, W. Ye, and J. Heidemann, Low-power acoustic modem for dense underwater sensor networks, *Proceedings of the ACM International Workshop on Underwater Networks*, 2006; doi: 10.1145/1161039.1161055.
44. J. Namgung, N. Yun, S. Park, C. Kim, J. Jeon, and S. Park, Adaptive MAC protocol and acoustic modem for underwater sensor networks, Demo Presentation, *Proceedings of ACM International Workshop on Underwater Networks*, November 2009, http://wuwnet.acm.org/2009/posters/poster6.pdf
45. I. Vasilescu, C. Detweiler, and D. Rus, AquaNodes: An underwater sensor network, *Proceedings of ACM International Workshop on Underwater Networks*, September 2007, pp. 85–88; doi: 10.1145/1287812.1287830.
46. Texas Instruments News, Radio Locman Media, Analog devices' new Blackfin family offers the fastest and most power-efficient processors for their class, March 31, 2003, http://www.radiolocman.com/news/new.html?di=475
47. M. Yovits, *Advances in Computers*. New York: Academic Press, 1993, pp. 105–107.
48. A. Amara, F. Amiel, and T. Ea, FPGA vs. ASIC for low power applications, *Microelectronics Journal*, 2006, 37: 669–677.
49. E. Rocha, Implementation trade-offs of digital FIR filters. *Military Embedded Systems*, Open Systems Publishing, September 2007, http://tnt.etf.rs/~ms1dps/FIR%20filter.pdf
50. R. Kastner, A. Kaplan, and M. Sarrafzadeh, *Synthesis Techniques and Optimizations for Reconfigurable Systems*. Boston: Kluwer Academic, 2004.
51. W. Mangione-Smith, B. Hutchings, D. Andrews, A. DeHon, C. Ebeling, R. Hartenstein, O. Mencer et al., Seeking solutions in configurable computing, *Computer*, 1997, 30(12): 38–43.
52. A. DeHon and J. Wawrzynek, Reconfigurable computing: What, why, and implications for design automation, *Proceedings 1999 Design Automation Conference, IEEE*, June 1999, pp. 610–615.
53. K. Bondalapati and V. K. Prasanna, Reconfigurable computing systems, *Proceedings of the IEEE*, 2002, 90: 1201–1217.
54. K. Compton and S. Hauck, Reconfigurable computing: A survey of systems and software, *ACM Computing Surveys*, 2002, 34: 171–210.
55. P. Schaumont, I. Verbauwhede, K. Keutzer, and M. Sarrafzadeh, A quick safari through the reconfiguration jungle, *Proceedings of the 38th Design Automation Conference (IEEE Cat. No.01CH37232), ACM*, 2001, pp. 172–177.
56. M. LaPedus, FPGAs can outperform DSPs, study says, *DSP DesignLine*, November 13, 2006.
57. BDTI, FPGAs vs. DSPs: A look at the unanswered questions, *DSP DesignLine*, January 11, 2007.

58. Z. Yan, J. Huang, and C. He, Implementation of an OFDM underwater acoustic communication system on an underwater vehicle multiprocessor structure, *Frontiers of Electrical and Electronic Engineering in China*, 2007, 2(2): 151–155.

59. E. M. Sozer and M. Stojanovic, Reconfigurable acoustic modem for underwater sensor networks, *Proceedings of ACM International Workshop on Underwater Networks*, Los Angeles, California, September 25, 2006.

60. Y. Li, B. Benson, X. Zhang, and R. Kastner, Hardware implementation of symbol synchronization for underwater FSK, *IEEE International Conference on Sensor Networks, Ubiquitous, and Trustworthy Computing*, Newport Beach, California, June 7–9, 2010, pp. 82–88.

61. Y. Li, B. Benson, L. Chen, and R. Kastner, Determining the suitability of FPGAs for a low-cost, low-power, underwater acoustic modem, *International Conference on Remote Sensing*, October 2010, http://cseweb.ucsd.edu/~b1benson/publications/icrs10.pdf

62. H. Li, D. Liu, J. Li, and P. Stoica, Channel order and RMS delay spread estimation with application to AC power line communications, *Digital Signal Processing*, 2003,13: 284–300.

63. W. Dally and J. Poulton, *Digital Systems Engineering*. Cambridge, UK: Cambridge University Press, 1998, pp. 280–285.

64. B. Benson, Y. Li, B. Faunce, K. Domond, D. Kimball, C. Schurgers, and R. Kastner, Design of a low-cost, underwater acoustic modem for short-range sensor networks, *IEEE Oceans Conference*, Sydney, 2010, http://ieeexplore.ieee.org/xpl/freeabs_all.jsp?arnumber=5603816

65. R. Jurdak, C. V. Lopes, and P. Baldi, Software acoustic modems for short range mote-based underwater sensor networks, *Proceedings of IEEE Oceans Asia Pacific*, Singapore, May 2006, http://www.ics.uci.edu/~lopes/documents/oceans%2006/oceans06.pdf

66. H. Yan, S. Zhou, Z. Shi, and B. Li, A DSP implementation of OFDM acoustic modem, *Proceedings of ACM International Workshop on Underwater Networks*, September 2007.

67. J. Namgung, N. Yun, S. Park, C. Kim, J. Jeon, and S. Park, Adaptive MAC protocol and acoustic modem for underwater sensor networks, demo presentation, *Proceedings of ACM International Workshop on Underwater Networks*, November 2009

68. Marport Deep Sea Technology, http://www.marport.com/

69. H. Krauss, C. Bostian, and F. Raab, *Solid State Radio Engineering*. New York: John Wiley & Sons, 1980.

70. J. Honda and J. Adams, Application Note AN-1071, International Rectifier, Class D Audio Amplifier Basics, www.irf.com/technical-info/appnotes/an-1071.pdf

71. Electronics Tutorial about Amplifiers, Electronics-Tutorials, http://www.electronics-tutorials.ws/amplifier/amp_1.html

72. C. Verhoeven, A. van Staveren, G. Monna, M. Kouwenhoven, and E. Yildiz, *Structured Electronic Design: Negative Feedback Amplifiers*. Boston/Dordrecht: Kluwer Academic, 2003.

73. S. Bregni, *Synchronization of Digital Telecommunications Networks*. New York: John Wiley & Sons, 2002.

74. E. M. Sozer and M. Stojanovic, Underwater acoustic networks, *IEEE Journal of Oceanic Engineering*, 2000, 25(1): 72–83.

75. C. Chien, *Digital Radio System on a Chip: A System Approach*. Boston/Dordrecht: Kluwer Academic, 2001.

76. B. Watson, FSK: Signals and demodulation, *W. J. Communications*, July 2004, http://www.wj.com/pdf/technotes/FSK_signals_demod.pdf

Part 2

Magnetic and Mechanical Sensors

8 Accurate Scanning of Magnetic Fields

Hendrik Husstedt, Udo Ausserlechner,
and Manfred Kaltenbacher

CONTENTS

INTRODUCTION

MOTIVATION

Magnetic sensors are used in versatile automotive applications such as the position detection of throttle valves, cam- and crankshafts, pedals, wipers, winders, and so on. Before assembling these sensors in a car, each device is extensively tested not only electrically but also magnetically. Consequently, dedicated test equipment is needed to generate magnetic reference fields emulating the fields during application. These reference fields are used to decide which parts are sorted out and which are free from defects. Thus, it is essential that the magnetic field vector at the position of the device under test (DUT) exactly meets the desired properties such as direction and absolute value of the magnetic field. Errors of the reference field affect the testing of millions of parts, and, therefore, the equipment is not assembled in the test environment before it is well analyzed. For this purpose, a highly accurate three-dimensional magnetic sensing system is needed to measure the spatial dependency of the magnetic field in the volume where the DUT is positioned.

APPROACH

The magnetic field is described with three degrees of freedom in every point in space which is represented by a vector field. Today, there are integrated magnetic sensors available that measure precisely the strength of the magnetic field in one direction. Combining three one-dimensional sensors on one silicon die or in separate packages allows for measurements of all three components of the magnetic field. Attaching such a three-dimensional magnetic field sensor to moving axes realizes the positioning of the sensor to arbitrary points near the field source. Furthermore, a probe is added to the moving axes so that the geometry of the field source can be measured. We call such a system a magnetic and coordinate measuring machine (MCMM) which is able to measure the magnetic field of an arbitrary field source in all three directions and relate the results to the geometry.

OUTLOOK

The rest of the chapter is organized as follows. First, a short introduction to magnetic sensors is given focused on Hall sensors that are used for the presented measurement setup. Second, the functional principle of a CMM is discussed, which includes moving axes and a probe to measure geometrical characteristics of an object. Thirdly, the scanning of magnetic fields is explained, and the approach of standard measurement systems is shown. Then, the measurement principle of an MCMM is presented, which allows for measurements of strongly inhomogeneous magnetic fields. In this system, the position and alignment of the magnetic sensor has to be calibrated. To this end, a calibration method is shown that uses a reference field generated by a straight conductor. Moreover, a method is presented that calibrates the alignment of a magnetic sensor with high precision. Finally, a measurement example is given that illustrates the working principle of an MCCM.

MAGNETIC SENSORS

OVERVIEW

There is a wide choice of magnetic sensors for a multitude of applications [1]. The majority of these sensors are used to detect secondary information, for example, digital reading heads for hard disks, magnetic switches in car doors, or magnetic sensors that measure contactless position, angle, or rotational speed. For our application, a sensor is needed that provides the primary information of the magnetic field, respectively, the components of the magnetic flux density vector.

Today, the magnetic sensor with the highest sensitivity is the superconducting quantum interference device (SQUID) [2]. This sensor consists of a superconductive ring and one or two Josephson contacts. A drawback of a SQUID sensor is the low temperature for the superconductive material. Another magnetic sensor is based on the fluxgate principle. This sensor consists of a soft magnetic core surrounded by an excitation and sensing coil. While the excitation coil periodically drives the core in saturation, the field component in the direction of the core can be measured with the output signal of the sensing coil. Furthermore, we have to mention magnetoresistive sensors that change their electric resistance under the influence of a magnetic field. Most magnetoresistive sensors are based on spin-dependent effects, for example, the giant magnetoresistance [3], the anisotropic magnetoresistance [4], the tunnel magnetoresistance [3], and the colossal magnetoresistance [5,6]. Other magnetoresistive sensors like the ordinary magnetoresistance and the extraordinary magnetoresistance [7] are based on the Lorentz force. In addition, there are non-magnetoresistive sensors also based on the Lorentz force such as magnetodiodes and magnetotransistors [8]. However, the prevalent magnetic sensor based on the Lorentz force is the Hall sensor [9].

For scanning magnetic fields, the sensing principle is unimportant as long as the following requirements are fulfilled: (i) the sensor has a unique, linear characteristic with no hysteresis; (ii) it has an adequate spatial resolution; and (iii) it is sensitive to the magnetic field range of the field source. In the following section, we focus on magnetic fields in the range of 10 μT to 200 mT which are typically used for automotive applications. With this in mind, Hall sensors are the best choice for the measurement system.

HALL SENSORS

In 1879, American physicist Edwin H. Hall discovered a voltage between two opposite contacts of a conductive plate caused by a magnetic field. On the other two opposite contacts the plate was supplied with a current I, and the direction of the magnetic field was perpendicular to the plate. The effect is called the Hall effect and such a conductive plate with four contacts is called the Hall plate. A schematic drawing of a simple Hall plate geometry is shown in Figure 8.1.

In an electromagnetic field a charge Q is influenced by the Lorentz force

$$\vec{F}_L = Q\vec{E} + Q(\vec{v} \times \vec{B}).$$ (8.1)

The first term in Equation 8.1 describes the electrostatic force $\vec{F}_{el}$ and the second term the magnetic force $\vec{F}_{mag}$. It is assumed that the charge transfer consists only of electrons that are charged with the negative elementary charge $Q = -e_0$. Without any magnetic field, there is only an electrical field in the direction of the current density which corresponds to the y direction (see Figure 8.1). With an external magnetic field, the magnetic part of the Lorentz force causes a force pushing the electrons in the x direction. This separation of charges generates an additional electrical field in the x direction which is called the Hall field, $\vec{E}_H$. In the steady state, the electrical force from the Hall field completely compensates for the magnetic part of the Lorentz force:

$$-e_0(\vec{v} \times \vec{B}) - e_0\vec{E}_H = 0.$$ (8.2)

In the center of the Hall plate, the current density has a component in the y direction only, which can be computed as

$$\vec{J} = \frac{I}{wt}\vec{e}_y$$ (8.3)

where $\vec{e}_i$ is the unit vector in the i direction with $i \in \{x, y, z\}$. With the smooth-drift approximation, the velocity of the electrons can be written with the number of free electrons N as

$$\vec{v} = \frac{\vec{J}}{-e_0 N}.$$ (8.4)

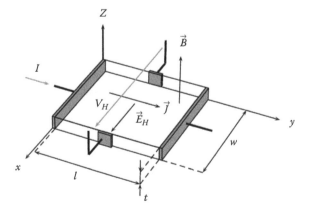

FIGURE 8.1   Schematic drawing of a Hall plate.

By summarizing Equations 8.2, 8.3, and 8.4, the Hall field can be calculated as

$$\vec{E}_H = -\vec{v} \times \vec{B} = \frac{1}{e_0 N}(\vec{J} \times \vec{B}) = \frac{IB}{e_0 Nwt}(\vec{e}_y \times \vec{e}_z) = \frac{IB}{e_0 Nwt}\vec{e}_x \qquad (8.5)$$

where B is the component of the magnetic induction orthogonal to the Hall plate. Finally, the electrical field has to be integrated between the two sensing contacts to calculate the Hall voltage

$$V_H = \int_0^w \vec{E} \cdot d\vec{s} = \int_0^w \vec{E}_H \cdot \vec{e}_x \, dx = \frac{1}{e_0 N}\frac{I}{t}B = K_H \frac{I}{t}B \qquad (8.6)$$

which depends on the thickness t of the plate, the current I, and the Hall coefficient K_H [9].

Most of the commercially available Hall plates are integrated sensors that allow for the implementation of peripheral circuits to improve the sensor performance. Moreover, the optimization of parameters, such as size, resistance, power consumption, and the reduction of parasitic effects lead to a multitude of different shapes of Hall plates. The calculation of the Hall voltage with Equation 8.6 holds only for the shape shown in Figure 8.1, where the sense contacts are small and the length is larger than the width. For all other shapes, Equation 8.6 has to be corrected with a geometrical factor. Besides standard Hall plates, which measure the magnetic field component normal to the surface of the silicon chip, there are vertical Hall plates available, which measure the magnetic field component parallel to the surface of the chip. The combination of standard and vertical Hall plates allows for fully integrated sensors measuring the magnetic field in three directions [10].

COORDINATE MEASURING MACHINE

WORKING PRINCIPLE

The manufacture of mechanical constructions with high precision requires quality checks of the dimensional accuracy. A real construction never exactly meets the ideal shape. It is only a question of what range are the tolerances. Usual measurement quantities are the size of features, the position of features relative to part coordinates, the distances between features, the form of features, such as flatness, circularity, and cylindricity, and angular relationships between features, such as perpendicularity. The coordinate measuring machine (CMM) was developed to measure these quantities flexibly, accurately, fast, and automatically [11,12]. To this end, a CMM detects several points on the surface of the DUT which results in a large amount of data (see Figure 8.2 [13]). Then, the measurement points are fitted to geometrical objects such as lines, planes, sphericals, cylinders, and so on. All objects together result in a reconstructed shape of the construction. Finally, a comparison of the ideal and the reconstructed shape provides the required quantities listed in the foregoing.

The detection of the measurement points is realized with moving axes including a frame of reference and a probe that detects the surface of the construction. This system

Real shape   Measuring points   Reconstructed shape   Ideal shape

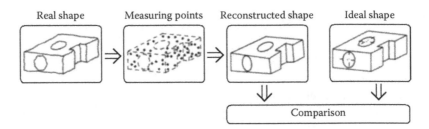

Comparison

FIGURE 8.2   Working principle of a coordinate measuring machine according to Keferstein and Dutschke (Adapted from C. P. Keferstein and W. Dutschke. *Fertigungsmesstechnik: Praxisorientierte Grundlagen, moderne Messverfahren*. Wiesbaden, Germany: B. G. Teubner Verlag, 2008. © Teubner 2008.)

is usually controlled by a computer that executes the measurement program and that can be used for the fitting and the postprocessing. The working range and the accuracy of CMMs reach from the nanometer scale up to a few meters (see Figure 8.3 [14]).[*]

MOVING AXES

The moving axes change the position of the geometrical probe relative to the DUT. There are versatile types of systems that move the geometrical probe, the DUT, or both relative to each other. At each position, the geometrical probe detects a point on the surface of the DUT. Common moving axes consist of three linear axes adjusted orthogonal to each other which provide a Cartesian coordinate system. Nevertheless, there are also non-Cartesian systems, for example, with two linear axes and a rotary

(a)   (b)

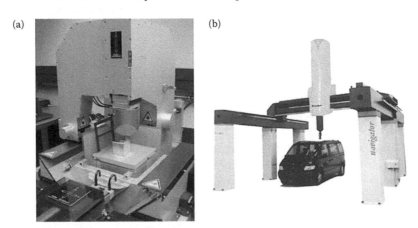

FIGURE 8.3   (a) Ultra precision coordinate measuring machine (CMM) with a measuring range of $100 \times 100 \times 40$ mm³ and an accuracy of 50 nm. (Adapted from T. Ruijl. *Ultra Precision Coordinate Measuring Machine*. PhD thesis, University of Delft, 2001.) (b) Navigator series CMM from the Leader Metrology Inc. with a measuring range of $5 \times 2.5 \times 2$ m³ and an accuracy of 30–100 μm.

[*] Leader Metrology, Inc., 2423 Computer & Space Building College Park, Maryland USA 20742. *Navigator Series CMM*, 2009.

stage that provide a cylindrical coordinate system, and so on. Furthermore, there are CMMs with more than three moving axes that define not only the position but also the alignment of the geometrical probe with respect to the DUT.* Moreover, the realization of a CMM with the same coordinate system and the same number of moving axes is versatile, which can be exemplarily seen in Figure 8.4.

The moving axes are connected to each other with structural elements usually made of granite or metal. Each axis is realized with a supporting bearing, a drive system and a feedback element. The supporting bearing defines the direction of the rotation or movement. It is important that this bearing is precise so that the desired degree of freedom is changed only, for example, for a linear axis the bearing should avoid a rotation, or a shift orthogonal to moving direction. The drive system changes the degree of freedom which is realized with DC or AC motors for high forces and long ranges, and with piezoelectric actuators for very small working ranges [15]. Since the drive system is not sufficiently accurate, there are feedback elements that measure the actual position of the moving axis. Common approaches for the measurement are laser interferometers, glass scales and capacitive systems. All feedback elements and the drive systems of all axes are connected to a controller that provides a comfortable interface to adjust the position of the moving axes. Furthermore, there are controllers available that provide additional features, for example, the transformation of the position in different coordinate systems, the movement on complex paths such as circles or ellipses, the definition of the acceleration and velocity during the movement, a direct coupling with the geometrical probe, and so on. In the following, the combination of the driving system, the feedback elements, and the controller is denoted as moving axes. However, in some literature, this combination is also denoted as CMM, although it does not include a geometrical probe.

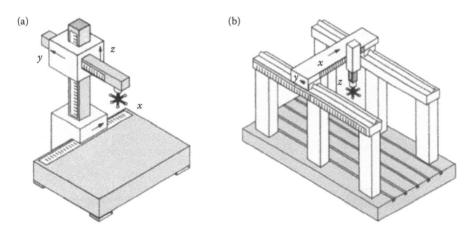

FIGURE 8.4 Example for two types of Cartesian coordinate measuring machines: (a) Horizontal arm; (b) Gantry. (Reprinted from C. P. Keferstein and W. Dutschke. *Fertigungsmesstechnik: Praxisorientierte Grundlagen, moderne Messverfahren.* Wiesbaden, Germany: B. G. Teubner Verlag, 2008. With permission. © Teubner 2008.)

---

* Physik Instrumente (PI) GmbH & Co. KG, Auf der Römerstr. 1 76228 Karlsruhe. M-810 *Miniature Hexapod*, 2008.

GEOMETRICAL PROBES

The geometrical probe is the connection between the surface of the DUT and the frame of reference of the moving axes. To use the full resolution of the moving axes, the CMM needs a suitable sensor. A wide variety of sensors with their own advantages and disadvantages are available.

An important group of sensors are tactile probes that consist of a probing pin and a contact tip that is moveable mounted. For the detection of one point, the moving axes of the CMM place the sensor next to the DUT so that the contact tip touches its surface. There are two different sensing principles (see Figure 8.5).

The dynamic principle consists of a switching probe that gives only information about contact and no contact. For detection, the probe is moved in the direction of the DUT until the signal state changes. This type of tactile probe has to be connected to the controller so that the exact position is saved and the movement is stopped when the probe touches the DUT. The second tactile sensing principle is the measuring probe that outputs the deflection of the contact tip. For the measurement, the probe has to be placed next to the DUT so that the probe touches its surface, and the probe is still inside its measuring range. Then, the touching point of the contact tip can be calculated by using the information of the deflection. A problem of tactile probes is the stress generated between the contact tip and the DUT. This may lead to surface damage of the DUT, and has to be compensated by reducing the probing force. Furthermore, the detection of elastic objects is difficult.

Another important group of geometrical probes are based on optical concepts. The main advantage is the contactless measurement and the measurement speed. A disadvantage is the susceptibility to different optical surface properties, for example, color, lighting, reflection characteristic, transparency, and so on. The variety of optical sensors reaches from a one-dimensional sensor up to full three-dimensional sensors with versatile measurement principles such as the autofocus sensor,[*] the triangulation laser [13], conoscopic holography [16], and the chromatic white light sensor [13]. An

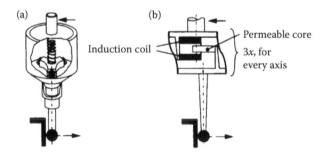

FIGURE 8.5 Two types of tactile probes: (a) Switching principle; (b) Measuring principle. (Reprinted from C. P. Keferstein and W. Dutschke. *Fertigungsmesstechnik: Praxisorientierte Grundlagen, moderne Messverfahren.* Wiesbaden, Germany: B. G. Teubner Verlag, 2008. With permission. © Teubner 2008.)

---

[*] Optische Präzisionsmesstechnik GmbH, Nobelstr. 7, 76275 Ettlingen, Germany. *AF16 Autofokussensor*, 2009.

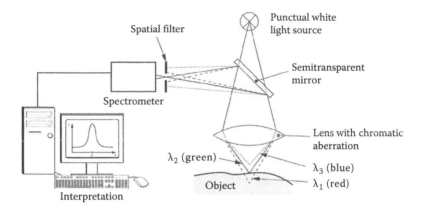

FIGURE 8.6 Working principle of a chromatic white light sensor. (Reprinted from Claus P. Keferstein and Wolfgang Dutschke, *Fertigungsmesstechnik: Praxisorientierte Grundlagen, moderne Messverfahren*, B. G. Teubner Verlag, 2008. © Teubner 2008.)

explanation of the sensing principle is given for the white light sensor only, because this is the sensor used in the measurement setup presented in Figure 8.6.

This sensing principle exploits the effect of chromatic aberration of the focus lens (see Figure 8.6). Chromatic aberration is the dependency of the focal length of a lens to the wavelength of the light. For the measurement a broadband light source (e.g., tungsten halogen lamp) is focused by a lens with chromatic aberration toward the surface of the DUT. The white light is separated along the optical axis in order of increasing wavelength. That means the focal length of blue light is shorter than the focal length of red light. Radiation reflected from the surface is collected by the very same object lens. This is the reverse process of the imaging described above, and also works best for the wavelength which is in focus. On the way back, the radiation is directed into a spectrometer. An external computer or an internal electronic determines the wavelength where the signal is maximal and calculates the distance to the measuring object.

For a geometric sensor, we selected an optical sensor because the contactless measurement avoids any damage of the DUT, it is faster, and the risk of a collision with the DUT is lower. From the different types of optical sensors the chromatic white light sensor is chosen, because it has a high axial resolution, a small spot size (a high lateral resolution), it is robust against variation of the optical properties, the measuring is coaxial with no shadowing in one direction, and the sensor head is totally passive and nonmagnetic.

MEASURING THE SPATIAL DEPENDENCY OF MAGNETIC FIELDS

STRAIGHTFORWARD APPROACH

A common approach to measuring the spatial dependency of magnetic fields is the combination of moving axes and a magnetic sensor.[*] For the measurement, the mag-

---

[*] Dr. Brockhaus Messtechnik GmbH & Co. KG, Gustav-Adolf-Str. 4, 58507 Lüdenscheid, Germany. *XYZ Field Scanner.*

netic sensor is positioned at several points where the magnetic field is measured. Such a scan provides roughly the field vector with respect to the geometry of the field source. Nevertheless, the main problem of such a system is that the exact position and alignment of the magnetic sensor and the field source are unknown. In detail, the magnetic sensor is usually assembled in a ceramic or plastic package. Inside this package, the position of the silicon die varies in the range of microns to ± 150 µm. Then, the package is mounted to an attachment that connects the sensor with the moving axes which also causes mechanical errors (see Figure 8.7). Overall, the position of the magnetic sensor with respect to the coordinate system of the moving axis can be determined to an accuracy of ± 300 µm, and typical alignment errors are in the range of $\pm 3°$. Furthermore, the positioning of the field source in the measurement setup is also accompanied by mechanical tolerances so that its orientation and exact position are also unknown. This reduces not only the absolute accuracy but also the repeatability. For homogeneous magnetic fields, these mechanical tolerances may be neglected. However, if strongly inhomogeneous fields have to be measured, for example, with a gradient of 1%/10 µm as used in automotive applications, mechanical tolerances are the crucial source of error.

MAGNETIC AND COORDINATE MEASURING MACHINE

To overcome problems caused by mechanical tolerances, an accurate system includes not only a magnetic sensor and moving axes but also a geometrical probe. We denote this system as MCMM, because it consists of a magnetic sensor and a CMM. A schematic drawing of an MCMM and a photograph of a real setup are shown in Figure 8.7.

Having a geometrical probe beside the magnetic sensor has the advantage that the geometry of the field source can be measured so that its exact position and alignment in the coordinate system of the moving axes are known. In the coordinate system of the moving axes, the reference axes of the DUT are represented by $\vec{x}_m$, $\vec{y}_m$, $\vec{z}_m$, and

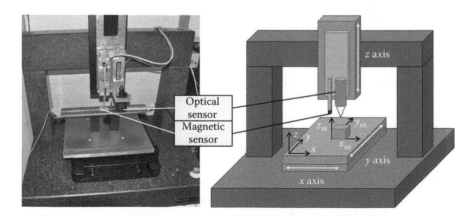

FIGURE 8.7 Photograph (left) and schematic drawing (right) of the setup of a magnetic and coordinate measuring machine. The reference frame of the moving system has the coordinates x, y, z, and the coordinates of the field source or the device under test are denoted as x_m, y_m, z_m.

the origin is given by $\vec{o}_m$. With this description, any point $\vec{r}$ of the coordinate system of the moving axes can be transformed to the frame of reference of the DUT

$$\vec{r}' = \mathbf{A}(\vec{r} - \vec{o}_m) \text{ with } \mathbf{A} = \begin{bmatrix} x_{m1} & x_{m2} & x_{m3} \\ y_{m1} & y_{m2} & y_{m3} \\ z_{m1} & z_{m2} & z_{m3} \end{bmatrix}. \tag{8.7}$$

In the next step, the magnetic sensor should be moved to points with respect to the axes of reference of the DUT. This requires that the distance between the magnetic and geometric sensor is known. Moreover, the measured field vectors can be transformed to the coordinate system of the field source only, if the alignment of the magnetic sensor in the measurement setup is known. Therefore, a calibration of the alignment and position of the magnetic sensor in the measurement system is necessary.

PARAMETERS OF CALIBRATION

The measurement of all three magnetic field components requires at least three one-dimensional sensing elements which are in this case Hall plates. The calibration of one sensing element requires the measurement of three degrees of freedom for the position vector $\vec{p}_B$ and two degrees of freedom for the alignment $\vec{n}_B$ (see Figure 8.8).

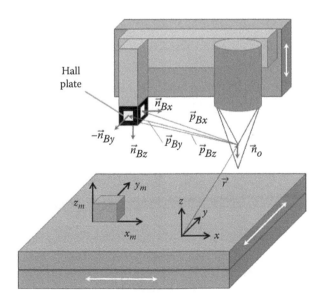

FIGURE 8.8 Schematic drawing of the measurements setup including the parameters of calibration.

So, in general, 15 degrees of freedom have to be calibrated for all three directions $(\vec{p}_{Bi}, \vec{n}_{Bi}$ with $i \in \{x, y, z\})$. However, if a multidimensional sensor is assembled, and the alignment and position of the sensing elements are exactly known relative to each other, not all distances and normal vectors have to be measured.

Besides the calibration of assembly tolerances, important parameters of magnetic sensors such as offset and sensitivity have to be calibrated. The offset can simply be measured without any field source whereas the calibration of the sensitivity S_i with $i \in \{x, y, z\}$ requires an external reference field.

In the following, a method is presented that determines all parameters of calibration such as alignment, position, and sensitivity of the Hall plates. Moreover, a method is shown that calibrates the alignment of a magnetic sensor with high precision.

MEASURING THE MAGNETIC FIELD WITH RESPECT TO THE GEOMETRY OF THE FIELD SOURCE

After the optical measurement, the orientation and position of the field source is known. Furthermore, the calibration provides the position and orientation of the magnetic sensor with respect to the geometrical probe. With this information, the position of the moving axes $\vec{r}_{Bi}$ with $i \in \{x, y, z\}$ can be calculated so that the magnetic sensor is placed at the position $\vec{r}'$ in the coordinate system of the field source

$$\vec{r}_{Bi} = \mathbf{A}^{-1} \vec{r}' + \vec{o}_m - \vec{p}_{Bi} \,. \tag{8.8}$$

Moving all three sensors to the position $\vec{r}'$ results in the three output signals

$$M_i(\vec{r}') = S_i \, \vec{n}_{Bi} \cdot \vec{B}(\vec{r}') \quad \text{with } i \in \{x, y, z\}. \tag{8.9}$$

Rewriting this equation and combining all three measurement values results in the magnetic flux density vector with respect to the coordinate system of the moving axes

$$\vec{B}(\vec{r}') = \begin{bmatrix} n_{Bx1} & n_{Bx2} & n_{Bx3} \\ n_{By1} & n_{By2} & n_{By3} \\ n_{Bz1} & n_{Bz2} & n_{Bz3} \end{bmatrix}^{-1} \begin{pmatrix} M_x(\vec{r}')/S_x \\ M_y(\vec{r}')/S_y \\ M_z(\vec{r}')/S_z \end{pmatrix} \tag{8.10}$$

Finally, the measured field vectors can be transformed to the coordinate system of the field source

$$\vec{B}'(\vec{r}') = \mathbf{A} \, \vec{B}(\vec{r}') \tag{8.11}$$

so that assembly tolerances have no influence anymore.

CALIBRATION OF AN MCMM

CONDUCTOR

Method

For the calibration of the position, alignment, and sensitivity of the magnetic sensing elements, an inhomogeneous reference field source is used which has a strong relation between its magnetic field and its geometry. Moreover, the magnetic reference field $\vec{B}_r(x, y, z) = \vec{B}_r(\vec{r})$ can be described analytically. First, the optical sensor measures the position and alignment of the field source so that the magnetic reference field is known. With this information, the output signal M_i of any sensing element can be calculated as a function of the sensitivity S_i, the alignment $\vec{n}_{Bi}$, and the position $\vec{p}_{Bi}$

$$M_i = S_i \, \vec{n}_{Bi} \cdot \vec{B}_r (\vec{r} - \vec{p}_{Bi}).\tag{8.12}$$

Then, the magnetic sensor scans the field which results in the output signal $\tilde{M}_i$. Finally, an optimization problem is formulated where the mean-squared error between the calculated and measured values should be minimized

$$\min_{S_i, \vec{n}_{Bi}, \vec{p}_{Bi}} \| \tilde{M}_i - M_i \|_2^2 .\tag{8.13}$$

Solving this minimization problem results in the unknown parameters of calibration.

Reference Field

The reference field is provided by a current-supplied straight conductor with a circular cross section. If all materials are nonpermeable and the conductor is assumed to be infinite, the magnetic induction outside of the conductor is given by

$$\vec{B}_c (a) = \frac{\mu_0 I}{2\pi a} \vec{e}_\phi ,\tag{8.14}$$

where a denotes the orthogonal distance to the axis of the conductor, and $\vec{e}_\phi$ the tangential unit vector with respect to the axis of the conductor [17]. In the coordinate system of the moving axes, the axis of the conductor is defined by the linear equation

$$\vec{g}_c (\lambda) = \vec{p}_c + \lambda \vec{d}_c ,\tag{8.15}$$

where $\vec{p}_c$ is the position vector and $\vec{d}_c$ the direction vector. With this description, the magnetic field of the conductor is represented by

$$\vec{B}_r (\vec{r}) = \frac{\mu_0 I}{2\pi} | \vec{d}_c | \frac{\vec{d}_c \times (\vec{r} - \vec{p}_c)}{| \vec{d}_c \times (\vec{r} - \vec{p}_c) |^2}.\tag{8.16}$$

Generating the Reference Field

To generate the reference field, a conductor made of copper is stretched and then glued to a frame made of aluminum (see Figure 8.9). The stretching supports the conductor axis to be straight and the cross section to be circular. On the frame, the position of the conductor is defined by a small guiding channel that is filled with adhesive. At the ends, the conductor is completely covered with adhesive, since the connection of the conductor to the power supply results in mechanical stress in these areas. Moreover, the adhesive is optimized for metallic surfaces and has a high thermal conductivity that is important for heat dissipation from the conductor to the attachment.

The frame is anodized to electrically isolate the conductor from the aluminum attachment so that it is surrounded by a hard coating. In the center, there are two openings so that the magnetic sensor can be positioned over and beside the conductor. Moreover, the frame can be attached with fixing pins and reamed holes in the measurement setup in several orientations (see Figure 8.4). Finally, optical measurements of the conductor surface in the range of ±70 mm around the center show that the error of the geometry compared to an ideal cylinder is in the range of ±10 μm.

Orientation of the Conductor

The orientation of the conductor has to fit to the orientation of the magnetic sensing element to assure an adequate measurement signal. If the axis of the conductor and the normal vector of the sensing element are in parallel no measurement signal occurs. Second, the surface of the conductor has to be measured optically, thus, an orientation of the conductor parallel to the z axis of the moving axes is not possible. Thirdly, the conductor is homogeneous along its rotation axis, and a measurement in one orientation restricts the position of the magnetic sensing element only to a line parallel to the conductor axis. Therefore, at least two measurements with different orientation of the conductor are necessary. Finally, after all considerations, four orientations of the conductor are enough to calibrate all three sensing elements (see Figure 8.10).

Realization

The calibration of one magnetic sensing element requires that the conductor is attached in two orientations. For each orientation the geometry is measured with the optical sensor first so that the position vector $\vec{p}_c$ and the direction vector $\vec{d}_c$ is known, and the magnetic reference field can be calculated according to Equation 8.16. Then, a scan with the magnetic sensing element around the conductor is performed and the optimization problem in Equation 8.13 is formulated. Finally, solving this problem leads to the desired calibration parameters such as the position $\vec{p}_B$, the alignment $\vec{n}_B$, and the sensitivity S.

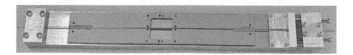

FIGURE 8.9 Conductor while it is stretched and glued to the aluminum frame.

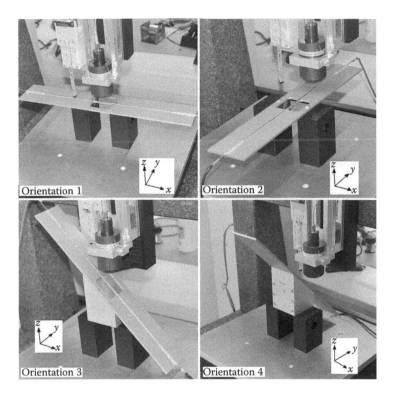

FIGURE 8.10   Possible orientations of the conductor in the frame of reference of the moving axes.

Accuracy Assumption

Firstly, the length of the conductor cannot be infinite. If the length is limited, the strength of the magnetic field is reduced, but the direction does not change [17]. During the measurement, the distance of the Hall plate to the axis of the conductor is small compared to the length of the conductor, and the reduction of the magnetic field is around −0.01%.

Secondly, the conductor is supplied with an appropriate current so that effects caused by a heating-up of the conductor can be neglected. Moreover, at each measurement point, the conductor is supplied with a positive and negative current. For this differential measurement, the output is the average of the measurement value at positive current and the negative measurement value at negative current. This calculation filters out all offsets caused by the magnetic sensor or by external fields. It should be mentioned that all even terms of a nonlinear sensor characteristic are also canceled by this differential technique, even though the characteristic of Hall probes is sufficiently linear within the field range applied during calibration.

The supply cables and the coupling of the reference field with permeable material around the setup generate an interference field at the position of the sensor which is not eliminated by the differential measurement technique. Although all attachments, screws, and so on, are made of nonmagnetic materials, there are some parts in the moving system, such as the guide rails, that are permeable. Both the supply cables and the permeable material are far away compared to the distance between the magnetic sensing element and the conductor. Therefore, the interference field is almost homogeneous at the position of the sensing element and can be approximated by a constant term. This term is additionally fitted

$$M_i = S_i \, \vec{n}_{Bi} \cdot \vec{B}_r(\vec{r} - \vec{p}_{Bi}) + M_{\text{offset}} \tag{8.17}$$

so that the impact of the supply cables and permeable materials is sufficiently suppressed.

Finally, all sources of errors are combined in one model. To this end, the magnetic fields of the conductor and the supply cables are calculated by means of the Biot Savart law. Moreover, in the range of ±70 mm, the axis of the conductor is represented by the data measured by the optical sensor which considers the tolerances of the conductor geometry. Averaging the magnetic field over several points around the center of the sensor takes the finite size of the Hall plate into account. Finally, noise is added to the averaged measurement signal and a Monte Carlo simulation is performed which results in probability distributions of the calibration parameters (position, alignment, and sensitivity). With the results of the model, the accuracy of the CMM, and a temperature stability of 0.5°C, the accuracy of the setup is estimated to be 10 μm for the position, 0.1° for the alignment, and 0.4% for the sensitivity. Moreover, if a higher accuracy for the sensitivity is needed, the sensor can be calibrated in a Helmholtz coil, and a higher accuracy for the alignment can be reached by using the calibration algorithm explained in the following section.

Accurate Angle Calibration

Method

For the accurate calibration of the alignment, an integrated magnetic sensor with multiple magnetic sensing elements on one silicon die is required. The position of the sensing elements on the die is precisely known from the layout of the chip. Moreover, all elements are located in a plane representing the surface of the die. During calibration, the distance between several sensing elements is detected with respect to the coordinate system of the moving axis. For this measurement, the sensor is moved on a line in the reference field generated by a magnetic core. Then, the zero crossings of the output signal of the sensing element are compared, which provides the distance in one direction. Furthermore, the orientation of the sensor can be uniquely determined by using the distances between the sensing elements. This method has the advantage that only the position of the die has to be known roughly in advance, and the calibration procedure directly provides the alignment of the magnetic sensor relative to the frame of reference with high precision.

Magnetic Sensor

The calibration method is demonstrated with a test chip that has 15 Hall plates with a size of $75 \times 75 \times 2 \ \mu m^3$ arranged in rectangular grid (see Figure 8.11). Each Hall

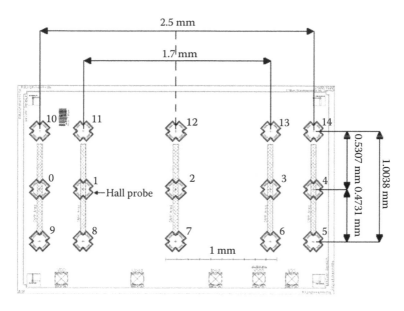

FIGURE 8.11 Test chip with 15 Hall plates provided for experimental purposes by Infineon Technologies AG (H. Husstedt, U. Ausserlechner, and M. Kaltenbacher. *Sensors Journal, IEEE*, 10(5), 984–990, 2010. © 2010 IEEE.)

plate can be operated in the spinning current mode [18,19] which reduces the offset to ±15 µT. The sensitivities of the Hall plates are calibrated with an accuracy of ±0.15%, and the absolute detection limit is 7 µT.

The Hall plates of the test chip are one-dimensional sensing elements that measure the component of the magnetic field orthogonal to the surface of the silicon die. Hence, a measurement of all three degrees of freedom of the magnetic field would require three of these test chips orientated orthogonal to each other, or a test chip with three-dimensional Hall sensors. However, for the demonstration of the calibration method one-dimensional Hall plates are sufficient.

Misalignment of the Silicon Die

The position and orientation of the Hall plates are described by vectors with respect to the frame of reference of the moving axes. The position of the center Hall plate no. 2 is defined by $\vec{p}$ (see Figure 8.12).

Furthermore, all Hall plates are located in the surface of the die, which is supposed to be plane. Any point in this surface can be expresses by a linear combination of two orthonormal vectors $\vec{n}_1, \vec{n}_2$. If the position of the ith Hall plate is denoted as $\vec{r}_i$ the vectors $\vec{n}_1, \vec{n}_2$ read as

$$\vec{n}_1 = (\vec{r}_5 - \vec{r}_9)/|\vec{r}_5 - \vec{r}_9|, \tag{8.18}$$

$$\vec{n}_2 = (\vec{r}_{10} - \vec{r}_9)/|\vec{r}_{10} - \vec{r}_9|. \tag{8.19}$$

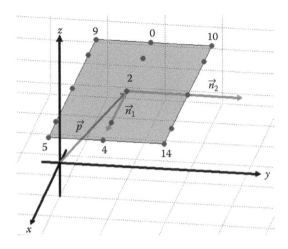

FIGURE 8.12 Die surface of the test chip perfectly aligned to the reference frame (x, y, z) of the moving axis.

The die is ideally aligned in the frame of reference, if $\vec{n}_1$ points in the x direction, and $\vec{n}_2$ in the y direction. For the orientation of the die plane a multitude of descriptions is possible, for example, with Euler angles [20]. In the following, a more intuitive description is applied where the angles are used, which an observer would see when s/he is looking from each direction of the reference frame (x, y, z) on the orthonormal vectors $\vec{n}_1$ and $\vec{n}_2$. The angle α_x is defined between the y axis and the projection of $\vec{n}_2$ on the yz plane

$$\tan \alpha_x = n_{2z}/n_{2y}, \tag{8.20}$$

and α_y denotes the angle between the x axis and the projection of $\vec{n}_1$ on the xz plane

$$\tan \alpha_y = -n_{1z}/n_{1x}. \tag{8.21}$$

For the angle α_z, the projection of either $\vec{n}_1$ or $\vec{n}_2$ on the xy plane can be used. We define this angle between the x axis and the projection of $\vec{n}_1$ on the xy plane as

$$\tan \alpha_z = n_{1y}/n_{1x}. \tag{8.22}$$

An example of a tilted test chip is shown in Figure 8.13 with $\alpha_x = 20°$, $\alpha_y = 20°$, and $\alpha_z = 30°$.

The tangent function is unique for misalignments in the range of $\pm 90°$ so that the vectors $\vec{n}_1$ and $\vec{n}_2$ can be represented only by the angles α_x, α_y, and α_z. To this end, the definitions of the angles and the orthonormal properties are used which results in

$$\vec{n}_1 = \frac{\vec{e}_x + \tan \alpha_z \vec{e}_y - \tan \alpha_y \vec{e}_z}{\sqrt{1 + \tan^2 \alpha_z + \tan^2 \alpha_y}}, \tag{8.23}$$

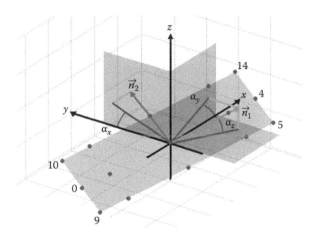

FIGURE 8.13 Example for a tilted die with $\alpha_x = 20°$, $\alpha_y = 20°$, and $\alpha_z = 30°$, where $\vec{p}$ is set to zero.

$$\vec{n}_2 = \frac{\left(\tan\alpha_x \tan\alpha_y - \tan\alpha_z\right)\vec{e}_x + \vec{e}_y + \tan\alpha_x\vec{e}_z}{\sqrt{1+\tan^2\alpha_x + \left(\tan\alpha_x \tan\alpha_y - \tan\alpha_z\right)^2}}, \tag{8.24}$$

where $\vec{e}_i$ denotes the unit vector in the i direction with $i \in \{x, y, z\}$. In the following, only three Hall plates (nos 5, 9, and 10) are used to derive all angles of orientation. To this end, the distances between Hall plate nos 5 and 9, and the distance between Hall plate nos 9 and 10 are expressed in terms of the orthonormal basis $\vec{n}_1$ and $\vec{n}_2$

$$\vec{r}_5 - \vec{r}_9 = 2.5\,\vec{n}_1, \tag{8.25}$$

$$\vec{r}_{10} - \vec{r}_9 = 1.0038\,\vec{n}_2. \tag{8.26}$$

For Equations 8.25 and 8.26, the distances between the Hall plates are given in millimeters taken from the layout of the sensor shown in Figure 8.11. If the positions of the Hall plates are written as

$$\vec{r}_j = x_j\vec{e}_x + y_j\vec{e}_y + z_j\vec{e}_z \quad \text{with } j \in \{0,1,...,14\} \tag{8.27}$$

the angles, representing the orientation of the die, are simply expressed by

$$\tan\alpha_x = (z_{10} - z_9)/(y_{10} - y_9), \tag{8.28}$$

$$\tan\alpha_y = -(z_5 - z_9)/(x_5 - x_9), \tag{8.29}$$

$$\tan\alpha_z = (y_5 - y_9)/(x_5 - x_9). \tag{8.30}$$

Thus, the alignment of the sensor plane can be calculated from differences of position of three Hall plates. For this calculation, five differences of position are used which are not independent from each other. All components of the vectors and all angles are completely defined only with the measurement of three distances by taking the orthonormal properties of $\vec{n}_1$ and $\vec{n}_2$ into account. Moreover, for small angles the nominators in Equations 8.23 and 8.24 may be approximated by 1 so that $n_{1x} = n_{2y} = 1$, and each angle can be calculated only with one distance

$$\tan \alpha_x \approx n_{2z}/1 = (z_{10} - z_9)/1.0038, \tag{8.31}$$

$$\tan \alpha_y \approx n_{1z}/1 = -(z_5 - z_9)/2.5, \tag{8.32}$$

$$\tan \alpha_z \approx n_{1y}/1 = (y_5 - y_9)/2.5. \tag{8.33}$$

For typical misalignments of ±5°, the error of this approximation is less than 1%.

Magnetic Reference Field

In the foregoing, the relation between distances of Hall plates and the exact orientation of the die with respect to moving axes are presented. Now, the measurement of distances between Hall plates with high accuracy is shown. To this end, a magnetic reference field should be designed whose flux density shows a marked pattern only in one direction while being homogeneous in the other two directions. If the die is moved in such a field in an inhomogeneous direction, all Hall plates measure the same field dependency. If the Hall plates are not at the same coordinate in the moving direction, the measurement curves are shifted against each other. This shift is equal to the distance in moving direction between the two involved plates.

To generate such a reference field, our approach is a combination of a permanent magnet with a remanence of 880 mT and steel parts that are precisely manufactured (±0.01 mm manufacturing tolerance). Although the geometry of a permanent magnet is inaccurate due to sinter shrinkage, the flux is precisely guided through milled steel parts whose surface was polished after assembly but before mounting the magnet. These L-shaped flux guides define a narrow gap similar to reading heads of audio tapes, and they avoid far-reaching stray fields (see Figure 8.14).

In the volume above and below the air gap, the field lines exit the core and stray field occurs (see Figure 8.15).[*] This stray field is highly inhomogeneous in the y direction and homogeneous in the x direction. In the symmetry plane, that is parallel to the xy plane, the field has no component in the x direction. Only near the borders this statement does not hold anymore.

Over the air gap, the field has almost semicircular field lines [21], and can be approximated by

$$B_y(y,z) = C \frac{z}{y^2 + z^2}, \tag{8.34}$$

---
[*] Simetris, Am Weichselgarten 7 91058 Erlangen. *Numerical Analysis of Coupled Systems*, 2009.

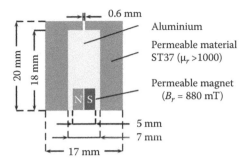

FIGURE 8.14 Dimensions of the magnetic core generating the reference field. The core is 10 mm thick in the direction perpendicular to the drawing plane. (H. Husstedt, U. Ausserlechner, and M. Kaltenbacher. *Sensors Journal,* IEEE, 10(5), 984–990, 2010. © 2010 IEEE.)

$$B_z(y,z) = C \frac{-y}{y^2 + z^2}. \tag{8.35}$$

The magnetic core has to be mounted in the coordinate system of the moving axes with several orientations to measure the distances between the Hall plates in each direction. In the following section, we confine the discussion to the measurement of α_z with the core mounted as shown in Figure 8.15.

A line plot of the y and z components of the magnetic induction in the y direction results in curves with a maximum for the y component and a zero crossing for the z component (Figure 8.16).

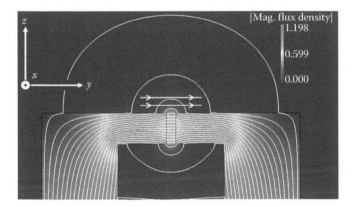

FIGURE 8.15 Field lines and magnetic flux density of the magnetic core generating the reference field. The results are obtained by a two-dimensional finite-element method simulation calculated using the software NACS with nonlinear permeability.[*]

---

[*] Calculated from Simetris, Am Weichselgarten 7 91058 Erlangen. *Numerical Analysis of Coupled Systems,* 2009.

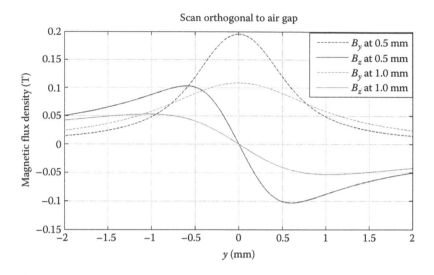

FIGURE 8.16 Two scans of the magnetic induction orthogonal to the air gap at a distance of 0.5 and 1 mm relative to the surface of the core. The results are taken from the two-dimensional finite-element method simulation of Figure 15 which also visualizes the two scanning paths.

For an ideal symmetric setup the maximum and the zero crossing are located in the center of the air gap. Along z direction it shows a marked decrease inversely proportional to distance; nevertheless, the locations of zeros and maxima do not change.

Measurement of Distances

If the test chip is moved in a normal direction to the air gap, a characteristic point like a zero or a maximum can be detected whose position is independent from the x and z coordinate. At a maximum, the curve is flat and the derivation with respect to the position is zero. For this reason, small errors in the magnetic flux density caused by noise result in large position errors. Therefore, it is more accurate to detect a zero crossing.

Exemplarily, the measurement of α_z is explained where the difference in y coordinates of the Hall plate nos 5 and 9 is required (see Equation 8.33). In a first step, the test chip is moved in the y direction while the magnetic core is attached as shown in Figure 8.15, until zero is detected with Hall plate no. 5. The x position is chosen in a way so that Hall plate no. 5 is approximately at $x = 0$, which is the mid-plane of the magnetic core. In a second step, the test chip is shifted 2.5 mm in a positive x direction so that Hall plate no. 9 is approximately in the mid-plane of the magnetic core. At this x position, a scan along the y direction is performed to detect the zero with Hall plate no. 9. If the die would be ideally aligned to the frame of reference, Hall plate no. 9 would be at exactly the same position during the second step as Hall plate no. 5 during the first step. Therefore, the flux density during the scan in the z direction is exactly the same for both sensor elements.

A tilt of the die has two main effects. Firstly, the orientation of the Hall plates is changed which causes a reduction of the z component measured. Moreover, if the die is tilted by α_x the Hall plates also measure field components in the y direction, and if the die is tilted by α_y the Hall plates measure components in the x direction. The reduction of the z component is small, and it does not change the position of the zeros. In x direction the field has no component that impairs the measurement whereas the effect of the mix of the y and z components caused by α_x has to be analyzed. To this end, the approximation from Equations 8.34 and 8.35 is used, and the flux density of the tilted Hall plate is computed as

$$B_H(y,z,\alpha_x) = C\,\frac{z\sin\alpha_x - y\cos\alpha_x}{y^2 + z^2}. \tag{8.36}$$

Since the position at zero field is detected, B_H is set to zero which results in the y coordinate of the position at zero field as a function of the z position:

$$y_{zero} = z\tan\alpha_x. \tag{8.37}$$

If the die is tilted, the second effect is a different position of Hall plate nos 5 and 9 during the scan in the y direction. The shift in the y direction is the parameter that should be measured, and the shifts in x and z directions should have no effect, since the position of the zeros is homogeneous in these directions. However, the combination of the different z coordinate caused by the tilt α_y, and the dependency of the zero crossing caused by the angle α_x results in a measurement error for the distance $y_5 - y_9$ which is used for the calculation of α_z. Consequently, the measurement of the angle α_z is impaired by the other two angles. The same holds for the measurement of the angles α_x and α_y, respectively. Typical misalignments of $\pm5°$ can be expected which results in an error of $0.5°$ for the measurement of one angle caused by the other two angles.

Once the alignment is measured, it is straightforward to compensate for it by some mechanical tilt mechanism so that the misalignment is reduced by one order of magnitude ($\pm0.5°$ instead of $\pm5°$). Then, a second measurement gives even higher accuracy. In addition, even if there is no possibility to correct the tilt mechanically, an iterative measurement may also cancel out the error caused by the other two angles. For this purpose, the distance from the previous measurement has to be used to adapt the position perpendicular to the moving direction of the two involved sensor elements. This iterative measurement also reduces the impact of small asymmetries and wrong orientations of the reference field. Thus, aside from parasitic effects and mechanical tolerances, the magnetic core theoretically allows one to measure differences of position without any errors.

Estimation of Accuracy

An analysis of errors that includes the impact of sensor offsets, sensitivity mismatches, noise, and a curved die surface results in an accuracy of $\pm0.027°$ for α_z of the central Hall element no. 2 [22]. So, Hall plate nos 5, 9, and 10 are used for the measurement of the alignment, and the central Hall plate no. 2 for the actual measurement. This reduces the impact of a curved die surface, because the die is almost symmetrically curved with respect to the center of the die [23].

MEASUREMENT EXAMPLE

MEASUREMENT SYSTEM

An example of an MCMM is shown in the left part of Figure 8.7. This setup consists of a moving system from Feinmess Dresden with a resolution of 0.1 μm, an absolute accuracy of ±2 μm, and a measurement range of ±75 mm in all directions. The alignment between the moving axes is accurate to ±0.004°.[*][†]

Secondly, the geometric sensor of the setup is the white light sensor CHRocodileE RB200031 from Precitec Optronik GmbH with a measuring range of 3300 μm, a resolution in z direction of 100 nm, and a resolution in lateral direction of 6 μm. The white light sensor is a one-dimensional sensor, and it is assembled in the setup so that it points in a negative z direction. For the geometrical measurement, the white light sensor is placed near the DUT so that the light beam is focused on the surface of the DUT. The position of the measurement point is then calculated by subtracting the optical measurement value o from the z coordinate of the actual position $\vec{r}$ that is provided by the moving axes (see Figures 8.7 and 8.8).

Thirdly, the magnetic sensor of the setup is a three-dimensional integrated Hall sensor from Senis GmbH with a measuring range of 200 mT, and an accuracy of 0.1% [10]. Hall probes are chosen for the measurement setup, because they are sensitive to magnetic fields used in automotive applications (10 μT–200 mT), and they have a linear characteristic without hysteresis.

FIELD SOURCE

As an example, the magnetic field of a neodymium permanent magnet is analyzed. According to the data sheet, the magnet has a remanence of 1.240 T, it is plated with nickel, and it has the shape of a cube with the dimensions of $2 \times 2 \times 2$ mm³. To easily refer to the faces of the magnet, they are numbered equally to the sides of dices (see Figure 8.17).

FIGURE 8.17 Photograph of the cubic permanent magnet and the magnetic sensor.

[*] Feinmess Dresden GmbH, Fritz Schreiter Str. 32, 01259 Dresden, Germany. *Linear Stage PMT 160-DC*, 2009.
[†] Feinmess Dresden GmbH, Fritz Schreiter Str. 32, 01259 Dresden, Germany. *Compact-XY-Stage KDT 380*, 2009.

For the measurement, an attempt is made usually to orientate the field source toward the axes of the moving system so that only tolerances of the positing have to be corrected. However, in this example, the magnet is totally misaligned toward the axes of the moving system to clearly show that an MCMM does not require any pre-alignments or exact positioning of the field source (see Figure 8.18).

OPTICAL MEASUREMENT

The optical scan results in several measurement points of the surface of the permanent magnet. Outliers due to measurement errors of the white light sensor are filtered out which cause holes in the surface plot of the measurement results (see Figure 8.18). To compare the real with the ideal geometry, it is necessary to define the way in which the reconstructed geometry is uniquely created from the measurement points. To this end, several points of each side are selected (see right part of Figure 8.18), and the least-square solution for the best-fitting plane is calculated for each set of points.

The intersection lines of these plane equations are also computed. These lines represent the edges of the reconstructed shape (plotted in Figure 8.18). Moreover, a unique coordinate system has to be defined for the field source, which also takes into account tolerances. Usually, it is desired that the frame of reference of the DUT consists of three orthonormal vectors. Therefore, it is not possible to define the axes of the reference coordinate system to be the edges of the magnet, because they are, in general, not orthogonal to each other. For this example, the coordinate system of the magnet is defined by the three orthonormal axes $\vec{x}_m, \vec{y}_m, \vec{z}_m$ and the origin $\vec{o}_m$ which

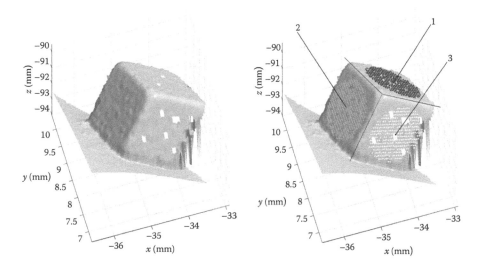

FIGURE 8.18 Optical measurement results of the scan of the permanent magnet. The holes in the surfaces occur since outliers of the measurement points are filtered out. On the right hand side, the points of each surface are plotted which are used for the fitting of the corresponding plane equation.

are defined as follows: (1) $\vec{z}_m$ is parallel to the normal vector of the plane equation of surface no. 1 and points away from the cube; (2) the origin $\vec{o}_m$ is the intersection point of the three planes 1, 2, 3; (3) $\vec{x}_m$ is parallel to the intersection line of planes 1 and 3 and points from the origin to the intersection point of planes 1, 3, and 5; (4) $\vec{y}_m$ is given by $\vec{z}_m \times \vec{x}_m$.

MAGNETIC MEASUREMENT

In this example, the magnetic field should be scanned in x_m and y_m directions in the range of 0–2 mm with steps of 0.5 mm, and for a height z_m of 1, 2, and 3 mm. Before the magnetic measurement is started, the system is calibrated by using a straight conductor as explained in the foregoing so that the parameters of calibration $(\vec{p}_{Bi}, \vec{n}_{Bi}, S_i$ with $i \in \{x, y, z\})$ are known. With this information, according to Equation 8.8, the magnetic sensor can be moved to points in the coordinate system of the field source. Then, according to Equation 8.10, the magnetic field vectors with respect to the coordinate system of the moving axes, visualized in the left part of Figure 8.19, can be calculated from the output signals of the three magnetic sensors.

Finally, the geometry and the magnetic field vectors are transformed into the coordinate system of the magnet shown in the right part of Figure 8.19. This plot demonstrates that the original orientation and position of the magnet and the magnetic sensor have no influence on the measurement, and, thus, assembly tolerances do not impair the measurement results. In addition, the absolute value of the magnetic flux density is plotted on the line $x_m = y_m = 1$ mm against z_m in the range of 1–3 mm, as shown in Figure 8.20 where the left axis indicates the absolute value and the right axis the relative change compared to the value at $z_m = 1$ mm. At this position,

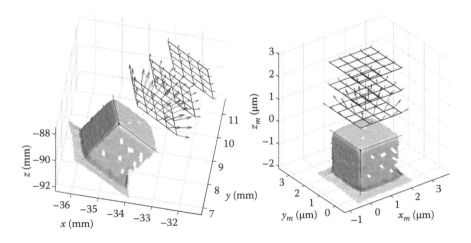

FIGURE 8.19 Magnetic field over surface 1 of the permanent magnet in the coordinate system of the moving axes (left) and in the coordinate system of the magnet (right). Moreover, in both plots, a black grid visualizes the measurement points of the scan.

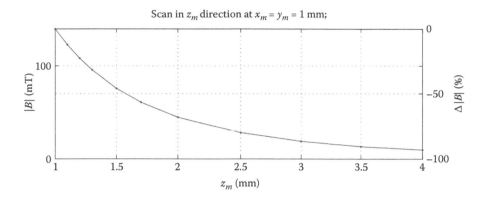

FIGURE 8.20 Scan of the absolute value of the magnetic flux density in the z_m direction.

the relative change is about 1.2%/10 μm which demonstrates the strong inhomogeneity of the magnetic field and the crucial impact of small positioning errors.

CONCLUSION

A measurement principle of scanning magnetic fields with respect to the geometry of the field source is explained. The corresponding measurement setup consists of a three-dimensional magnetic sensor and CMM, and, therefore, the system is known as a magnetic and coordinate measuring machine (MCMM). An introduction to magnetic sensors and to the working principle of a CMM is given which includes components such as the geometric sensor and the moving axes. Then, a straightforward approach of scanning magnetic fields is shown consisting of moving axes and a magnetic sensor only. Such a setup is affected by assembly tolerances that are the crucial source of error, if inhomogeneous magnetic fields are analyzed. An MCMM suppresses the impact of assembly tolerances because the geometry is measured with the optical sensor first so that the magnetic field vectors can be measured at points relative to the field source and the field vectors can be transformed into the coordinate system of the measurement object. This measurement principle requires calibration of several parameters of magnetic sensors such as sensitivity, alignment, and position. To this end, a method is shown that uses the magnetic field of a current-supplied straight conductor as reference and calibrates all parameters. Moreover, another method is discussed that uses several sensing elements on a silicon die to measure the alignment of a magnetic sensor with high accuracy. Finally, a real measurement setup of an MCMM is presented that consists of an integrated three-dimensional Hall sensor, a white light sensor, and moving axes. With this setup, the magnetic field of a cubic permanent magnet is analyzed to demonstrate the working principle of an MCMM.

To conclude, an MCMM is a tool to measure the geometry and the magnetic field of an arbitrary field source. After calibration of the setup, the magnetic field can be measured in the coordinate system of the field source so that assembly tolerances do not impair the measurement. Moreover, if the results do not conform to the expectations, it is possible to distinguish between deviations due to changes in the geometry or in the magnetic material parameters such as permeability and magnetization.

REFERENCES

1. P. Ripka (ed.). *Magnetic Sensors and Magnetometers.* Norwood, MA: Artech House Publishers, 2001.
2. J. C. Gallop. *SQUIDs, the Josephson Effects and Superconducting Electronics.* New York: Taylor & Francis, 1991.
3. E. Hirota, H. Sakakima, and K. Inomata. *Giant Magneto-Resistive Devices.* Berlin: Springer-Verlag, 2002.
4. Klemens Ginter. *Ein Sensor auf Basis des anisotropen magnetoresistiven Effekts.* PhD thesis, Universität Erlangen-Nürnberg, 1999.
5. A. P. Ramirez. Colossal magnetoresistance. *Journal of Physics: Condensed Matter,* 9:8171–8199, 1997.
6. R. C. O'Handley. *Modern Magnetic Materials: Principles and Applications.* New York: John Wiley & Sons, 1999.
7. O. Kronenwerth. *Extraordinary Magnetoresistance Effekt: Metall-halbleiter-Hybridstrukturen in homogenen und inhomogenen Magnetfeldern.* PhD thesis, Universität Hamburg, 2004.
8. C. S. Roumenin. *Handbook of Sensors and Actuators, Volume 2: Solid State Magnetic Sensors* (S. Middlehoek, series ed.). Amsterdam, The Netherlands: Elsevier Science B.V., 1994.
9. R. S. Popovic. *Hall Effect Devices,* 2nd edn. Bristol, UK: Institute of Physics Publishing, 2004.
10. D. R. Popovic, S. Dimitrijevic, M. Blagojevic, P. Kejik, E. Schurig, and R. S. Popovic. Three-axis teslameter with integrated Hall probe. *IEEE Transactions on Instrumentation and Measurement,* 56:1396–1402, 2007.
11. J. A. Bosch (ed.), *Coordinate Measuring Machines and Systems.* New York: Marcel Dekker, Inc., 1995.
12. A. H. Slocum. *Precision Machine Design.* Prentice Hall, NJ: Society of Manufacturing Engineers, Prentice-Hall College Divison, 1992.
13. C. P. Keferstein and W. Dutschke. *Fertigungsmesstechnik: Praxisorientierte Grundlagen, moderne Messverfahren.* Wiesbaden, Germany: B. G. Teubner Verlag, 2008.
14. T. Ruijl. *Ultra Precision Coordinate Measuring Machine.* PhD thesis, University of Delft, 2001.
15. A. Weckenmann, S. Büttgenbach, O. Tan, J. Hoffmann, and A. Schuler. Sensors for geometric accurate measurements in manufacturing. In *Sensors + Test Conference.* Wunstorf, Germany: AMA Service GmbH, 2009, pp. 131–138. ISBN 978-3-9810993-5-5.
16. Gabriel Y. Sirat. Conoscopic holography. I. Basic principles and physical basis. *Journal of the Optical Society of America,* 9:70–83, 1992.
17. H. Hofmann. *Das elektromagnetische Feld.* Berlin: Springer-Verlag, 1974.
18. S. G. Taranov, V. V. Braiko, A. D. Nizhensky, V. P. Belousov, D. V. Kovalchuk, I. P. Grinberg, E. E. Laschuk, J. M. Panchishin, and J. T. Chigirin. Method of and apparatus for eliminating the effect of non-equipotentiality voltage on the hall voltage, United States Patent No. 4037150, July 19, 1977.
19. U. Ausserlechner. Limits of offset cancellation by the principle of spinning current Hall probe. *Proceedings of IEEE Sensors 2004,* 3:1117–1120, 2004.
20. H. Goldstein, C. P. Poole, and J. L. Safko. *Classical Mechanics.* Cambridge, MA: Addison-Wesley, 2001.

21. J. C. Mallinson. *The Foundations of Magnetic Recording*. London: Academic Press, 1993.
22. H. Husstedt, U. Ausserlechner, and M. Kaltenbacher. Precise alignment of a magnetic sensor in a coordinate measuring machine. *Sensors Journal, IEEE*, 10(5):984–990, 2010.
23. H. Husstedt, U. Ausserlechner, and M. Kaltenbacher. In-situ measurement of curvature and mechanical stress of packaged silicon. In *Proceedings of the 9th IEEE Conference on Sensors*, pp. 2563–2568, 2010.

9 Artificial Microsystems for Sensing Airflow, Temperature, and Humidity by Combining MEMS and CMOS Technologies

Nicolas André, Laurent A. Francis, Bertrand Rue, Denis Flandre, and Jean-Pierre Raskin

CONTENTS

INTRODUCTION

Since the 1990s, the construction of structures at the micrometer and nanometer scales, in silicon and in its homologous materials, has become common. In fact, microstructures such as beams, bridges, plates, and so on, which are either fully or partially anchored onto a silicon substrate, provide adequate topologies to sense airflow or humidity. For example, out-of-plane silicon beams with a length of 200 μm, a width of 10 μm, and a thickness of only 1 μm deflect under airflow just as natural structures such as spider hairs (*Cupiennus salei*, approximately 10 μm in diameter at their base and 0.1–1.4 mm in length [1]). On the other hand, in-plane micromachined beams, coated with hydrophilic materials, are used as humidity- and dew-based sensors. However, a structure such as a silicon beam is merely a constitutive part of a fully functional miniaturized sensor. Additionally, signal processing circuits would be required. Thus, building a totally embedded smart sensing system incurs the challenge of selecting materials and methods that are fully compliant for their direct cointegration.

THREE-DIMENSIONAL MICROBEAMS AS AIRFLOW SENSORS

In this section, we introduce a method to fabricate three-dimensional (3D) cantilevers using both microfabrication techniques and mechanical stress in multilayered thin films. The cointegration of a mechanical sensing component and the complementary metal–oxide–semiconductor (CMOS) circuit on the same silicon chip is also presented. This will be supported by measurements of the components under various stimuli.

THREE-DIMENSIONAL MULTILAYERED CANTILEVERS

Microcantilever-based sensors are silicon beams that are typically less than 1 μm thick. The term "three-dimensional" (3D) is used here to describe a structure presenting a nonflattened geometry, that is, an out-of-plane curvature when the reference plane is the silicon substrate wafer. Microcantilevers are widely used in atomic force microscopy, mass sensing, contact sensing, and force measurements. Classically, with microelectromechanical system (MEMS)-based cantilevers, a change in surface tension or surface stress, due to interfacial interactions between the surface and the environment or intermolecular interactions on the surface [2], is detected electrically by a piezoresistive gauge or a piezoelectric resonance frequency shift.

For building 3D MEMS, several techniques have been proposed in the literature such as: projection microstereolithography [3], plastic deformation under magnetic field [4], reflow of solder hinges [5], and multistack silicon-direct wafer bonding [6]. In the study by Kolesar et al. [7], a probe tip under the released structures is even incorporated to reach such a shape. Finally, in the studies by Iker et al. [8] and Kao et al. [9], a novel miniaturization technique, based on appropriate use of built-in stresses, is used to produce the 3D MEMS components. This, in turn, allows for an increase in fabrication portability while also reducing cost of production.

Movable 3D cantilevers offer detection as a result of a stimulus changing their deflection. This change in deflection (due to an airflow, for example) bends the cantilevers downward or upward, respectively increasing or decreasing their capacitance. These microcantilevers have therefore an out-of-plane movable part that is sensitive to airflow impinging on it or to temperature variations. Flow and temperature sensors, built in such a way, are basically composed of two interdigitated combs of electrodes, one series of cantilevers that are released and anchored at their extremity form the first electrode (3D movable electrode) and the other electrode is composed of metallic fingers that are attached (unreleased) at the top surface of the silicon substrate. Ensuring CMOS compatibility by the use of classical materials and fabrication techniques, these microcantilevers can therefore be cointegrated with an oscillating circuit, which would be appropriate for signal processing.

FABRICATION OF COINTEGRATED CMOS 3D SENSORS

Since the fabrication of our MEMS sensors is based on techniques and materials widely used in the CMOS industry for building digital and analog integrated circuits, that is, oxidation, silicon doping, or metallization, we can easily extend our process to the fabrication of cointegrated CMOS-integrated circuits with 3D MEMS.

For the construction of these, a silicon-on-insulator (SOI) wafer is commonly considered as the starting material to facilitate the cointegration with our CMOS SOI process, though a classical bulk wafer can also be used. The SOI wafer consists of a 100-nm-thick monocrystalline silicon thin film on top of a 400-nm-thick buried oxide lying on a 780-μm-thick bulk silicon handling substrate. The fabrication of 3D capacitive sensors requires the deposition of two layers on top of the silicon film forming the active layer of the starting SOI wafer, as illustrated in Figure 9.1. A 250-nm-thick layer of silicon nitride is coated using low-pressure chemical vapor deposited (LPCVD) silicon nitride. Then, a process known as "loading" is implemented which consists of cooling the system to room temperature (20°C) from the thin-film deposition temperature (i.e., 800°C for silicon nitride). Third, a layer of 900-nm-thick aluminum is then evaporated at 150°C which represents a new loading

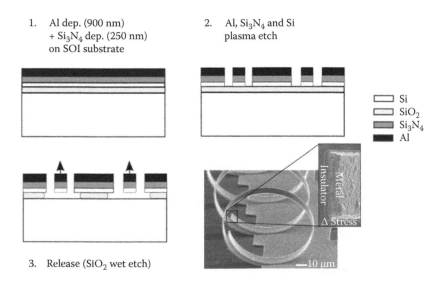

1. Al dep. (900 nm)
    + Si$_3$N$_4$ dep. (250 nm)
    on SOI substrate

2. Al, Si$_3$N$_4$ and Si
    plasma etch

☐ Si
☐ SiO$_2$
▨ Si$_3$N$_4$
▪ Al

3. Release (SiO$_2$ wet etch)

—10 μm

FIGURE 9.1 Process flow for curved three-dimensional beams obtained by use of stacking a trilayer (silicon/silicon nitride/aluminum—Si/Si$_3$N$_4$/Al) presenting a gradient of stress over its thickness.

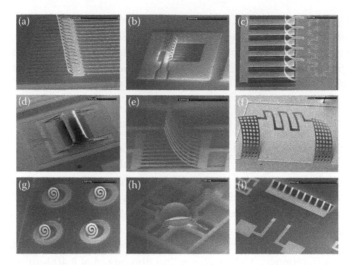

FIGURE 9.2 Microphotograph gallery of processed three-dimensional sensors: (a) spiral beams, (b) capacitive plate, (c) thermal actuator (zoom), (d) rectangular bolometer, (e) inter-digitated capacitive flow sensor, (f) meander inductor, (g) spirals, (h) circular bolometer, and (i) thermal actuator.

phase. During these two cooling processes, stress builds up in the layers as a result of the thermal expansion coefficient mismatch between the different materials.

After a single photolithographic step to pattern the structures as desired, the aluminum film is first etched by a chlorine-based plasma. Subsequently, a sulfur hexafluoride (SF_6) and silicon tetrachloride ($SiCl_4$)-based plasma is used to etch, respectively, silicon nitride (Si_3N_4) and silicon (Si), while aluminum is then used as a masking material. The release of the microstructure must then be performed by withdrawing the buried silicon dioxide (SiO_2) located below the thin silicon layer by using a mixture of concentrated hydrofluoric acid (HF 73%) and isopropanol 1:1. After further rinsing in pure isopropanol, the samples are dried in a critical point dryer machine (Tousimis 915B) in order to avoid adhesion of the beams to the substrate. Finally, a thermal annealing step is performed where the multilayered structure is brought to 432°C in a forming gas (95% nitrogen, 5% hydrogen) for 30 min. During this annealing step, changes in the mechanical stress state in the upper plastic aluminum layer lead to the self-assembling of the microstructures that is, the fabrication of the 3D MEMS structure itself. This process is presented in detail in Reference [10]. Micrographs of several out-of-plane microstructures obtained by this process are shown in Figure 9.2, demonstrating the design flexibility offered by this process.

Design of 3D Sensors

The trilayer process described in Section "Fabrication of Cointegrated CMOS 3D Sensors" to fabricate the MEMS component is not restricted to silicon/silicon nitride/aluminum ($Si/Si_3N_4/Al$) stacking and applies to producing out-of-plane beams with different layer stacking. Specifically, this can pertain to the interdigitated capacitor structures (Figure 9.3). Since the CMOS circuits and the 3D MEMS are made with the same materials, the focus is now to integrate both on the same chip and therefore to benefit from pre-existing stacked layers of the SOI substrate. The thin silicon layer can be used as the sacrificial layer and etched with SF_6, leading to bilayered out-of-plane beams (i.e., SiO_2/Al). The cointegration of the CMOS electronics and 3D microbeams requires one extra lithographic step in comparison with the complete CMOS process. The additional photolithography is required to protect the integrated circuit when processing only the MEMS devices (etching and release). In this case, half of the fingers are out-of-plane, while the others remain in the substrate plane. The first half of the structure is therefore able to move under airflow or temperature stimuli and transduces into a capacitance change (see Sections "Flow Sensors" and "Thermal Sensors").

The trilayer process can also be applied to build a 3D Lorentz force-based magnetometer integrating piezoresistive transducers at the anchors of the released 3D beams (see Section "Magnetic Sensors"), or to build thermally actuated structures that can be of interest to design on-wafer microrobots (see Section "Thermal Actuators").

Flow Sensors

The proposed design of flow detectors offers a list of advantages such as CMOS compatibility (see Figure 9.3 where CMOS inverters with variable interdigitated capacitors were designed to sense flow), SOI process, simple-to-build 3D microstructures, no direct current consumption due to capacitive detection leading to extremely low power consumption of the order of a few tenths of microwatts (see Table 9.1), large sensing range, as well as small occupied chip area (1.25 mm² including CMOS circuits and MEMS).

For a range of flow velocities from 0 to 120 m/s, there is a measurable change in capacitance over a range of 500–550 fF, leading to a 10% change in oscillation

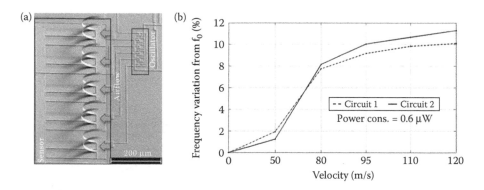

FIGURE 9.3 Three-dimensional flow capacitive anemometer and its associated complementary metal–oxide–semiconductor ring oscillator in the 1-µm fully depleted silicon-on-insulator process (a); relative oscillation frequency variation for different applied air flow velocities and two different flow anemometers (b).

TABLE 9.1

Review of Some Published Flow Sensors Results[a]

Reference	Technology	Static Power Concentrations	Flow Range (ms⁻¹)	Integrated Circuit Compatibility
[11]	Silicon substrate	Yes	2–18	Yes
[12]	Membrane	Yes	0.01–200	No
[13]	Bridge	Yes	0–4	No
[14]	Membrane	Yes	0–8	Yes
Our work	Surface machining	No	0–120	Yes

[a] The first three are on bulk micromachining techniques and composed of a microheater and thermopiles.

frequency (decreasing). Further to this, lower- or higher-speed flow can be sensed by decreasing or increasing, respectively, the cantilever stiffness as the artificial crickets hairs, developed in Reference [10].

Magnetic Sensors

The out-of-plane magnetic flux is converted into a mechanical force by Lorentz force F on the M-shaped Si_3N_4/Al-bilayered cantilever (Figure 9.4):

$$F = I \cdot L \cdot B \cdot \sin(\theta) \tag{9.1}$$

where I is the half-loop current, L is the top beam length, B is the magnetic field flux density across this beam, and θ is the angle between the magnetic field and the current flowing into the top beam. A piezoresistive gauge located at the M-shaped cantilever anchor converts the bending motion of the beam into a resistance change. The piezoresistor is incorporated into a Wheatstone structure with three other fixed resistances. As shown in Figure 9.4, the response is linear over a high range of magnetic field (from 1 to 1000 Gauss) with a small offset at 0 V. The output voltage signal for

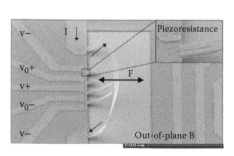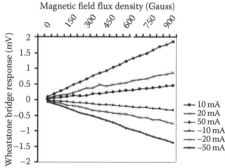

FIGURE 9.4 Silicon-Lorentz magnetometer with its integrated Wheatstone bridge. Sensor response in millivolts for different rectangular currents at 4 kHz.

FIGURE 9.5   Silicon nitride (Si_3N_4)–nickel (Ni) microtweezers.

the Lorentz force-based device for different values of current through the bridge shows a sensitivity of 0.2 mV per Gauss (see Reference [15] for details).

Thermal Sensors

Under a substantial increase in temperature, the out-of-plane component of these interdigitated capacitive thermal sensors (bolometers in Figure 9.2) bends downward because this part is made up of three different layers (Si, Si_3N_4, and Al, from bottom to top). Each layer is characterized by its own thermal expansion coefficient (around 2×10^{-6}, 3×10^{-6}, and $20 \times 10^{-6}°C^{-1}$, respectively). As aluminum expands 10 times more than silicon and is 1 μm thick, a large bending motion downward occurs, resulting in temperature detection over a certain range of temperatures.

Thermal Actuators

Under substantial and controlled local heating by Joule effect in a thin conductive layer, the out-of-plane cantilevers bend downward because of different thermal expansion coefficients. Building circularly assembled cantilevers (in place of straight in line), an Si_3N_4/Ni bilayered microtweezer for the release or capture of microcompounds (cells, proteins, microbes, yeasts, etc.) is shown in Figure 9.5.

TWO-DIMENSIONAL MICROELECTRODES AS HYDROPHILIC SENSORS

Two-Dimensional Multilayered Microsystems

Taking advantage of miniaturization offered by microelectronic fabrication techniques, a humidity microsensor based on water adsorption of aluminum oxide is developed as a human breath analyzer. It demonstrates a sensitivity of 7.5 fF/%RH/mm² and an increase in capacitance by up to two orders of magnitude upon condensation due to breathing. The fundamental mechanism responsible for the breath detection is the presence of a chemisorbed layer of hydroxyl ions on which physisorption of water molecules can easily occur.

Physisorbed water alters impedance by increasing conductance and capacitance, respectively, because of ionic conduction (Grotthus chain reaction [16]) and higher permittivity of water. Thermal oxide between the microelectrodes fingers as well as aluminum oxide anodized fingers permit high sensitivity to humidity.

Also, in view of further miniaturization, we previously reported the successful cointegration of the driving and interface electronic circuits with a flow sensor on the same silicon chip (see Section "Flow Sensors") and the same approach is currently successful for the humidity sensor. Furthermore, combining several sensors with their associated CMOS electronics, such a hybrid circuit platform will act as a mechanical transducer for airflow and temperature sensing and as a chemical transducer for water concentration sensing.

FABRICATION OF COINTEGRATED CMOS HUMIDITY SENSORS

For this microsensor a thin ceramic material (aluminum oxide) is used as a humidity-sensitive layer (Figure 9.6). Starting from an SOI wafer, a wet thermal oxidation of the thin silicon film produces a 500-nm-thick insulating layer. A 900-nm-thick aluminum film is evaporated and patterned to define the contact pads and the interdigitated metallic fingers (microelectrodes) with a width of 2 μm, a length of 450 μm, and separated by a spacing of 2 μm. The last fabrication step concerns the formation of the thin humidity-sensitive layer that covers the interdigitated fingers. This thin hydrophilic layer is a dense aluminum oxide film with a thickness of 100 nm obtained by anodizing the top and sidewall surfaces of each of the aluminum interdigitated fingers [17,18].

ELECTRICAL CHARACTERIZATION OF HUMIDITY SENSORS

Condensation due to humidity can be viewed as an added layer with a very high conductivity and permittivity. Depending on the microsensor design, a change of one

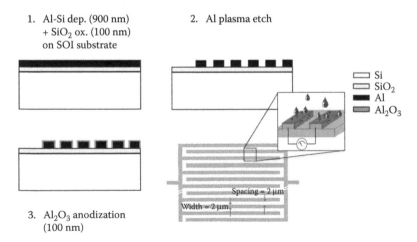

1. Al-Si dep. (900 nm)
   + SiO$_2$ ox. (100 nm)
   on SOI substrate

2. Al plasma etch

$\square$ Si
$\square$ SiO$_2$
$\blacksquare$ Al
$\square$ Al$_2$O$_3$

Spacing = 2 μm
Width = 2 μm

3. Al$_2$O$_3$ anodization
   (100 nm)

FIGURE 9.6 Process flow for hydrophilic surfaces obtained by means of anodization.

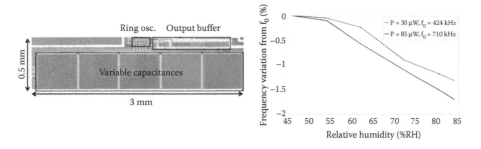

FIGURE 9.7 Humidity sensor associated with an output buffer in a 1 μm fully depleted silicon-on-insulator process.

or even two orders of magnitude can be reached between the simulated capacitance under dry air condition and under full immersion, that is, a several micrometer-thick water layer covering the whole area of the microelectrodes.

In dry air, the equivalent capacitance of the microsensor approximates the air capacitance. When relative humidity increases, the capacitance of the microsensor slightly increases at the rate of 7.5 fF/%RH/mm², leading to a 2% variation of the oscillation frequency between 45% and 85%RH (Figure 9.7).

However, when water molecules condense at the sensor surface, the equivalent resistance of the medium above the interdigitated electrodes drastically decreases whereas the capacitance largely increases. The aluminum oxide capacitance, hundreds of time higher than air capacitance, is more and more observed. The design of the electrode spacing and aluminum oxide thickness dimensions as well as relative permittivities of the fluids and materials give rise to a capacitance variation of two orders of magnitude [18].

Table 9.2 shows a comparison between the performance of humidity sensors reported in the literature. We focus especially on capacitive sensing improved by hydrophilic materials (sensitive to water vapor and other polar organic molecules) based on absorption and diffusion such as polyimide, and adsorption and capillary condensation such as aluminum oxide. In this table, the absorption time is defined as the time taken by the transient curve for a change of the capacitance from 10% to 90% of its maximum as the desorption time is defined as the time for a change of the capacitance from 90% to 10% of its maximum. On the one hand, most humidity sensors based on aluminum oxide use the porosity obtained by acid anodization to take advantage of capillary condensation inside pores [22]. However, such porosity contributes to a drift in capacitance characteristics [26] and leads to slow response times (2 s at least) [24]. On the other hand, high moisture absorption of polyimide leads to very high sensitivity; nevertheless, important drifts are observed beyond 70%RH [19].

High-speed responses are mainly limited by humidity diffusion into the sensing area and evaporation. By thinning and heating polyimide, the time response improves at the expense of sensitivity and power consumption. In Table 9.2, the sensitivity of our sensor demonstrates small but measurable values with regard to comparable capacitive sensors. One must nevertheless consider the simplicity of the process requiring no heater in contrast to References [19–21] and one-step anodization with no sulfuric acid

TABLE 9.2

Comparison of Sensitivity for Some Humidity Sensors Presented in the Literature[a]

Reference	Best Sensitivity	Sensitive Layer	Absorption/ Desorption	Comments
[19]	6.8 pF/%RH/mm²	Polyimide	—	—
[20]	23 fF/%RH/mm²	Polyimide	1 s/unknown	30%–90%RH
[21]	120 fF/%RH/mm²	Polyimide	200 ms/11 s	Human breath
[22]	0.4 pF/%RH/mm²	Porous aluminum oxide	5 s/5 s	40%–100%RH
[23]	≅140 fF/%RH/mm²	Porous aluminum oxide	25 s/30 s	2–45%RH
[24]	2.2 µA/%RH/mm²	Porous aluminum oxide	2 s/10 s	0–100%RH
[25]	8.8 pF/%RH/mm²	Porous aluminum oxide	—	—
Our work	7.5 fF/%RH/mm²	Dense aluminum oxide	250 ms/3 s	Human breath

Notes: RH, relative humidity.

[a] All based on capacitive transduction excepted drain current of a transistor for Reference [24].

in contrast to References [23–25]. Furthermore, our dense aluminum oxide layer also adequately protects the electrodes (here, made of low-cost CMOS-compatible aluminum) from chemical aggressions, and desorption time as small as 300 ms is measured for sensors drying forced by human inspiration (see application in Section "Application for Respiratory Rate Detection"). The microsensor desorption time is indeed greatly improved when a forced evaporation occurs due to the patient's inspiration.

APPLICATION FOR RESPIRATORY RATE DETECTION

Condensation sensing is shown to be very sensitive and finely correlated with the breathing cycle. However, a dedicated packaging (i.e., the protection and the connection of the sensor chip) has been set in order to avoid saturation of the capacitive sensor by large condensation (Figure 9.8). The interdigitated microelectrode has been fabricated on a silicon substrate by the same techniques as described in Figure 9.6. The difference lies in the capacitance-to-frequency converter, which is assured by a Schmitt trigger and a resistance (surface-mounted devices) onto the surface of a printed circuit board.

CONCLUSION

Planar and 3D SOI sensors were built and characterized under various stimuli such as airflow, temperature, magnetic field, and humidity, illustrating both the high flexibility of the surface micromachining and the 3D stressed-cantilever concept. Such

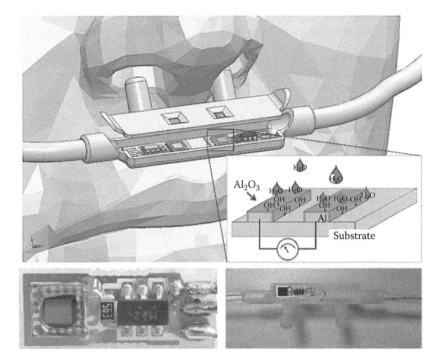

FIGURE 9.8 Human breath monitoring with embedded humidity sensor copy of oxygen supply canula with upward openings (module size: $30 \times 8 \times 8$ mm^3).

microintegrated sensors can easily be built incorporating their associated electronics on the same chip. Using simple fabrication techniques, the presented technology can then be seen as a technology for the fabrication of highly integrated low-power MEMS sensors, built with cointegrated metal–oxide–semiconductor-integrated circuits.

ACKNOWLEDGMENTS

This research is supported by the project CAVIMA of the Walloon region of Belgium and by the project MINATIS cofunded by the European program FEDER and the Walloon region. The authors would like to thank P. Simon for the RF measurements of the microsensors, C. Emmerechts for his skills with rapid prototyping, and Pat Chambers for his kind advice about redaction.

REFERENCES

1. Barth, F. G. 2004. Spider mechanoreceptors. *Current Opinion in Neurobiology* 14, 415–422.
2. Wee, K. W., Kang, G. Y., Park, J., Kang, J. Y., Yoon, D. S., Park, J. H., and Kim, T. S. 2005. Novel electrical detection of label-free disease marker proteins using piezoresistive self-sensing micro-cantilevers. *Biosensors and Bioelectronics* 20, 1932–1938.

3. Sun. C., Fang, N., Wu, D. M., and Zhang, X. 2005. Projection micro-stereolithography using digital micromirror dynamic mask. *Sensors and Actuators A: Physical* 121, 113–120.
4. Wilson, S., Jourdain, R., Zhang, Q., Dorey, R., Bowen, C., Willander, M., Wahab, Q. et al. 2007. New materials for micro-scale sensors and actuators: An engineering review. *Materials Science and Engineering* 56, 1–129.
5. Dahlmann, G. W., Yeatman, E. M., Young, P., Robertson, I. D., and Lucszyn, S. 2002. Fabrication: RF characteristics and mechanical stability of self-assembled 3D micro-wave inductors. *Sensors and Actuators A: Physical* 97–98, 215–220.
6. Miki, N., Zhang, X., Khanna, R., Ayon, A., Ward, D., and Spearing, S. M. 2003. Multi-stack silicon-direct wafer bonding for 3D-MEMS manufacturing. *Sensors and Actuators A: Physical* 103, 94–201.
7. Kolesar, E. S., Ruff, M. D., Odom, W. E., Howard, J. T., Ko, S. Y., Allen, B. A., Wilken, J. M., Wilks, R. J., Bosch, J. E., and Boydston, N. C. 2001. Three-dimensional structures assembled from polysilicon surface micro machined components containing continuous hinges and microrivets. *Thin Solid Films* 398–399, 566–571.
8. Iker, F., Andre, N., Pardoen, T., and Raskin, J.-P. 2006. Three-dimensional self-assembled sensors in thin-film SOI technology. *Journal of Microelectrochemical Systems* 15, 1687–1697.
9. Kao, I., Kumar, A., and Binder, J. 2007. Smart MEMS flow sensor: Theoretical analysis and experimental characterization. *IEEE Sensors Journal* 10, 713–722.
10. Dagamseh, A. M. K., Lammerink, T. S. J., Kolster, M. L., Bruinink, C. M., Wiegerink, R. J., and Krijnen, G. J. M. 2010. Dipole-source localization using biomimetic flow-sensor arrays positioned as lateral line system. *Sensors and Actuators A: Physical* 162, 355–360.
11. Makinwa, K. A. A. and Huijsing, J. H. 2002. A smart wind sensor using thermal sigma-delta modulation techniques *Sensors and Actuators A: Physical* 97–98, 15–20.
12. Kohl, F., Fashing, R., Keplinger, F., Chabicovsky, R., Jachimowicz, A., and Urban, G. 2003. Development of miniaturized semiconductor flow sensors. *Measurement* 33, 109–119.
13. Fürjes, P., Legradi, G., Ducso, Cs., Aszodi, A., and Barsony, I. 2004. Thermal characterisation of a direction dependent flow sensor. *Sensors and Actuators A: Physical* 115, 417–423.
14. Laconte, J., Raskin, J.-P., and Flandre, D. 2006. *Micromachined Thin-Film Sensors for SOI-CMOS Co-integration.* Dordrecht, The Netherlands: Springer Science, p. 186.
15. Sobieski, S., Andre, N., Raskin, J.-P., and Francis, L. A. 2009. Temperature effect on Lorentz based magnetometer. *Sensor Letters* 7, 1–4.
16. Igreja, R. and Dias, C. J. 2004. Analytical evaluation of the interdigitated electrodes capacitance for a multi-layered structure. *Sensors and Actuators A: Physical* 112, 291–301.
17. Moreno-Hagelsieb, L., Lobert, P. E., Pampin, R., Bourgeois, D., Remacle, J., Flandre, D. 2004. Sensitive DNA electrical detection based on interdigitated Al/Al_2O_3 microelectrodes. *Sensors and Actuators B: Chemical* 98, 269–274.
18. André, N., Druart, S., Gerard, P., Pampin, R., Moreno-Haglesieb, L., Kezai, T., Francis, L. A., Flandre, D., and Raskin, J.-P. 2010. Miniaturized wireless sensing system for real-time breath activity recording. *IEEE Sensors Journal* 10, 178–184.
19. Dokmeci, M. and Najafi, K. 2001. A high-sensitivity polyimide humidity sensor for monitoring hermetic micropackages *Journal of Microelectrochemical Systems* 10, 197–204.
20. Kang, U. and Wise, K. D. 2000. A high-speed capacitive humidity sensor with on-chip thermal reset. *IEEE Transactions on Electron Devices* 47, 702–709.

21. Laville, C. and Pellet, C. 2002. Interdigitated humidity sensors for a portable clinical microsystem. *IEEE Transactions on Biomedical Engineering* 49, 1162–1167.
22. Chen, Z. and Chin, M. C. 1992. An alpha-alumina moisture sensor for relative and absolute humidity measurement. *IEEE Industry Application Society Conference* 2, 1668–1675.
23. Varghese, O. K. and Grimes, A. G. 2003. Metal oxide nanoarchitectures for environmental sensing. *Journal of Nanoscience and Nanotechnology* 3, 277–293.
24. Chakraborty, S., Hara, K., and Lai, P. T. 1999. New microhumidity field-effect transistor sensor in ppm$_v$ level. *Review of Scientific Instruments* 70, 1565–1567.
25. Juhasz, L., Vass-Vamai, A., Timar-Horvath, V., Desmulliez, M. P. Y and Dhariwal, R. S. 2008. Porous alumina based capacitive MEMS RH sensor. *IEEE Design, Test, Integration & Packaging Conference* (Nice, France), pp. 381–385.
26. Dickey. E. C., Varghese, O. K., Ong, K. G., Gong, D., Paulose, M., and Grimes, C. A. 2002. Room temperature ammonia and humidity sensing using highly ordered nanoporous alumina films. *Sensors* 2, 91–110.

10 Microelectromechanical System-Based Micro Hot-Plate Devices

Jürgen Hildenbrand, Andreas Greiner, and Jan G. Korvink

CONTENTS

Microelectromechanical system (MEMS)-based devices with a thermally decoupled region are in wide use. Radiation detectors based on a temperature change due to absorbed light are a typical example for a device showing a low temperature change. For these kinds of micro hot-plates the focus is restricted to the thermal decoupling and/or the reduction of the heated thermal mass.

Devices with an integrated heater element—micro hot-plates—form another family of these MEMS devices. Typically, the required temperatures are of the order of several hundreds of degree Celsius. The need for the micro hot-plate approach for

257

sensor or actuator integration can have different motivations. Low power consumption and fast transient operation are the crucial reasons in most applications, but direct—monolithic or hybrid—integration with additional electronic components, or the advantage of minimization and reduction of fabrication costs, can influence the decision for micro hot-plates.

Besides the thermal decoupling of the hot-plate region, the mechanical stability of the devices themselves as well as the stability of the functional structure commonly deposited onto the hot-plate platform are the challenging aspects of development and fabrication. Two important micro hot-plate devices, which strongly influenced the development in this field, are metal-oxide gas sensors and thermal emitter infrared gas spectrometers. The micro hot-plate variants of these types are quite similar, but the infrared emitter has a typical operation temperature range of 600–800°C, and more would be better, where for metal-oxide-based gas sensors, temperatures of less than 400°C are sufficient for most applications.

STATE OF THE ART

Since the late 1980s, micro hot-plates have been investigated as substrates for metal-oxide gas sensors. These devices are typically operated at 350–400°C. In the beginning, closed membranes were realized. Sometimes a heat spreader was integrated beneath the functional materials and structures [1–8]. Later, suspended membranes* were fabricated [8–14].

Considering thermal emitters, the developing course seems similar. Up to now, there are some micro hot-plate thermal emitters commercially available. These are based on a closed-membrane design. However, a suspended-membrane micro hot-plate for use as a thermal emitter in miniaturized gas sensor systems was reported in Reference [15]. Typically, the suspended-membrane micro hot-plates are based on a silicon-on-insulator (SOI) substrate.

Scitec Instruments Ltd offers thermal emitters for fast modulation. These emitters can be considered as a lateral thin-film filament emitter. The infrared-50 [16] thermal emitter is based on a very low thermal mass diamond-like carbon thin-film element. The filament suspension looks like a closed-membrane micro hot-plate. Details about the fabrication itself and the use of MEMS technologies are not known. Closed-membrane-based micro hot-plate emitters are provided by Intex and Axetris. The key component of the Intex emitter MIRL17-900 is a 2 µm thin amorphous carbon nanocomposite which forms a closed membrane together with the SiO_2 and Si_3N_4 layers. The low thermal mass allows a good modulation behavior [17]. The Axetris emitter infrared source—also a closed-membrane type—achieves black-body-like emission characteristics with a black platinum layer. The maximum temperature is restricted to 450°C [18,19]. This emitter is well suited for implementation in a spectroscopic system in the spectral range from 4 to 6 µm. Due to the low operation temperature, the optical power is often not sufficient at wavelengths greater than 6 µm. A higher thermal emission in the fingerprint region is desired. Besides these commercially available emitters, several research groups are working on this topic.

---

* Membranes structured in the area between the active region and the rim.

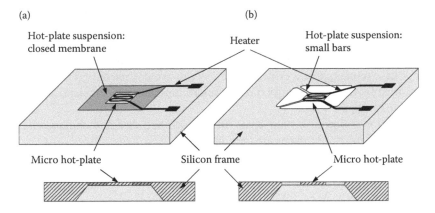

(a)

Hot-plate suspension:
closed membrane

Heater

Hot-plate suspension:
small bars

(b)

Micro hot-plate

Silicon frame

Micro hot-plate

FIGURE 10.1 Two different concepts for micro hot-plates. (a) The closed membrane concept typically uses a dielectric suspension layer for the hot-plate suspension. (b) The suspension bar concept is an improvement of the closed membrane concept. The suspension bars are typically made of silicon.

Here, it seems that the trend goes toward structured membranes, so that only some suspension bars connect the hot-plate with the frame. Figure 10.1 shows the main difference between the closed-membrane concept and the suspension-bar concept.

Spannhake et al. [20,21] presented interesting investigations on new high-temperature realizations of micro hot-plates. Approaches of a direct heating of the SOI hot-plate using the thin silicon layer itself as a heater as well as the deposition of an antimony-doped tin-oxide heater element onto the micro hot-plate were investigated. Temperatures up to 1000°C were achieved. This approach is a possibility for increasing the operating temperatures of such emitters. Probably, the reproducibility of the electrical resistance and its sensitivity to ambient oxygen concentration and temperature coefficients are two disadvantages in contrast to platinum heaters. Further, the electrical resistance of tin oxide is also sensitive to oxygen coverage at its surface like a metal-oxide gas sensor.

DESIGN PROCESS FOR MICRO HOT-PLATES

The design process for micro hot-plates can be divided into a coarse and a fine design. In the coarse design, the relations between the required hot-plate temperature, the hot-plate size, and the thermal resistance between the hot-plate and the frame and the thermal resistance for a membrane or suspension bar are determined. This can be done using a simple lumped element model, which represents at least the thermal resistances between the hot-plate and the frame and the hot-plate and the ambient by a rough estimation of conduction, convection, and radiation. If the thermal decoupling between the hot-plate and the frame is quite low, the thermal resistance between the hot-plate frame and the mounting device or element, which probably has a constant temperature, should be taken into account. For a rough estimation of the time transient operation characteristics, the thermal masses of the parts of the lumped element model can be added to the model.

The fine design process deals with the optimization of the distribution of the temperature field. For this purpose, simulation software based on the finite element method (FEM) is ideal to analyze the problem. In most cases, a homogeneous temperature distribution on the hot-plate is required, the current density in the heater does not have to overcome a critical value (e.g., to avoid electromigration), and the stress caused by thermal expansion of the micro hot-plate and the deposited structures has to be kept low too.

In the following section some basic information and design techniques for the development of micro hot-plates are introduced.

THERMAL ENERGY TRANSFER IN MICRO HOT-PLATES

The design of the heater element, the mounting in the housing, or, in the case of micro hot-plate devices, the substrate design itself have to be developed considering the heat transfer from the heated region to the ambient region.

Heat transfer is the transition of thermal energy. It is a compensation process from an item with temperature $T > 0$ K to one or more cooler items. The heat transfer is based on three processes: thermal conduction, convection, and thermal radiation. The time derivative of the total heat quantity Q_{tot}—also the total power consumption— is given by the sum of the time derivatives of the corresponding heat quantities Q_{cond}, Q_{conv}, and Q_{rad}:

$$\dot{Q}_{tot} = \dot{Q}_{cond} + \dot{Q}_{conv} + \dot{Q}_{rad}.$$

The consideration of the overall power consumption is necessary for the dimensioning of the heater. In general, the aim is to reduce the thermal conduction between the heated region and the environment in order to reduce the power consumption of the devices. For pulsed emitters it can be advantageous to increase the thermal coupling in order to improve their time constants in the cooling process compared with a design, which is tuned for minimal power consumption.

Thermal Conduction

Thermal conduction is the transport of thermal energy in matter by electron diffusion or phonon vibrations. The basic equation for heat flux q_{heat}, which describes the static heat energy flow for a given specific thermal conductivity κ and the temperature profile T, is defined by Fourier's law:

$$\bar{q}_{heat} = -\kappa \cdot \text{grad } T$$

For transient analysis, the specific thermal capacity has to be introduced. This leads to the heat conduction equation

$$q_{gen} = c \cdot \rho_m \cdot \dot{T} - div(\kappa \cdot \text{grad } T)$$

where ρ_m is the mass density and q_{gen} is the heat generation rate per unit volume.

For good thermal decoupling of the heated region it is necessary to use items with low thermal conductivity. The thermal conductivity is determined by a combination of the specific thermal conductivity and its geometry. In most cases the main heat sink is given by the socket. Good thermal decoupling is reached using suspended approaches for the heated regions. In that case, the suspension elements require a high mechanical stability. Higher specific thermal conductivities are tolerable if the mechanical properties allow smaller cross sections.

If the optimization of the transient characteristics of such suspended devices is required, the minimization of the thermal mass of the heated region is the main strategy. Additionally, it is possible to improve the time constants—given by the product of the thermal resistance and the thermal mass—using numerical optimization algorithms for a defined operation point (e.g., modulation frequency and temperature hub).

Convection

In the case of convection, the heat energy is transported by particle flow. This implies that convection only occurs in fluids or gases. There are two kinds of convection: natural convection and forced convection. Natural convection means that the reason for particle flow is a temperature gradient only. Forced convection implies another source of particle flow, for example, a ventilator.

The natural convection is the dominant convection type in thermal emitters, because the emitters are typically mounted in housings and protected from the ambient flow field. Thermal flow resulting from natural convection can be estimated by the heat-transfer coefficient α_{trans} and a temperature difference between two different materials. T_1 is the temperature of material 1 and T_2 is the temperature of material 2.

$$q_{conv} = \alpha_{trans}(T_2 - T_1)$$

For an estimation, α_{trans} can be considered as a constant (i.e., $\alpha_{trans} = \max(\alpha_{trans}(T))$). The forced convection is given by

$$q_{conv} = \alpha_{trans}(T_2 - T_1) \cdot \sqrt{q_m}$$

where q_m is the mass flow rate of the forcing medium. The heat transfer coefficient results from a linearization of the convection at a specific operation point (temperature, temperature difference, mass flow rate, etc.) and is only valid in a limited region around this operation point.

The thermal energy loss caused by convection can be minimized using a sealed housing (e.g., sealed with a BaF_2 window at the top for thermal emitters). Then, the free convection process is limited to the housing inside. The heat flow from the inner surface to the outer surface of the housing is a heat conduction process.

A minimization of the convection by reducing the lateral surface area is counterproductive, because the desired high thermal radiation decreases with this surface area too. The influence of the sidewalls is insignificant due to the relatively small surface areas.

Thermal Radiation

The thermal radiation of any matter can be referred to as black-body radiation, which defines the maximum possible thermal radiation emitted by matter depending on its temperature. According to its name, the black body is an object which absorbs any photon over the complete spectral region (or in the region where it is defined as a black body). A practical example of a black body is a small pinhole in a box with highly absorbent inner surfaces. The probability of an incoming photon of hitting the pinhole again after many reflections is very low. On the other hand, a black body with a uniform temperature $T > 0$ K emits electromagnetic radiation according Planck's distribution:

$$u_{bb}(\lambda, T) = \frac{8\pi hc}{\lambda^5} \frac{1}{e^{hc/(k_B \lambda T)} - 1}$$

The spectral energy density u of an arbitrary thermal emitter corresponds to the thermal emission of a black body the spectral emissivity factor ⊠ to read

$$u = \varepsilon(\lambda, T) \cdot u_{bb}$$

The spectral emissivity of a black body is defined as ⊠ = 1 over the complete spectrum. Emitters having the same relative spectral distribution but a lower energy density (⊠ < 1 and constant) are named gray emitters. For a worst-case estimation, the heat flow of a micro hot-plate device can be estimated using an emissivity of 1.

Kirchhoff's law states that the absorptance α of a body is equal to its emission ratio ⊠. This implies that there is a steady energy transfer with the ambient objects at a temperature $T > 0$ K. Typically, micro hot-plates are operated in an environment that is around room temperature. In that case, it is sufficient to consider the energy transfer only from the micro hot-plate to the ambient region and not vice versa. In special cases, like vacuum packaging sealed with caps, which can reach a temperature higher than the ambient temperature, this may be taken into account.

In most cases, a rough estimation of the energy transfer caused by radiation is sufficient and mostly detailed information of the spectral emissivity at different temperature is not available. In this case, the energy transfer can be estimated with the Stefan–Boltzmann law, to read

$$q_{rad} = \varepsilon \cdot \sigma_{SB} \cdot T^4$$

where σ_{SB} is the Stefan–Boltzmann constant. The Stefan–Boltzmann constant is defined by

$$\sigma_{SB} = \frac{2\pi^5 k_B^4}{c^2 h} = 5.670400 \times 10^{-8} \frac{J}{sm^2 K^4}$$

The thermal resistor R_{th} and the thermal mass C_{th} required for making lumped models for rough estimations in the beginning of new micro hot-plate types are used equivalent to its electrical counterparts [10]. The temperature equals the voltage and the heat flux equals the electrical current.

For a rough estimation, it is sufficient to consider only the main heat flow path as a one-dimensional problem. Like every numerical solution method, it is necessary to discretize the model. This means that the geometric representation of the sensor has to be divided into i cuboids along the heat flow path. The transformation of heat conduction equation leads to

$$C_{\mathrm{th}\,i} \cdot \dot{T} - \frac{1}{R_{\mathrm{th}\,i}} \Delta T = h$$

where h is the heat generation rate, $R_{\mathrm{th}\,i}$ is the thermal resistance of the i-th cuboid, and $C_{\mathrm{th}\,i}$ is the thermal mass of the i-th cuboid.

A thermal quadripole shown in Figure 10.2a is one possibility for the representation of the cuboid. For example, a suited representation for a suspended-type hot-plate is a serial connection of one quadripole for the hot-plate, one for the suspension bars, one for the frame, and one for the bottom side of the device to the ambient region (Figure 10.2b). The resistor $R_{\mathrm{th}\,i} = R_{\mathrm{th}\,i\,1} + R_{\mathrm{th}\,i\,2}$ depends on the area A_i of the cuboid cross section, the length $l_{i\,x}$, and the specific resistivity $\rho_i \cdot R_{\mathrm{th}\,i\,x}$ and $C_{\mathrm{th}\,i\,x}$ can be calculated by

$$R_{\mathrm{th}\,i\,x} = \frac{1}{2} \cdot \frac{l_{i\,x}}{A_i} \cdot \rho_i \text{ and } C_{\mathrm{th}\,i} = c_{\mathrm{th}} \cdot m_i.$$

For a more complex model, the thermal losses caused by radiation and convection can be also added in the form of and further resistance to the ambient region. For a fixed working point, a linearization of this thermal resistance could be used.

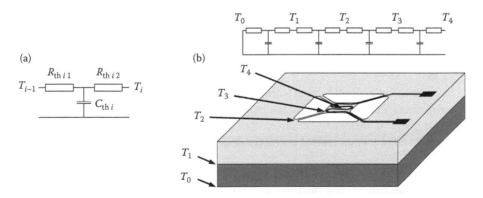

FIGURE 10.2 (a) Quadripole network consisting of two thermal resistors and a thermal capacity. The arrangement of the symmetrically placed resistors around the capacity results in mean temperature applied to the thermal mass. (b) Micro hot-plate structure and the corresponding thermal circuit for quarter of the device. For devices with extreme high thermal decoupling, the estimation of the convection and radiation also has to be considered.

Considering the transient characteristic, this would be a best-case estimation in the cooling process, because the energy transfer is considered for its maximum value.

Besides the possibility of a fast estimation of the power consumption and the transient characteristic the lumped model, it is easy to understand which part of the hot-plate has to be changed in order to achieve the requirements or to make a best-case study with minimum effort.

Hot-Plate Design

The geometrical size and shape of the hot-plate are determined by the individual application. The heating and cooling times are closed coupled with the thermal mass of the hot-plate and thus with the hot-plate area. The hot-plate suspension is the thermal decoupling element of the active area. Its thermal resistor influences the power consumption and also the heating and cooling times.

Dielectric membranes based on SiN_3, SiO_2, or a composite of these are widely used as a basis for the closed-membrane micro hot-plates. Suspension bars are typically fabricated in silicon, often in combination with a dielectric layer that separates the hot-plate silicon from the bulk silicon and acts also as etch barrier during fabrication.

The material properties are fixed values, but the static and transient characteristics of the hot-plate device can be adjusted with its geometrical design. For low power consumption in a static mode a high thermal decoupling is required, but the thermal mass is irrelevant. For fast transient operation, the thermal mass has to be reduced to a minimum, but thermal decoupling by the suspension bars also influences transient characteristics. For a fast heating process, a high thermal isolation of the hot-plate is the ideal configuration, but this will result in a slow cool-down characteristic. If the thermal resistance is too low, the complete device—especially at the mounting points of the suspension bars—will experience an increase in temperature during the heating process and the thermal capacity of the device will also influence the transient characteristic in the hot-plate area. Therefore, it is not possible to make a general design strategy for the thermal resistor of a suspension bar, because the transient characteristic also depends on the thermal capacity of the hot-plate and the amount and division of its heat loss (conduction, convection, radiation).

In order to ensure mechanical stability during fabrication and later in the operation, designers also have to look at the mechanical characteristic in the thermal optimization step. Considering a micro hot-plate platform with several hundred micrometer edge length, typical geometry data are:

- Membrane thickness of around 1 μm for closed-membrane types.
- Thickness of several micrometers, width of several tens of micrometers, and a length of some hundreds of micrometers for suspension bars.

A possible mechanical fatigue mechanism during operation of the thermal emitter could be caused by buckling. Buckling can occur due to the thermal expansion of the hot-plate and the suspension bars, which are fixed at the more stable silicon frame. The suspension bars and the hot-plate are under compression in this state. Depending on the geometrical design, buckling can occur at a defined temperature difference between the hot-plate and the silicon frame. Additionally, the combination of different thermally mismatching materials causes thermomechanical stress. Considering long-term stability, thermomechanical fatigue of the silicon suspension bars can occur, caused by pulsed operation. A further experimentally observed defect mechanism, which is also related to the bending of the microstructure, is the mechanical rupture of the platinum heater. The heater is placed outside the neutral axis of the bending part of the emitter. Bending or buckling of the structure yields a high stress in the interface of two materials. This could cause a lift-off of the platinum structure from the silicon or a crack or void formation, superposed with electromigration effects. A minimization of the bending of the suspended microstructure will result in a decrease of these effects.

A strain deformation of the suspension bars and the hot-plate is not possible for standard designs (Figure 10.1). In case of a membrane-type micro hot-plate, there is really no room for improvement. May be a deposition with a negative prestress could shift the critical temperature for buckling a little bit. In case of the suspension-bar micro hot-plates, designers have more freedom to shift the critical temperature for buckling. Two main concepts, which can be adjusted by tuning the shape of the suspension beams, are important:

1. The suspension beams can be shaped like a meander forming a spring. If the hot-plate area expands during the heating process, the spring suspension beams can be compressed without much force.
2. The suspension bars can be connected to the hot-plate area in such way that a thermal expansion will result in a rotational movement of the hot-plate [22].

Figure 10.3 shows two example layouts for the suspension-beam concept, which introduce a rotational movement of the hot-plate, caused by the thermal expansion of the hot-plate and the suspension bars. Figure 10.3b shows a variant with an additional 90° arc, which also acts as a spring structure. This variant allows larger temperature differences between the hot-plate and the frame until buckling occurs. The black structures sketched in Figure 10.3 are the ohmic heater, which goes on the suspension bars to the hot-plate area and back and a meander-shaped temperature sensor on the frame. The hot-plate temperature can be estimated by measuring the electrical resistance of the ohmic heater or by an additional temperature sensor on the hot-plate. The rim temperature is used to evaluate the thermal decoupling of the device.

HEATER AND TEMPERATURE SENSOR LAYOUT

Material Considerations

The choice of materials for the heater and temperature sensor structures depends mainly on the compatibility with the temperature range, the bulk material used, and the fabrication processes.

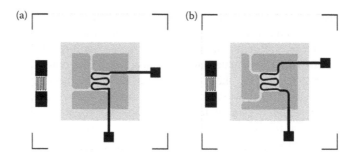

FIGURE 10.3 Two variants for suspension bars, which introduce a rotational movement during the thermal expansion of the hot-plate and its suspension. The variant (b) has an additional arc in the suspension beams. This will also contribute a spring characteristic to the suspension beams. An improved mechanical stress characteristic during operation is the result.

For operation temperatures up to 600°C sputtered thin-film platinum is well suited for this purpose. It is possible to deposit thin-film platinum using a tantalum adhesion layer on silicon. The main advantage of platinum over silicon or metal oxides with high-temperature stability is the precise temperature characteristics. Platinum temperature sensors are standard devices for temperature monitoring. Also for the heater a precise and reproducible thin-film material is advantageous even if the accuracy is not as important as for the temperature sensor.

Polysilicon is also used sometimes as heater material. The main advantage here is the availability and experience of the process in almost any clean room and the complete compatibility with complementary metal–oxide–semiconductors. Unfortunately, the thermal conductivity of polysilicon is strongly nonlinear and the electrical conductivity at room temperature is low. The use of doped polysilicon improves these, but it could also give rise to new problems, for example, for long-term operation at a higher temperature.

For the realization of high-temperature heater structures operating at 1000°C or even more, semiconducting metal oxides could be an alternative to thin-film platinum.

Heater and Temperature Sensor Design

For the development of a heater design, it is necessary to calculate the heater electrical resistance R_{el}. Under the assumption that the electrical power P_{el} will be completely transformed into the heating power P_{heat} the following equation can be used:

$$P_{el} = P_{heat} = U_{el} \cdot I_{el} = \frac{U_{el}^2}{R_{el}}$$

where I_{el} is the electrical current. Hereby, electrical resistivity is a function of temperature itself:

$$R_{el}(T) = R_0 \cdot (1 + \alpha T + \beta T^2 + \cdots)$$

where R_0 is the resistance at temperature T_0 and α and β are temperature coefficients.

The resistance of a one-dimensional conductor is defined by length l, width w, height h, and electrical specific resistance ρ_{el} as

$$R_{el}(T) = \rho_{el} \cdot \frac{l}{h \cdot w}$$

In addition to the total resistance, which determines the heating power, the lateral heater structure itself is important with regard to temperature distribution and local current density. A uniform temperature distribution can be achieved with a meander structure. Here, the size of the radii must be sufficiently high in order to ensure that the electrical current density does not increase too much at the inside of the curves. If the temperature difference in the heated region has to be kept nearly constant—that means in the magnitude of some degrees Celsius—the meander structure should be optimized with regard to the position-dependent heat generation. In the case of the thermal emitter, this optimization process is not required.

FEM ANALYSIS OF MICRO HOT-PLATES

The two mostly required simulation types in the micro hot-plate design process are electrothermal and thermal expansion. In principle, all this could be done with one geometrical model, but in order to reduce the simulation effort for a buckling analysis, it is sufficient to use a reduced model for this, which only contains the suspension beams and the hot-plate. In most cases, the microstructured hot-plate and suspension region can be geometrically modeled with two-dimensional elements, because there are no big temperature differences between the top and the bottom side of the hot-plate.

In the following discussion, the approach for electrothermal FEM analysis of micro hot-plate devices is exemplarily shown for some devices with an edge length of 3 mm, a hot-plate size of 500 μm, and a distance of 450 μm.

The simulations are performed for the suspension designs shown in Figures 10.3b and 10.4. The considered devices are based on an SOI waver, which consist of 400 μm bulk silicon, 1 μm SiO, and further 15 μm silicon. There are deposits of 400 nm Si_3N_4 at the top and the bottom of the waver. The heater structure consists of 200 nm platinum and a 20-nm-thick adhesion layer.

In order to keep the simulation effort at an acceptable complexity the micromachined part of the emitter is considered as one silicon object. This object is completely modeled in three dimensions and consists of the 400-μm-thick bulk and the 15-μm-thin silicon plate. The platinum heater with the tantalum adhesion layer is embedded as a two-dimensional structure on the surface of the silicon plate. The CAD files of the photolithographic masks can be used for the model generation, sometimes requiring a manual finishing of the geometry—especially at the curves.

The dielectric thin films at the top and the bottom as well as the one between the silicon substrate and the silicon plate—which are necessary for the emitter fabrication—are not considered in the model. The device is mounted using a

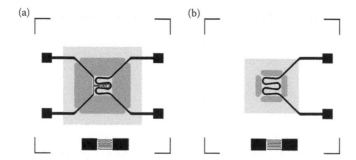

FIGURE 10.4 Two variants for standard suspension bar design. (a) Standard design, which has the same distance between micro hot-plate and frame, the variant shown in Figure 10.3. (b) Standard design, which has a strong reduced distance between micro hot-plate and frame, in order to have an example for a bad thermal decoupling.

ceramic adhesive. The adhesive is modeled as a 50-μm-thick volume beneath the silicon substrate. The thermal conductivity of air is implemented in the space between the hot-plate and the adhesive. The thermal radiation is also considered at the surfaces of the emitter. An emissivity of 1 (black body) is used as a worst-case estimation with regard to power consumption. Additionally, a heat transfer coefficient for the linear approximation of the free convection is implemented at the outer surfaces of the emitter. Room temperature is defined at the bottom of the ceramic adhesive and the heater voltage is varied. Literature values are used for the temperature-dependent thermal conductivity of silicon and platinum. In case of the tantalum/platinum sandwich structure, the electrical conductivity of test structures were measured at different temperatures and implemented in the model. Figure 10.5a shows exemplarily the temperature distribution of a hot-plate with standard suspension operated at a voltage of 10 V. A significant temperature decrease occurs at the silicon suspension bars and the intermediate air. The silicon rim is almost at room temperature.

The hot-plate temperature as a function of the heater resistance is shown in Figure 10.5b. This dependence is used to determine the temperature of the micro hot-plates. Even if there is a considerable difference in power consumption and hot-plate temperature between simulation and measurements, temperature distribution over the platinum structure for a defined hot-plate temperature and thus the electrical resistance of the heater will be nearly independent of this difference.

All simulated temperature–heater resistance curves show nonlinear characteristics. The reason for this property is that the heater is partly on the rim, on the suspension bar, and on the hot-plate. The temperature changes in these parts are nonlinear with respect to the hot-plate maximum temperature. The reasons for this are the temperature-dependent thermal conductivities, convection, and radiation. The curves of the variants with the same distance between the hot-plate and the frame seem only to shift by the difference of the base resistance of the heater. Apart from this, the curves show no noticeable differences. The variant with a short distance between hot-plate and frame shows clearly different characteristics. Compared

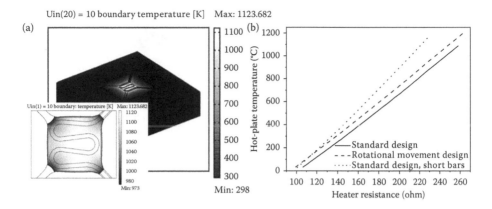

FIGURE 10.5 (a) Temperature distribution of a micro hot-plate emitter operated at 10 V obtained by a finite element method simulation. The simulated maximum temperature is 850°C. The micro hot-plate is well thermally decoupled from the silicon frame. Only at the fixing of the suspension bars is a larger decrease of the temperature observable (small picture). (b) Simulated hot-plate temperatures as a function of the electrical resistance of the heater for three different hot-plates. The curve is just shifted by the difference of the electrical resistance. The variant with the smaller distance between the hot-plate and the frame shows a clearly stronger increase, caused by the worth thermal decoupling.

with the others, the thermal decoupling is decreased and the required heating power is increased.

The stress–strain behavior of micro hot-plates operating at several hundred degrees Celsius can also be studied using FEM analysis. The implementation of prestress caused by the deposition of the thin-film layers is in principle possible, but quite involving. The prestress of the single layers had to be investigated previously using test structures fabricated with the planned fabrication process. The sensitivity for buckling of micro hot-plates can also be estimated by FEM analysis. In the case of micro hot-plates based on SOI substrates with suspension bars, the main influence on buckling is due to the thermal expansion of the hot silicon platform and suspension. There is also a contribution of the thin-film layers on it, but to obtain a rough value for the first buckling mode, the model can be simplified to the main material of the hot-plate and the suspension bars.

Figure 10.6 shows the simulated shapes of the first buckling mode for three different suspension bar designs. This linear buckling analysis shows only the critical temperature for the respective buckling mode. Information on the absolute displacement of the hot-plate from these simulations is not possible. Buckling analyses were applied on the silicon suspension bars and the silicon part of the hot-plate. The position of the ends of the suspension bars—the connection to the silicon frame—is fixed. The standard suspension type, shown in Figure 10.6a, having an axis-symmetric design has the lowest critical temperature difference of 379 K. The suspension design shown in Figure 10.6b has a significant improvement of the buckling problem, because of its possible degree of freedom in translation. This reduces the comprehensive stress caused by thermal expansion. Its critical temperature of

FIGURE 10.6 First buckling mode shapes of a linear buckling analysis for different support layouts caused by the thermal expansion. The critical temperature change that leads to the buckling is calculated to $\delta T = 379$ K for (a), $\delta T = 2507$ K for (b), and $\delta T = 8163$ K for (c). The displacement scale reaches from 0 to 1 depending on the eigenvalue calculation.

buckling is 2507 K. The design in Figure 10.6c also has the rotational degree of freedom and suspension bars with spring characteristics. Even if both supporting points of such bars are at a fixed position, a thermal expansion is possible. The critical temperature for buckling here is 8163 K. The critical temperature for buckling of the variants with a rotational movement of the hot-plate during operation is significantly above the melting point of silicon and thus above the possible maximal operation temperature. The length of the suspension bars also has a strong influence on the critical temperature for buckling.

FABRICATION

Typically micro hot-plates are fabricated using bulk micromachining from the front and the back side. Most of the material is removed using fast wet-etching processes from the back followed by a structuring process of the suspension beams from the thin front side. Micro hot-plate fabrication from only one side is also possible, but then the hot-plate region and its suspension have to be made etch resistant, for example, by ion implantation of silicon. Using two-side structuring, a dielectric layer acts in most cases as an etch stops. SOI substrates are well suited and becoming a standard starting material for micro hot-plate devices. Therefore, in the following discussion the fabrication of a micro hot-plate device is explained on the basis of an SOI substrate.

The thermal emitter hot-plate is based on a common 4-in. SOI wafer with 15 μm Si, 1 μm SiO_2, and 380 ± 15 μm Si. The SiO_2 layer acts as an etch stop for the KOH etch process. The bulk silicon has a (100) orientation and is p-doped with boron. On both sides a 400-nm-thick low-pressure chemical vapor deposition Si_3N_4 layer is deposited as a passivation for further processes. Figure 10.7a shows the cross section of the wafer.

The first part of the process comprises the front-sided structuring of the platinum heater elements and the platinum temperature sensors. An aluminum layer with a thickness of 280 nm is deposited and acts as a sacrificial layer in the platinum structuring process. The layout of the platinum structure is transferred in the following photolithographical process in the photoresist. The resulting photoresist structure is used for wet etching of the adjacent aluminum layer. Wet etching of the aluminum yields an undercut of approximately 1 μm. This undercut is important for creating a homogeneous border area of the sputtered platinum elements. The aluminum layer has the negative shape of the final platinum structures. Before the deposition of the

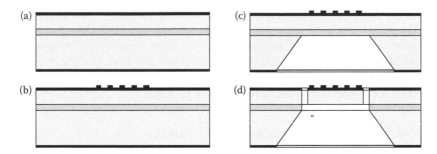

FIGURE 10.7 Fabrication of the thermal hot-plate (cross-section view). (a) Silicon on insulator substrate with a thickness of 386 μm (380 μm silicon, 1 μm SiO_2, and 15 μm silicon. (b) Both sides have a Si_3N_4 layer with platinum heater and platinum T-sensor elements on the top. (c) The bottom has a KOH etch. (d) Top side structuring of the suspension bars.

200-nm-thick platinum layer, a thin 20 nm-thick tantalum layer is deposited by sputtering in order to increase the adhesion of the platinum. After the deposition of platinum, the photoresist structure is stripped with an acetone–propanol cascade and the aluminum layer is removed by wet etching. The cross section of the structured platinum elements is shown in Figure 10.7b.

For the backside definition, the Si_3N_4 layer is photolithographically structured. The remaining photoresist acts as a negative mask for the following reactive-ion etch step. After stripping of the photoresist with an acetone–propanol cascade the wafer is prepared for the KOH etch of the backside cavity. The bare silicon is anisotropically etched in a KOH etch solution at a temperature of 80°C (Figure 10.7c).

In order to protect the front side, a special wafer holder that allows sealing one side of the wafer is used. The KOH etch has a much slower etch rate at the Si_3N_4 mask, at the (111) surfaces in the silicon bulk material and at the SiO_2 layer. This results in cavities with the shape of truncated pyramids. The angle between the backside surface and the sidewalls is 54.7°. The SiO_2, which is the etch stop for the KOH etch, is finally removed by a backside reactive-ion etch process. This is done uniformly without any additional passivation.

In the current state, silicon membranes with platinum structures are fabricated. Now, these membranes have to be structured. Once again this is done with an aluminum sacrificial layer process (thickness 500 nm). The aluminum is structured using a photolithographic process and a wet etch. The aluminum hard mask protects the hot-plate region with the platinum structures and the suspension bars from the dry-etch process. Using reactive ion etching the 400 nm Si_3N_4 as well as the 15 μm Si between the hot-plate, the rim and the suspension bars are removed. Finally, the remaining aluminum hard mask is removed by a wet-etching process (Figure 10.7d).

These fragile devices are diced using a wafer saw. The wafer is applied with its bottom side to an adhesion foil. The top side is protected with a viscous photoresist that also fills the cavities next to the hot-plates. After an out-gasing of 2 h the photoresist is baked for 15 min at 50°C. This leads to an improved adhesion between the wafer and the foil. The wafer is cut along dicing marks, which were deposited together with the platinum heaters and sensors on the front side. After the dicing process, the adhesion foil with the diced chips and the photoresist is soaked in acetone. The photoresist is removed and the chips are removed from the adhesion foil. After cleaning in propanol and deionized water the chips are prepared for packaging. Now, the chips are ready for packaging. A good possibility is to glue the chips with a ceramic adhesive in suite socket (e.g., TO5). Standard wire-bonding techniques can be used for the connection of the device and the package.

CHARACTERIZATION OF MICRO HOT-PLATES

The most important data of micro hot-plates are the hot-plate temperature as a function of the heater voltage or power and the modulation heater temperature as a function of the modulation frequency. This section discusses the experimental approach for the determination of these data and their problems.

STATIC ELECTRIC INVESTIGATIONS

The basis of the electrical characterization is the determination of the V-I characteristics. The results of the standard design are shown—exemplarily, one curve per design—in Figure 10.8a. Due to the inhomogeneous temperature along the heater, the V-I characteristics are clearly nonlinear. The platinum regions at the rim are at or close to room temperature, while the platinum regions on the hot-plate have a temperature of several hundred degrees Celsius and the platinum structures on the suspension bars have a temperature distribution between rim and hot-plate temperature. The electrical characteristics of the platinum heater can be modeled as a serial connection of resistors changing their resistance in dependence of the hot-plate temperature, but not uniformly.

Measurements in a vacuum chamber are performed in order to get information on the influence of the convection and conduction of air. These results can be also used for the FEM model in a redesign. In general, the influence of the position emitter in the vacuum chamber should not influence its own temperature. In order to assure one that there is no relevant position dependence in the chamber, a series of

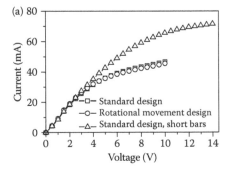
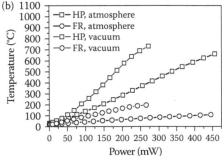

FIGURE 10.8 (a) *V–I* characteristics of the three different micro hot-plate emitters in normal atmosphere. (b) Hot-plate temperature (HP) and frame temperature (FR) as a function of the heating power in vacuum and in a normal atmosphere.

test measurements can be performed at different positions. Figure 10.8b shows the hot-plate temperature and frame temperature of the emitter as a function of the heating power. A double heating power (400 mW) is required in normal atmosphere to achieve a hot-plate temperature of 580°C. In both cases, the hot-plates are heated to ~700°C. The micro hot-plate operated in vacuum shows almost twice the frame temperature ($T_{frame,vac} = 197°C$; $T_{hot-plate,vac} = 733°C$) compared with the micro hot-plate operated in a normal atmosphere ($T_{frame,atm} = 109°C$; $T_{hot-plate,atm} = 662°C$).

TRANSIENT INVESTIGATIONS

The micro hot-plates are designed for a fast modulation. The response time of the hot-plate temperature—the time span from room temperature to operation temperature in steady-state condition—gives information of the modulation behavior during operation. Typically, a voltage step is applied to the heater. This implies that the heater current—and thus, the heating power—is changing during the transient process, because of the temperature dependence of the electrical resistance of the heater. The response characteristics have an exponential shape. It is not sufficient to determine the end value during the transient temperature measurements, because the temperature change at the end of the heating or cooling process is relatively small.

For a better comparison of different emitters, it is common to use the t_{90} time. The t_{90} time is the time span during the heating process from the static ambient temperature to 90% of the static operation temperature and the time span during the cooling process from the static operation temperature down to 10% of it, respectively.

For the t_{90} time measurements it is useful to have hot-plate devices with an integrated temperature sensor on the hot-plate. A constant current of 0.1%–1% of heating current has to be applied to the sensor and the voltage can be measured with an oscilloscope during the heating and cooling processes, respectively. The sensor current has to be chosen taking into consideration that the power loss through the temperature should not yield in hot-plate heating of several degrees Celsius. With the FEM results of the electrothermal simulations (Figure 10.5b) it is possible to estimate the hot-plate temperature on the basis of heater resistance. The steady-state voltage of the temperature sensor at a constant current was determined and used as 800°C reference in the modulated operation. Figure 10.9 shows the transient heating and cooling characteristics of one micro hot-plate with a standard suspension. During the heating process the hot-plate reaches 90% of the steady-state operation temperature after 39 ms. The cooling process down to 10% of the operation temperature lasts for 45 ms.

Alternatively, or even additionally, the frequency dependence of the hot-plate maximum and minimum temperatures can be determined using an optical detector by measuring the radiated power. Ideally, the detector has a linear characteristic with the incoming power. The dependence of the optical power and the hot-plate temperature can be measured in static operation. Now, the amplitude–frequency response can be determined by modulating the heater current and evaluating the detector signal. The interesting frequency span of micro hot-plates today is around 0.1–100 Hz.

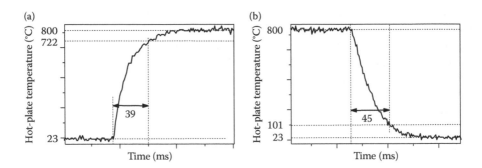

FIGURE 10.9 Determination of the transient characteristics of the hot-plates during (a) heating process up to 800°C and (b) cooling down to room temperature.

FURTHER RECOMMENDED INVESTIGATIONS

The measurement of the radiation power can be quite interesting for further modeling or better understanding of the different heat losses in order to optimize the micro hot-plate. The key element for this is the integrating sphere. The highly diffuse reflective inner surface of an ideal integrating sphere ensures that each segment of its surface is illuminated with the same light intensity, independent of where the light source is and how it radiates. The amounted of emitted power can be calculated on the basis of the surface and the measured optical power density.

The most temperature-limiting part of the micro hot-plate is its heater. In principle, it is possible to operate the micro hot-plate for a specified time at a fixed operation point, monitoring its heater current and checking the heater structure using microscopy after this period. But if the heater degradation is much more than expected, it is helpful to check the temperature influence on the heater structure without a heating current. Here, an oven, which allows the control of the nitrogen, oxygen, and hydrogen mixture in it, is useful for testing the influence of temperature at various ambient conditions.

The investigation of mechanical failure is important for micro hot-plates operating at high temperatures, and especially the types with low critical temperatures for buckling. An external acceleration of the hot-plate vibration with significantly larger amplitudes can be used to make highly accelerated life tests.

MICRO HOT-PLATES FOR METAL-OXIDE-BASED GAS SENSORS

Gas sensors based on metal oxides are quite sensitive to oxidizing or reducing gases such as CO, NO_x, and hydrocarbons. Typical metal oxides are SnO_2, ZnO, TiO_2, and WO_3, but the most investigated and commercially used material is SnO_2 [9]. The sensing mechanism is based on the availability of free electrons of the metal oxide. Used metal oxide has free oxygen vacancies because of its nonstoichiometric bond. In atmosphere, the metal oxide surface is covered with adsorbed oxygen. At higher temperatures (>>100°C), most of the adsorbed oxygen exists as O^-. This single molecule takes an electron from the metal oxide material for its bond. Thus, the adsorbed O^- atoms at the surface resulting in a depletion zone in the metal oxide.

Oxidizing and reducing gases can react at the surface and take or leave an O$^-$ molecule. If a CO molecule reacts to CO_2 at the surface, an electron will be given back to the metal oxide and its conductivity will increase. This also means that the sensing principle is an oxygen sensor. Under the assumption that the partial pressure change in the ambient is constant or at least small compared to the change of the aim gas, the sensor response is a function of the change in the gas concentration. In order to get a high signal change, small metal-oxide grains or thin films are used. Here, the fraction of the depletion zone related to the overall thickness is high or it even covers the complete material.

Because of the last cross-sensitivity to other gases and the strong sensor drift, it is not possible to use the sensors for quantitative analytical measurement. Its advantage is the extremely low price and its high sensitivity. Applications, that require the detection of a high change of a gas concentration, as in smoke gas concentration in buildings or the control system for fresh air quality in cars, are typical for these sensor types.

Metal-oxide gas sensors are typically operated at 200–400°C.

The first industrial available sensors were based on a heater embedded in an alumina ceramic tube. The metal oxide was mounted around this tube which had two printed electrodes (Taguchi-type sensor) [23–25]. Nowadays, the metal oxides are deposited mostly with a screen printing process (thick film) or sputtering (thin film) on a substrate. In order to achieve a high surface compared with the thin thickness of a material, nanostructured metal oxides are a major concern for research.

Micro hot-plates have two big advantages in metal-oxide gas sensing. The realization of operation temperatures up to 400°C with "normal" filament-like heat-able structures or the use of bulk substrates requires a high amount of electrical power. This allows the reduction of the power consumption from several Watts to some 100 mW or even less, depending on the required size of the heated area. Due to its fast heating and cooling times, power consumption can be reduced further. In many applications, it is sufficient to have a measurement each minute. The sensor can be turned on for a time span of some seconds to take the measurement and afterwards turned off. In general, the metal-oxide sensors have a large cross-sensitivity. But there is a temperature- and material-dependent sensitivity available, which allows getting some information on the gas composition. Micro hot-plate substrates allow a fast change of the operation temperature. The transient response of the temperature scan can be used as a gas-type-dependent characteristic. An integration of several different metal-oxide elements on one hot-plate is possible. Micro hot-plate concepts can also be used to build sensor arrays operating at different temperatures in one chip. Then, one micro hot-plate per chip is required.

Due to low electrical conduction, finger electrodes are used for connecting the sensitive layers. It is possible to design the heater and electrodes in such way that it can be fabricated together in one process step. Concerning the design of micro hot-plates for metal-oxide gas sensors, it is important to ensure a low temperature difference in the metal oxide. It should not exceed some degrees Celsius. In case of the screen-printed and sputtered approaches, the problem is mainly the lateral temperature distribution. If the micro hot-plate has a quite low thermal conductivity, an additional heat-spreading layer has to be deposited beneath the sensing region. In the case of nanostructured metal oxides, temperature distribution along the vertical

(a)

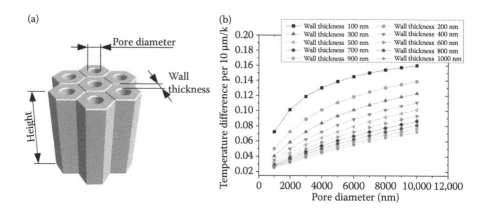

Pore diameter

Wall thickness

Height

(b)

Wall thickness 100 nm · Wall thickness 200 nm
Wall thickness 300 nm · Wall thickness 400 nm
Wall thickness 500 nm · Wall thickness 600 nm
Wall thickness 700 nm · Wall thickness 800 nm
Wall thickness 900 nm · Wall thickness 1000 nm

Temperature difference per 10 µm/k

0.20
0.18
0.16
0.14
0.12
0.10
0.08
0.06
0.04
0.02

0   2000   4000   6000   8000 10,000 12,000
Pore diameter (nm)

FIGURE 10.10   (a) Cut-out of a possible nanostructured metal oxide. (b) Estimation of the temperature difference from the top to the bottom of different wall thickness and pore diameters of the basis of Al_2O_3 operated at 300°C (bottom temperature). Pore heights of around 100 µm can be realized without critical temperature difference along the vertical direction.

direction may also become important. Here, making some calculation in order to obtain same "Figure of Merits" is recommended. Figure 10.10a shows an example cut-off from a pore array. The temperature difference between the bottom and the top for different pore diameters and wall thickness is shown in Figure 10.10b. Structures with very thin walls and large pore diameters could reach some critical values. A pore with a wall thickness of 100 nm, a diameter of 10 µm, and a height of 300 µm will have a vertical temperature difference of around 4.8°C. Together with a lateral temperature difference over the metal oxide of, for example, a further 5°C, this will result in a total temperature difference of almost 10°C. This value is too high for a metal-oxide gas sensor. The height of the structure had to be minimized and/or the lateral temperature distribution had to be improved.

MICRO HOT-PLATES FOR THERMAL EMITTERS

An increasing number of processes and safety and environmental applications require measurement systems for gas detection as well as for contamination monitoring of liquids. Examples of such applications are the monitoring of toxic gases and early detection of leakages. Another field is the chemical industry, which needs sensor systems for process control. In these applications, absorption measurements are an important detection technology combining high sensitivity, fast response time, and high reliability.

In particular, infrared spectroscopy facilitates the selective and sensitive measurement of various molecules by their specific absorption. It uses the characteristic absorption of the molecules in the mid–infrared region and allows the determination of the species and its concentration. Especially by absorption at longer wavelengths between 8 and 12 µm, the so-called fingerprint region, molecules can be measured with the highest selectivity. In the last years, infrared detection and measurement technologies have gained increasing importance.

Small thermal emitters are a key component in nondispersive infrared (NDIR) systems. In contrast to dispersive infrared systems that use a wavelength-selecting element like a Michelson interferometer or a grating, NDIR systems have a detection unit (channel) for a small specific spectral band. Typical is the combination of an optical filter with a broadband detector or the use of a photoacoustic detector. The main advantage of an NDIR system is the significantly low fabrication costs compared to other infrared spectroscopic systems.

Gas molecules have a defined and typical light interaction characteristic, which allows the determination of a gas concentration by measuring the amount of transmitted light through a gas volume with a known optical path length dependent in the wavelength. The basic setup of such systems contains a light source emitting light in the specific spectral range—where the target gas is absorbing—with the intensity I_0, a defined length of the optical path d, and a detection unit for measuring the transmitted light I_t. Knowing the spectral molar absorption coefficient $\varepsilon_{molabs}(\lambda)$ of the gas in the considered spectral range, the molar gas concentration c_{gas} can be calculated using the Beer–Lambert law [26]. Figure 10.11 shows the basic setup for spectroscopic transmission gas measurement.

Conventional thermal emitters for use in NDIR spectroscopic measurement systems are operated in a static mode. However, the modulation of the optical emission allows techniques for noise reduction or the use of pyroelectric detectors. Mechanical modulation with a chopper is a possibility, but it has the disadvantage of moving mechanical components, and at least it is an additional component, which increases the system costs. In the case of medium-priced systems, the cost of this optical chopper can be in the magnitude of 10–30% of the system costs—depending on the system. Thermal emitters with a low thermal mass are the right choice for this task. One possibility is to form filaments with a low thermal mass and a large surface area [16,27–30]. The use of MEMS-based micro hot-plates allows a further improvement of the modulation property of thermal emitters. Modulation frequencies of around 10 Hz are typical for this application.

MEMS-based commercially available emitters for pulsed operation use thin dielectric membranes to achieve thermal isolation of the micro hot-plate (the emitting area) from the supporting frame of the silicon chip. Nevertheless, the performance of

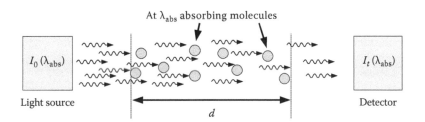

FIGURE 10.11 Basic setup for spectroscopic transmission gas measurement. The transmitted light intensity $I_t(\lambda)$ is measured with an detector after the light has passed a volume with absorbing molecules. On the basis of the absorbing coefficient at the wavelength λ_{abs}, the amount of light before gas interaction I_0 and the spatial length of the light/gas interaction d, the gas concentration can be determined by the Beer–Lambert law.

these "standard" MEMS-based emitters suffers from inadequate mechanical robustness and a low emission of radiation in the upper part of the mid-infrared region, due to the use of a closed dielectric membrane for thermal insulation. In contrast, micromachined infrared emitters based on the suspension-bar concept enable a fast transient temperature operation, higher operation temperatures, and, as a consequence, better emission characteristics in the mid-infrared spectral region.

Besides the high operation temperature, an improved spectral emissivity of the micro hot-plate is required in order to increase thermal emission. Emission coatings with low emittance at shorter wavelengths contribute toward saving heat energy. This is explained by Planck's distribution, where the maximum of the emitted power is shifted to shorter wavelengths with increasing temperature. At 800°C the emitted power of a black body at 10 μm is very low compared to the emitted power at 3 μm. If it is possible to reduce the emissivity in the unused spectral region, the efficiency factor of the emitter will be improved. Typically, an emissivity enhancement of a surface is achieved by increasing its roughness. In terms of MEMS devices, black silicon or black platinum are typical examples for such an emissivity enhancement.

The size of the hot-plate is directly related to its thermal mass and to its transient characteristic. Multichannel detectors are often used in NDIR systems in order to monitor several gas concentrations and a reference value simultaneously. Ideally, the used emitter has a similar area, size, and shape as the detector array, and the optical system images the emitter surface directly on the detector. If the emission area is smaller than the overall size of the detector array, the detector array has to be positioned out of the system's focal point and the intensity on the detector surface is reduced. Due to the fact that thermal emitter-based spectroscopic systems are in general limited by detector noise, this arrangement should be avoided.

ACKNOWLEDGMENTS

We heartily thank the Volkswagen Foundation for the generous financial support of the project that yielded a part of the reported results. We gratefully acknowledge the excellent cooperation and discussions with the teams at Fraunhofer IPM in Freiburg and Fraunhofer IWM in Halle, where the remainder of the work was performed. We also thank the "Der Andere Verlag," who generously allowed us to use copyrighted material from the PhD thesis of Jürgen Hildenbrand. Jan Korvink acknowledges support through the Excellence Initiative of the German Federal and State Governments.

REFERENCES

1. V. Demarne and A. Grisel. An integrated low-power thin-film CO gas sensor on silicon. *Sensors and Actuators*, 13(4):301–313, 1988.
2. U. Dibbern. A substrate for thin-film gas sensors in microelectronic technology. *Sensors and Actuators B: Chemical*, 2(1):63–70, 1990.
3. M. Gall. The Si planar pellistor: A low-power pellistor sensor in Si thin-film technology. *Sensors and Actuators B: Chemical*, 4(3–4):533–538, 1991.

4. H. S. Park, H. W. Shin, D. H. Yun, H.-K. Hong, C. H. Kwon, K. Lee, and S.-T. Kim. Tin oxide micro gas sensor for detecting CH_3sh. *Sensors and Actuators B: Chemical*, 25(1–3):478–481, 1995.
5. J. W. Gardner, A. Pike, N. F. De Rooij, M. Koudelka-Hep, P. A. Clerc, A. Hierlemann, and W. Göpel. Integrated array sensor for detecting organic solvents. *Sensors and Actuators B: Chemical*, 26(1–3):135–139, 1995.
6. V. Guidi, G. C. Cardinali, L. Dori, G. Faglia, M. Ferroni, G. Martinelli, P. Nelli, and G. Sberveglieri. Thin-film gas sensor implemented on a low-power-consumption micromachined silicon structure. *Sensors and Actuators B: Chemical*, 49(1–2):88–92, 1998.
7. D. Briand, A. Krauss, B. van der Schoot, U. Weimar, N. Barsan, W. Göpel, and N. F. de Rooij. Design and fabrication of high-temperature micro-hotplates for drop-coated gas sensors. *Sensors and Actuators B: Chemical*, 68(1–3):223–233, 2000.
8. S. Astié, A. M. Gué, E. Scheid, and J. P. Guillemet. Design of a low power SnO_2 gas sensor integrated on silicon oxynitride membrane. *Sensors and Actuators B: Chemical*, 67(1–2):84–88, 2000.
9. I. Simon, N. Barsan, M. Bauer, and U. Weimar. Micromachined metal oxide gas sensors: Opportunities to improve sensor performance. *Sensors and Actuators B: Chemical*, 73(1):1–26, 2001.
10. J. Hildenbrand, J. Wöllenstein, E. Spiller, G. Kühner, H. Böttner, G. A. Urban, and J. G. Korvink. Design and fabrication of a novel low-cost hotplate micro gas sensor. In *Proceedings of Design, Test, Integration, and Packaging of MEMS/MOEMS 2002*, Cannes, France, May 6–8, 2002, *SPIE Proceedings Series* 4755:191–199, 2002.
11. J. Hildenbrand. Mikrostrukturierte Halbleitergassensoren. Leistungsreduzierung durch den Einsatz von mikrostrukturierten "Micro-Hotplates". *MessTec and Automation*, 11(5):34–35, 2003.
12. D. Barrettino, M. Graf, M. Zimmermann, C. Hagleitner, A. Hierlemann, and H. Baltes. A smart single-chip micro-hotplate-based gas sensor system in CMOS-technology. *Analog Integrated Circuits and Signal Processing*, 39(3):275–287, 2004.
13. D. Briand, S. Heimgartner, M.-A. Gretillat, B. van der Schoot, and N. F. de Rooij. Thermal optimization of micro-hotplates that have a silicon island. *Journal of Micromechanics and Microengineering*, 12(6):971–978, 2002.
14. I. Elmi, S. Zampolli, E. Cozzani, M. Passini, G. Pizzochero, G. C. Cardinali, and M. Severi. Ultra low power MOX sensors with ppb-level VOC detection capabilities. In *Proceedings of the IEEE Sensors Conference 2007*, Atlanta, Georgia, October 28–31, 2007, pp. 170–173.
15. J. Spannhake, O. Schulz, A. Helwig, G. Müller, and T. Doll. Design, development and operational concept of an advanced MEMS IR source for miniaturized gas sensor systems. In *Proceedings of the IEEE Sensors Conference 2005*, Irvine, California, October 30–November 3, 2005, 4pp.
16. Scitec Instruments. Series 50, Thin film 0.9 watt infra-red emitter, data sheet, issue 1.2, www.scitecinstruments.de/irsources, 2007.
17. T. S. Skotheim, G. G. Kirpilenko, V. K. Dmitriev, P. Ohlckers, and J. Kunsch. Nanoarmophous carbon miniature thermal infrared source. *VDI-Berichte* 2047:161, 2008.
18. Axetris. IR source, data sheet, leaflet F25/09.2003/05.06, www.leister.com/axetris, 2007.
19. Axetris. IR source with reflector, data sheet, leaflet F29/03.2006/05.06, www.leister.com/axetris, 2007.
20. J. Spannhake, O. Schulz, A. Helwig, A. Krenkow, G. Müller, and T. Doll. High-temperature MEMS heater platforms: Long-term performance of metal and semiconductor heater materials. *Sensor*, 6(4):405–419, 2006.

21. J. Spannhake, A. Helwig, A. Müller, G. Faglia, G. Sberveglieri, T. Doll, T. Wassner, and D. Eickhoff. SnO_2:Sb—A new material for high-temperature MEMS heater applications: Performance and limitations. *Sensors and Actuators B: Chemical*, 124(2):421–428, 2007.

22. J. Hildenbrand. *MEMS-Based Thermal Emitters and Beyond—Components and Investigations for Miniaturised NDIR Systems. Microsystem Simulation, Design and Manufacture,* 2:58–61, 2010.

23 N. Taguchi. Japanese Patent No. 45-38200.

24. N. Taguchi. Japanese Patent No. 47-38840.

25. N. Taguchi. U.S. Patent No. 3 664 795.

26. P. W. Atkins. *Physikalische Chemie*, 3rd edn. Weinheim: Wiley-VCH, 2001.

27. ICx Photonics. Broadband pulsed infrared light sources, data sheet, http://photonics.icxt.com, 2007.

28. ICx Photonics. Pulsir™ high power, multi-element devices, product specifications sheet, http://photonics.icxt.com, 2007.

29. Scitec Instruments. Series 40, Thin film 1.2 and 4 watt infra-red emitters, data sheet, issue 1.4, www.scitecinstruments.de/irsources, 2007.

30. Laser Components. Cal-Source™ infrared emitters, pulsable IR emitters: SVF series, data sheet, www.lasercomponents.com, 2007.

11 Vibration Energy Harvesting with Piezoelectric Microelectromechanical Systems

Marcin Marzencki and Skandar Basrour

CONTENTS

WHY AMBIENT ENERGY HARVESTING?

Imagine that a miniature electronic device never needs battery recharging. Impossible? Not, if the energy source in the device is constantly replenished in an efficient and unobtrusive way. For decades, harvesting of the energy of ambient light has been a common method of making electronic devices operate longer. If we can use the energy of light, why not use other forms of ambient energy? Pressure variations, structural deformations, or mechanical vibrations are readily available in many environments. It is only recently, however, that the decreasing energy consumption of electronic devices enabled the creation of complex wireless systems powered solely from the energy present in their immediate surroundings. Furthermore, advancements in the microelectromechanical system (MEMS) technology have allowed miniaturization of the ambient energy harvesters to accompany the already small electronic systems, thus opening ways to the creation of fully autonomous miniature systems.

Use of ambient energy to extend the lifetime of electronic devices is especially appealing in two scenarios. The first one concerns devices that operate in difficult-to-access environments: mountains, forests, bridges, contamination zones, sealed structures, and also inside common machinery. In such cases, battery replacement or recharging is impossible or very inconvenient. The other scenario of interest applies when it is not economically viable to replace or manually recharge the battery in a device, for example, when a multitude of miniature devices compose a larger system. Those two cases usually apply to wireless sensor nodes that are used to continuously measure key parameters in a system. These devices would greatly benefit from additional energy to extend their operation time.

System Architecture

Figure 11.1 presents a block diagram of an example system, where the energy required for its operation is provided by an ambient energy-harvesting device. The ambient energy is converted into an electrical signal by an electromechanical transducer—the energy harvester. The electrical signal generated is rectified by an AC/DC circuit* and then regulated by a DC/DC circuit to efficiently recharge a small local energy storage unit (microbattery or supercapacitor). The client electronics uses the energy stored in the intermediate energy storage unit. Therefore, based on the amount of

---

* An AC/DC circuit is needed only if an AC signal is initially generated. For example, it is needed for mechanical vibration harvesting, but not for semiconductor solar panels.

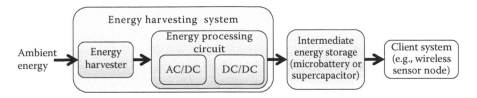

FIGURE 11.1 Block diagram showing the principal components of an ambient energy powered system.

energy available in the environment, the duty cycle of operation of the client electronics can be adjusted allowing the energy-harvesting system to recharge the battery, even if the average power available in the environment is lower than the active power consumption of the client electronics.

Recent studies show that current technology enables the creation of wireless sensor nodes that use sufficiently little power to be entirely powered from harvested ambient energy [1]. Hempstead et al. [2] analyze a sensor system that consumes on average only 2 µW when operated at 10% duty cycle and Hui Teo et al. [3] propose a sensor system for health monitoring that requires only 7.5 µA when operated at 0.9 V. A commercial wireless smart sensor platform Toumaz Sensium [4] is claimed to operate for over one year from a single 30 mAh battery at 1 V, which translates to a mean current consumption of less than 3.5 µA. All these examples prove that continuous power generation in the order of several microwatts would suffice to power a wireless sensor node operating at a very low duty cycle.

SIZE MATTERS

It has been proven that it is possible to use energy extracted from ambient vibrations to power wireless sensor nodes. Nevertheless, most of the current energy-harvesting devices are macroscopic, which negates one of the principal advantages of their use, that is, miniaturization of the device through reduction of size or complete elimination of the battery. Miniaturization of wireless sensor nodes is crucial not only to reducing their cost but also to minimizing their impact on the monitored environment. Imagine a wireless sensor that monitors motion of a shaft in a machine. If the sensor is too big, in the first place, it could be impossible to fit it into the best measurement location. Furthermore, if it is heavy compared with the monitored part, its presence would affect the part behavior negating the validity of the acquired data. Therefore, if the implementation of an energy-harvesting device in a system resulted in an increase in the overall dimensions over a similar system equipped with a battery containing enough energy to power it through its lifetime, the interest in using energy harvesting would be questionable. Therefore, miniaturization of energy harvesters is crucial to proliferate their use. To this end, MEMS technologies can be used. MEMS structures are created using manufacturing technologies originating from the standard batch microelectronic fabrication processes and thus provide a means of producing large quantities of inexpensive miniature devices, similar to electronic chips that revolutionized the electronic industry.

Ambient Mechanical Vibrations

Up to now, solar radiation has been the most common source of ambient energy, widely examined and successfully used [5]. Nevertheless, its application is limited to environments where direct light is available, which is rarely the case for miniature sensor nodes. As an example, let us consider possible ambient energy sources that could be used for powering wireless sensor nodes for industrial machinery health surveillance. Solar energy is to be excluded as in very few cases ambient light is available inside machinery. Furthermore, given the harsh and dirty nature of such environments, the efficiency of solar panels would quickly deteriorate. Thermal energy can be considered, but implementation of such a solution is very challenging due to high-temperature gradients required that are not practically realizable in miniature systems [5]. On the other hand, mechanical vibrations are often present with high power densities in industrialized environments [6]. Furthermore, mechanical vibrations can be transferred to the harvester device by means of a simple mechanical coupling. Due to these advantages a lot of research has been done in this field [7] and three main methods for converting mechanical energy into electrical energy have been identified: capacitive, electromagnetic, and piezoelectric. Each method has its advantages and drawbacks. A device using the capacitive method can be miniaturized relatively easily using a complementary metal–oxide–semiconductor (CMOS)-compatible process, but it requires very high polarization voltages for efficient operation [8]. On the other hand, the electromagnetic method provides high power densities, but relies on high-quality magnets and coils, which currently excludes easy miniaturization and integration with CMOS electronics [9]. Finally, the piezoelectric method offers elevated power densities from mechanically simple structures, but requires high-quality piezoelectric materials for efficient operation [6]. It has already been shown that the energy of mechanical vibrations can be successfully used to power wireless sensor nodes using the piezoelectric effect [10,11]. Additionally, the recent advancements in the piezoelectric thin-film deposition opened ways to further miniaturization of piezoelectric devices [12]. Given the fact that a wireless sensor node incorporating an array of MEMS energy harvesters could result in a truly miniature autonomous system, the piezoelectric energy conversion method seems to be the most promising one to use.

Research presented by Mitcheson et al. [13] analyzes various types of generic electromechanical energy converters with respect to their frequency characteristics. As the output power of a device converting mechanical vibrations into electricity increases with deformation of the device, obviously the highest power is generated at the resonance frequency. In case of the analyzed application (industrial machinery), a dominant frequency can usually be identified; therefore, a device with its resonance frequency matching this characteristic ambient vibration frequency can be created.

This chapter discusses the use of piezoelectric MEMS energy-harvesting devices to supply power to a miniature wireless sensor node. We start our considerations with a generic unidimensional model of an energy harvester followed by more detailed analytical models of actual geometrical structures common in MEMS implementations of piezoelectric energy harvesters. We also discuss various methods of implementing the presented models, ranging from the finite-element method

(FEM) to VHDL-AMS behavioral models. The most important factor discussed is the influence of various device parameters (e.g., material properties, damping, layer thickness) on the overall efficiency of energy conversion. Finally, all the modeling results are contrasted with experimental data obtained with actual piezoelectric MEMS energy-harvesting devices.

For clarity, an appendix containing all the symbols used in the presented models is provided at the end of this chapter.

GENERAL MODEL

The simplest representation of a resonant mechanical vibration energy harvester is a mass spring system, as schematized in Figure 11.2. This type of assembly was originally analyzed in the aspect of energy harvesting by Williams and Yates [14] where the fact of energy extraction was simplistically represented by a viscous damper. More precise models adapted for the piezoelectric transduction were introduced by duToit et al. [15] and by Lefeuvre et al. [16].

UNIDIMENSIONAL MODEL

A piezoelectric mechanical energy-harvesting system that is constrained to move in one dimension only is composed of a frame in movement $y(t)$ relative to a motionless reference base and a seismic mass m connected to the inside of the frame by a means of a piezoelectric element. When an external acceleration is applied on such system through movement $y(t)$, inertial forces act on the seismic mass m and induce its displacement $w(t)$ relative to the frame, which in turn results in compression of the piezoelectric element. The piezoelectric element is represented by a mechanical stiffness component (a spring with stiffness k_0) and a piezoelectric transduction element connected to an electrical load R. Furthermore, three types of losses are considered: viscous (proportional to the displacement speed) represented by the viscous damping coefficient λ, structural (proportional to the displacement amplitude) expressed by the structural damping ratio γ, and dielectric represented by the tangent of the loss angle ($\tan \delta$). Assuming that all components except for the seismic mass

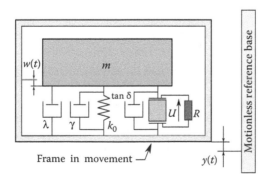

FIGURE 11.2   General one-dimensional model of a piezoelectric resonant power generator.

are weightless, this system can be described by the differential Equation 11.1 derived from the force equilibrium principle.

$$m\ddot{w} + \lambda\dot{w} + kw + F_p + m\ddot{y} = 0 \qquad (11.1)$$

The component $m\dot{w}$ represents the force of inertia of the seismic mass relative to the frame. The component $m\ddot{y}$ represents the force of inertia of the entire frame relative to the reference base. The component $\lambda\dot{w}$ represents the force of viscous damping and the component kw represents the force related to the mechanical stiffness of the piezoelectric element, where k is the complex stiffness incorporating the structural damping ratio γ (Equation 11.2, where $j = \sqrt{-1}$). Finally, F_p is the force introduced by the piezoelectric effect, which also includes the influence of the dielectric losses present in the piezoelectric element.

$$k = k_0 \left(1 + j\gamma\right) \qquad (11.2)$$

In the simplest case, the output power is evaluated on a resistive load R connected directly to the electrodes of the piezoelectric element. In a real implementation though, the power conversion system would be much more complicated, that is, composed of a rectification AC/DC circuit and a DC/DC converter for voltage regulation. Modeling of an entire energy-harvesting system including the piezoelectric transducer and the power conditioning circuit will be discussed in Section "Complete System Modeling."

In order to determine the value of force F_p, the constitutive equations of piezoelectricity, Equations 11.3 and 11.4, for a one-dimensional system are used. In the general model, it is assumed that the force is applied on the piezoelectric element along its polarization axis (piezoelectric axis 3). Therefore, following the IEEE standard on piezoelectricity [17] the 33 mode coefficients are used.

$$T_3 = c_{33}^E S_3 - e_{33} E_3 \qquad (11.3)$$

$$D_3 = e_{33} S_3 + \varepsilon_{33}^S E_3 \qquad (11.4)$$

T_3 and S_3 are respectively the stress and strain components along the thickness of the piezoelectric element, E_3 and D_3 are respectively the electric field and the electric displacement field components, c_{33}^E is the elastic stiffness constant measured at constant electric field, e_{33} is the piezoelectric constant, and ε_{33}^S is the permittivity component measured at constant strain. Equations 11.3 and 11.4 can be represented using the notation detailed in Table 11.1. Thus, after the introduction of structural and dielectric losses, Equations 11.5 and 11.6 are obtained, where C is the capacity of the piezoelectric element, linked with the static (lossless) capacity of the piezoelectric element C_0 by Equation 11.7.

$$F_p = kw + \alpha U \qquad (11.5)$$

TABLE 11.1

Conversion between Macroscopic and Microscopic Values

Quantity	Description
$F_p = T_3 A$	Relation between stress in the piezoelectric material and the external force applied, where A is the cross-sectional area of the piezoelectric element.
$U = -E_3 L$	Relation between the electric field intensity and the potential difference between the electrodes, where L is the length of the piezoelectric element.
$\dot{q} = A\dot{D}$	Relation between the electric displacement field variation and the charge variation on the electrodes (current between the electrodes).
$k_0 = c_{33}^E A/L$	Equivalent spring stiffness related to the piezoelectric element dimensions and material stiffness in the direction of pooling.
$C_0 = \varepsilon_{33}^S A/L$	Capacity of the piezoelectric element related to its dimensions and the electrical permittivity in the direction of pooling of the piezoelectric material.
$w = S_3 L$	Relation between the strain in the piezoelectric material and the external displacement of the seismic mass.
$\alpha = e_{33} A/L$	Piezoelectric force factor.
$\tau = RC$	Time constant of the circuit created by the piezoelectric capacity and the load resistance.
$\omega_0 = \sqrt{\frac{k_0}{m}}$	Angular resonance frequency of a purely mechanical system.
$\Omega = \omega/\omega_0$	Circular frequency ratio.
$\zeta = \dfrac{\lambda}{2m\omega_0}$	Unitless viscous damping ratio (percentage of critical damping).
$k_e^2 = \dfrac{\alpha^2}{C_0 k_0}$	Effective coupling factor of the system.

$$U = \alpha R\dot{w} - RC\dot{U} \tag{11.6}$$

$$C = C_0\left(1 - j\tan\delta\right) \tag{11.7}$$

The system of Equations 11.5 and 11.6 can be solved in the Laplace domain, leading to the system of Equations 11.8 and 11.9, where variables in the Laplace domain are indicated in bold and p is the Laplace variable.

$$F_p = kw + \alpha^2 R\frac{p}{1+\tau p}w \tag{11.8}$$

$$U = \alpha R\frac{p}{1+\tau p}w \tag{11.9}$$

Equation 11.8 indicates that force F_p applied on the seismic mass is composed of a purely mechanical component kw and a component due to the piezoelectric coupling dependent on the value of the electrical load R.

Furthermore, for the permanent state of the system and with a purely sinusoidal excitation, the frequency domain can be used. Therefore, the Laplace variable p is replaced by $j\omega$, where $j = \sqrt{-1}$ and ω is the angular frequency of vibrations. The relation in Equation 11.10 is obtained, which describes the complex displacement of mass $\hat{w}$ relative to the frame. Equation 11.11 describes the complex voltage $\hat{U}$ generated on the piezoelectric element. A_{in} is the amplitude of applied acceleration.

$$\hat{w} = \frac{(1 + j\tau\omega)mA_{in}}{(\lambda\tau + m)\omega^2 - k + j\left[m\omega^3\tau - (k\tau + \lambda + \alpha^2 R)\omega\right]} \qquad (11.10)$$

$$\hat{U} = \frac{j\omega\alpha RmA_{in}}{(\lambda\tau + m)\omega^2 - k + j\left[m\omega^3\tau - (k\tau + \lambda + \alpha^2 R)\omega\right]} \qquad (11.11)$$

OUTPUT POWER

Given that the ambient vibrations are abundant and that the harvesting device is sufficiently small not to influence its surroundings, the output power is the best criterion of harvester efficiency. In the simplest case, when the excitation is a sinusoidal acceleration and a purely resistive load R is connected to the piezoelectric element, the mean power dissipated on this load can be calculated following Equation 11.12.

$$P = \frac{|\hat{U}|^2}{2R} \qquad (11.12)$$

By using Equation 11.11 in Equation 11.12, it can be deduced that in the absence of damping, the output power is proportional to the seismic mass value and to the square of the input acceleration amplitude, and inversely proportional to the square of the resonance frequency of the system. This result is specific for the piezoelectric transduction method and differs from the generic one initially proposed by Williams and Yates [14].

OPTIMAL RESISTIVE LOAD

Structures using high-quality piezoelectric materials are strongly coupled, which means that the nature of the electrical load connected to the electrodes encompassing the piezoelectric material influences the mechanical behavior of the system. For very strongly coupled systems, two peak values of output power can be distinguished. One corresponds to the resonance ($\omega = \omega_0$), when impedance of the piezoelectric element is minimal, and the other to the antiresonance ($\omega > \omega_0$), when impedance of the piezoelectric element is maximal. There are, therefore, two optimal sets of excitation frequencies and the corresponding optimal load resistance values leading to the highest power output from an energy harvester. In an ideal lossless case, the optimal resistance would tend to zero at resonance and to infinity at antiresonance, with the corresponding power output values being identical. Furthermore, for each excitation frequency, an optimal resistive load value can be found that provides the highest output power.

INFLUENCE OF DAMPING

The presented model takes into account three types of damping: structural (proportional to deformation), viscous (proportional to deformation speed), and dielectric losses in the piezoelectric element. Figure 11.3 presents power dissipated on an optimal resistive load (a specific value is calculated for each excitation frequency) versus excitation frequency represented as circular frequency ratio for different damping levels. Structural and viscous damping influence both resonance and antiresonance peaks, tending toward a single peak for higher damping levels. The dielectric losses, however, influence only the antiresonance peak. It can be explained by modeling the dielectric losses as a parallel leakage resistor. The optimal load value at resonance being very small, the power dissipated on it is not influenced by a high-value parallel resistor. In the case of antiresonance, the optimal resistance value is very high (tending to infinity for a lossless case); therefore, any parallel resistance would cause a power drop. Piezoelectric materials often exhibit high dielectric losses up to 3% for soft lead zirconate titanate (PZT) [18] which significantly reduce the power generated at antiresonance. Figure 11.4 presents the influence of losses on the value of the optimal resistance. For viscous and structural damping, both values for resonance and antiresonance are influenced. For high damping ratios, these values tend to a single value, equal to $(\omega_0 C_0)^{-1}$. In the case of dielectric losses, the optimal load value corresponding to the antiresonance is much more influenced, in the same way as for the power output.

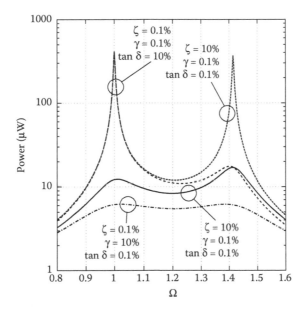

FIGURE 11.3 Analytical modeling results for the power dissipated on an optimal load resistance versus the circular frequency ratio for four sets of the three types of losses: viscous (ζ), structural (γ), and dielectric ($\tan \delta$).

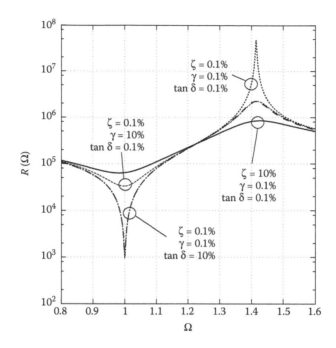

FIGURE 11.4 Analytical modeling results for the optimal resistance value versus the circular frequency ratio for four sets of the three types of losses: viscous (ζ), structural (γ), and dielectric (tan δ).

CRITICAL COUPLING

The principal parameter of an energy conversion system is the electromechanical coupling coefficient. This parameter describes the efficiency at which the input mechanical energy E_m is transformed into the output electrical energy E_e, as depicted in Equation 11.13. Its value is different when defined for materials used in the structure and for the complete structure. The coupling coefficient of the entire structure, called the effective coupling coefficient k_e, corresponds to the part of the mechanical energy present in the system that is converted into the output electrical energy.

$$k_e^2 = \frac{E_e}{E_m} \tag{11.13}$$

Figure 11.5 shows the evolution of power dissipated on an optimal resistive load as a function of the excitation frequency represented as the circular frequency ratio for four different values of k_e. It can be seen that the peak output power increases with the coupling coefficient but only up to a certain level. After a critical point, corresponding to the critical value of the effective coupling coefficient (in this case critical $k_e = 0.075$), the electrical power generated no longer increases and two peaks in the frequency domain appear (and therefore also two optimal resistive load values). Figure 11.6

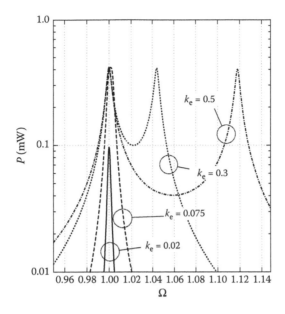

FIGURE 11.5 Analytical modeling results for the power generated on an optimal load resistance versus circular frequency ratio for four values of the effective coupling k_e.

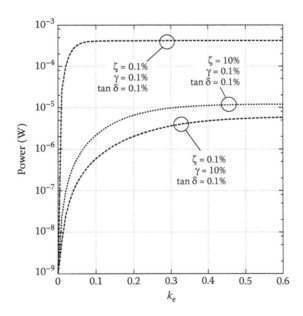

FIGURE 11.6 Analytical modeling results for the power dissipated on a matched resistive load at resonance versus the value of the effective coupling coefficient k_e for different levels of damping. Two zones can be identified, one where the output power increases rapidly with the increase of k_e, and second where the increase of k_e implies almost no increase of the output power.

presents the evolution of power dissipated at resonance on an optimal resistive load versus the coupling coefficient of the system at different damping levels. Two zones can be identified. The first one is where the increase in the coupling factor is followed by a rapid increase in the output power. The second one is where the output power is almost insensitive to the coupling coefficient change. A conclusion can therefore be made that when a system is strongly coupled (when there are two power peaks in the frequency domain and two optimal resistance values), the output power can only be increased by decreasing the losses present in the system. For example, for strongly coupled systems, it can be beneficial to use a material with inferior coupling but with lower losses, for example hard PZT versus soft PZT. It also proves that the use of very highly coupled materials, but with low mechanical quality factors (e.g., PZN-PT single crystals) does not guarantee higher output power.

COMPARISON OF PIEZOELECTRIC MATERIALS

In order to present the importance of the coupling coefficient and material quality on the performance of a piezoelectric energy generator, properties of four common piezoelectric materials (Table 11.2) are used in the model to estimate the output power generated on a matched resistive load. The angular resonance frequencies of all assemblies are set equal to $\omega_0 = 10^3$ rad s^{-1} by adjusting the geometrical dimensions of the piezoelectric element and keeping the mass size constant. The viscous damping ratio (independent of the piezoelectric material used) is fixed at 0.1%. Figure 11.7 presents the results and Table 11.3 summarizes the key values. It can be seen that all materials induce strong coupling in the system (two power peaks in the frequency domain) and, therefore, it is the level of losses introduced by these materials that determine the output power and not the coupling coefficient value. Even though the PZT-4 material does not present the highest coupling, it provides the highest output power.

The simple model described in this section is useful for analyzing the influence of general properties of the assembly on its performance. However, in order to precisely explore the behavior of a real system, it is necessary to create detailed analytical models of complete geometrical structures, as discussed in the next section.

TABLE 11.2
Properties of Four Common Piezoelectric Materials

	k_{33}	ε_{33}^s	Y (GPa)	Q	tan δ	Reference
PZN-PT	0.92	1386	120	61	1%	[19]
PZT-5H	0.75	3800	62	32	2%	[20]
PZT-4	0.7	1450	70	500	0.5%	[21]
AlN	0.3	11	300	250	0.1%	[22,23]

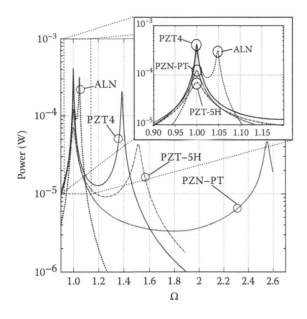

FIGURE 11.7   Analytical modeling results for the output power versus excitation frequency (expressed as circular frequency ratio) for devices using the four common piezoelectric materials, whose properties are listed in Table 11.2.

TABLE 11.3

Summary of the Modeling Results for the Four Common Piezoelectric Materials at the Resonance Frequency (res) and at the Antiresonance Frequency (ares)

	PZN-PT	PZT-5H	PZT-4	AlN
$P_{opt}@res$	123 µW	71 µW	417 µW	312 µW
$R_{opt}@res$	1.81 kΩ	2.52 kΩ	1.77 kΩ	12.45 MΩ
ω_{res}	2.55 ω_0	1.51 ω_0	1.39 ω_0	1.05 ω_0
$P_{opt}@res$	47 µW	43 µW	207 µW	312 µW
$R_{opt}@res$	30 MΩ	3.5 MΩ	43.3 MΩ	6.9 GΩ

CANTILEVER BEAM MODEL

The simple model presented in the previous section is often insufficient for precise performance evaluation of realistic energy-harvesting structures. Researchers proposed various shapes for vibration energy-harvesting devices, but a simple cantilever beam with a mass at the end proves to be the most promising one [7]. Additionally, a cantilever beam structure with an active piezoelectric layer can be fairly easily implemented as an MEMS [24].

Modeling of strongly coupled piezoelectric structures presents a significant challenge. Even though several models have been reported in the literature [7], very few of them focus on ambient energy harvesting. Furthermore, a structure fabricated using an MEMS process significantly differs from macroscopic structures and should be represented appropriately. This section introduces a precise analytical model of a piezoelectric cantilever beam energy harvester and compares the results with FEM simulations and experimental results.

SPECIFICITY OF MEMS

Miniaturization of energy-generating devices is one of the main goals in the field of ambient energy harvesting. In order to take full advantage of reducing the size of the energy reservoir present in an autonomous system, the energy-harvesting structure has to be miniaturized as well. MEMS technology can be used to create microscopic electromechanical structures, while maintaining low per-unit cost due to the batch fabrication process. Furthermore, it is possible to directly integrate MEMS devices with CMOS electronics to create self-powered Systems-on-Package or even self-powered Systems-on-Chip [25].

Structures fabricated using MEMS processes significantly differ from their macroscopic counterparts. The nature of the microfabrication process implies sequential deposition of layers of various materials interleaved by selective etching, which results in very specific geometries. The devices are usually made on a monocrystalline silicon wafer presenting a very high mechanical quality factor around 2×10^5 [26]. Additionally, the current piezoelectric material deposition processes allow for the creation of piezoelectric layers on one side of the device only, contrary to macroscopic structures, where multiple layers on both sides of the device are often present [27]. Furthermore, the mechanical influence of the metallic electrodes can no longer be neglected in MEMS as their thickness is comparable with the thicknesses of other layers. Also, large process variations induce high relative structure nonuniformity in the material properties, layer thicknesses, and geometrical dimensions. Finally, MEMS devices are often characterized by very high geometrical aspect ratios not practically realizable as macroscopic structures, for example, a very big mass on the tip of a very slender beam.

THIN-LAYERED PIEZOELECTRIC MATERIALS

Multiple piezoelectric materials are known, but only very few are available as thin layers to be used in MEMS. The materials most commonly considered for integration in MEMS devices are aluminum nitride (AlN), lead zirconate titanate (PZT), zinc oxide (ZnO), lithium niobate (LiNbO$_3$), and lead magnesium niobate-lead titanate (PMN-PT). Currently, AlN and PZT demonstrate the best combination of coupling factor, mechanical quality, and ease of deposition. For a more complete description of thin-layered piezoelectric materials and the methods of their deposition, the reader is referred to a review performed by Muralt [28].

An example structure of a cantilever beam MEMS piezoelectric generator created from a Silicon-On-Insulator (SOI) wafer is shown in Figure 11.8 and an SEM image of a manufactured device is shown in Figure 11.9 [25]. This structure has been fabricated using an SOI wafer to obtain a very uniform and slender beam from the top silicon layer. The seismic mass (proof mass) is made out of the bulk silicon and its thickness is equal to the wafer thickness.

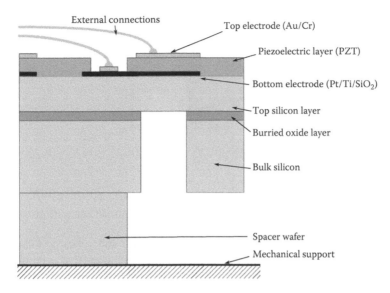

FIGURE 11.8 An example layer structure of a piezoelectric energy harvester built on a silicon-on-insulator wafer with lead zirconate titanate (PZT) piezoelectric thin layer (Adapted from M. Marzencki et al., *Proceedings of Nanotech 2007*, Santa Clara, California, May 20–24, 2007.)

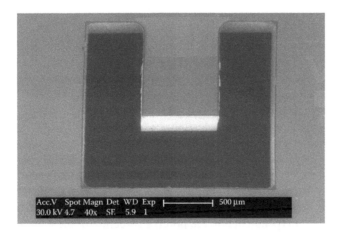

FIGURE 11.9 A piezoelectric microelectromechanical system energy harvester fabricated within the European Project VIBES. (Adapted from M. Marzencki, Y. Ammar, and S. Basrour, *Sensors and Actuators A: Physical*, 145–146, 363–370, 2008.)

GEOMETRY OF THE MODELED DEVICE

The piezoelectric cantilever beam structure that can be used to convert ambient mechanical vibrations into electricity is schematically represented in Figure 11.10. It is composed of a cantilever beam that carries a big seismic mass at its end. When excited with external acceleration, the force of inertia of the seismic mass deforms

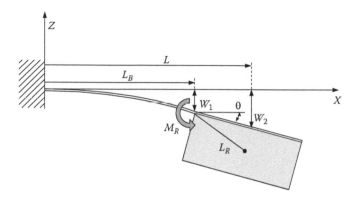

FIGURE 11.10   Schematic representation of the modeled cantilever beam structure.

the structure. The piezoelectric layer on top of the beam undergoes periodic cycles of extension–compression and thus generates electrical charges that accumulate on the metallic electrodes. As the dominating frequency values in the considered environment are around 1 kHz [30], the resonance frequency of the harvester should fall in the same range. In the case of MEMS this is not evident, taking into account the very small dimensions of these structures. In order to achieve a low resonance frequency, the size of the seismic mass carried at the beam end should be very significant compared with the rest of the device. As a result, the following assumptions are taken into account during creation of the model:

- The beam mass is neglected as it is much smaller than the seismic mass value.
- Only the first mode of resonance corresponding to the out-of-plane movement is modeled.
- The rotary inertia of the mass is taken into account.
- Mechanical clamping is considered to be perfect.
- The seismic mass is considered to be perfectly rigid.

A generic layer structure of the cantilever beam is shown in Figure 11.11. In the case of MEMS, there is usually only one piezoelectric layer sandwiched between electrode layers deposited on top of a silicon beam. In the case considered here, the electrodes are laid perpendicular to the pooling direction of the piezoelectric layer, and so piezoelectric axis 3 corresponds to geometrical axis z, piezoelectric axis 1 corresponds to geometrical axis x, and piezoelectric axis 2 corresponds to geometrical axis y. The placement of electrodes imposes that no electric field can develop in directions 1 and 2. On the other hand, during flexing of the beam, the charge density on the electrodes must be nonuniform in direction 1 in order to keep the electrode surfaces equipotential. The cantilever beam modeled is slender (very thin compared to its length) so the Euler–Bernoulli representation is used. Both the shear stresses and strains are neglected and Kirchhoff's and Love's assumptions are used, where the normal to the neutral surface remains normal and unstretched during deformation.

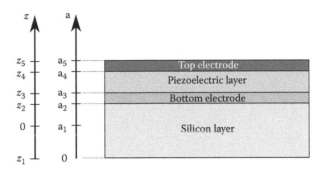

FIGURE 11.11 Layer composition of the beam, with the mechanical silicon layer, the bottom metallic electrode, the active piezoelectric layer, and the top metallic electrode. Location of the extremities of each layer on the two coordinate axes used in the model is presented as well.

BOUNDARY CONDITIONS

Definition of boundary conditions is very important for proper modeling of this structure. An assumption of "uniform strain" has been made for the beam, which implies that the beam width remains constant. It can be justified by the presence of both beam clamping at one end and attachment of the beam to the seismic mass at the other end. It is also assumed that the mechanical variables are independent of the y coordinate. Therefore, the independent mechanical variables comprise S_1 and S_2 horizontal strains and T_3 vertical stress. The effective mechanical and electrical properties of materials (denoted by superscript ef) are derived from boundary conditions of the structure (Table 11.4), as proposed by Muralt et al. [31].

The neutral axis location from the bottom of the support layer is calculated with Equation 11.14, where $Y_s^{ef}, Y_{eb}^{ef}, Y_{et}^{ef}$, and Y_p^{ef} represent the effective rigidity of the silicon support layer, bottom electrode, top electrode, and the piezoelectric layer,

---

TABLE 11.4
Definition of Electrical and Mechanical Boundary Conditions Used in the Model

Boundary Condition	Description
$w(x = 0, t) = 0$	The beam is mechanically clamped at $x = 0$.
$\left[\dfrac{\partial w(x,t)}{\partial x} \right]_{x=0} = 0$	Beam curvature at the clamping equals zero.
$T_3 = 0$	The layers are free to expand in the thickness direction.
$S_2 = 0$	The beam width is constant.
$E_1 = E_2 = 0$	Electrical field components in the horizontal directions equal zero.
$D_3 = f(x,t)$	The electric displacement field is a function of the x coordinate and time, but is uniform in the layer width and thickness.
$E_3 = f(z,t)$	The electric field is a function of the z coordinate and time, but is uniform in the layer width and length.

respectively; while h_s, h_{eb}, h_{et}, and h_p are the thicknesses of these layers, respectively. Equation 11.14 is obtained from the force equilibrium in the length of the beam (assuming that all layers are purely mechanical) and provides distance a_1 from the bottom of the beam to the neutral axis [32].

$$a_1 = \frac{Y_s^{ef} h_s^2 + Y_{eb}^{ef} h_{eb}\left(h_{eb}+2h_s\right) + Y_p^{ef} h_p\left(h_p+2h_s+2h_{eb}\right) + Y_{et} h_{et}\left(h_{et}+2h_p+2h_{eb}+2h_s\right)}{2\left(Y_s^{ef} h_s + Y_{eb}^{ef} h_{eb} + Y_p^{ef} h_p + Y_{et}^{ef} h_{et}\right)}$$

(11.14)

PIEZOELECTRIC COUPLING

Piezoelectric coupling of the structure has been derived from the piezoelectric constitutive equations [17]. Boundary conditions applied on the piezoelectric layer and the geometry of the device are used to generate Equations 11.15 and 11.16 [33,34].

$$\begin{bmatrix} T_1 \\ T_2 \\ S_3 \end{bmatrix} = \begin{bmatrix} c_{11}^{ef} \\ c_{12}^{ef} \\ -c_{13}^{ef} \end{bmatrix} \cdot S_1 - \begin{bmatrix} e_{31}^{ef} \\ e_{31}^{ef} \\ e_{33}^{ef} \end{bmatrix} \cdot E_3$$

(11.15)

$$D_3 = e_{31}^{ef} \cdot S_1 + \varepsilon_{33}^{ef} \cdot E_3$$

(11.16)

Parameters T_i and S_i are respectively the stress and strain vector components, D_3 and E_3 are the electric displacement field and electric field components, c_{ij}^{ef} represents the effective mechanical stiffness matrix coefficients, e_{31}^{ef} is the effective stress piezoelectric coefficient, and ε_{33}^{ef} is the effective electrical permittivity coefficient.

DAMPING TYPES

The behavior of the coupled electromechanical structure is greatly influenced by the presence of damping as discussed in Section "Boundary Conditions." The main types and the corresponding quality factors calculated for the analyzed structure are:

- Viscous damping that introduces a force proportional to the speed of displacement of the structure relative to the surrounding medium. Level of this damping depends mostly on the ambient gas pressure and the structure shape [26]. For simplification, we assume that only the seismic mass is affected by this type of loss, as its maximum speed of vertical movement is higher than that of the beam.
- Structural (hysteretic) damping that introduces a force proportional to the deformation amplitude. In the presented model, the quality factors of all materials are defined separately and result in a global quality factor of the structure when combined. The quality factor of monocrystalline silicon is equal to $2 \cdot 10^5$ [26] and that of the PZT layer is equal to 135 [35].

- Compression losses linked with compression of air between the mobile structure and fixed surfaces. In the analyzed case, for atmospheric pressure and the closest surfaces 500 μm away, the associated quality factor is equal to 49×10^3 [26].
- Clamping losses linked with energy radiation into substrate through clamping. The calculated quality factor of the structure linked with this type of loss is equal to $1.8 \cdot 10^6$ [36].
- Thermoelastic losses linked with heating of the structure and heat-related strains. In this case, the time of thermal relaxation of the structure is much smaller than the period of vibration. Using the Zener model [37] the quality factor linked with this type of loss is equal to 26×10^6.
- Dielectric losses in the piezoelectric material linked to electrical dipole mobility. This type of loss is usually very important in piezoelectric materials [18].

Due to small influence, the clamping, compression, and thermoelastic losses have been neglected in further modeling.

SYSTEM DYNAMICS

Equilibrium of bending moments $M_y(x,t)$ acting on the beam is used to determine the expression of the beam curvature $\kappa(x,t)$, defined in Equation 11.17.

$$\kappa(x,t) = \frac{\partial^2 w(x,t)}{\partial x^2} \tag{11.17}$$

The external forces acting on the beam are induced by the vertical movement of the seismic mass m (mass of the beam is neglected) and by the mass moment of inertia of the seismic mass J_0, as shown in Equation 11.18.

$$M_y(x,t) = \frac{m}{B_B}\left(a_{in}(t) - \ddot{w}_2(t)\right)\left(L - x\right) - \frac{J_0^2}{B_B}\ddot{\theta}(t) \tag{11.18}$$

As depicted in Figure 11.10, $w_2(t)$ is the displacement of the seismic mass center of gravity, $a_{in}(t)$ is the applied acceleration, and $\theta(t)$ is the angle of rotation of the seismic mass. Both $\theta(t)$ and J_0 are relative to the attachment of the seismic mass to the beam. The bending moment of the beam, relative to the neutral axis, is given by Equation 11.19.

$$M_y(x,t) = \kappa(x,t)\left[Y_s^{ef}\int_{z_1}^{z_2}z^2\,\mathrm{d}z + Y_{be}^{ef}\int_{z_2}^{z_3}z^2\,\mathrm{d}z + Y_{te}^{ef}\int_{z_4}^{z_5}z^2\,\mathrm{d}z\right]$$
$$+ Y_p^{ef}\kappa(x,t)\int_{z_3}^{z_4}z^2\,\mathrm{d}z + e_{31}^{ef}\int_{z_3}^{z_4}E_3(z,t)z\,\mathrm{d}z \tag{11.19}$$

Equation 11.19 can be simplified using the constitutive Equation 11.16 to obtain Equation 11.20.

$$M_y(x,t) = \kappa(x,t)D_G - \frac{\beta \varepsilon_{33}^{ef}}{h_p} \cdot u(t) \tag{11.20}$$

The equivalent rigidity of beam D_G is defined in Equation 11.21, and $u(t)$ is the voltage generated between the electrodes.

$$D_G = \frac{Y_s^{ef}}{3}\left(z_2^3 - z_1^3\right) + \frac{Y_{be}^{ef}}{3}\left(z_3^3 - z_2^3\right) + \frac{Y_{te}^{ef}}{3}\left(z_5^3 - z_4^3\right) + \frac{z_4^3 - z_3^3}{3}\left[Y_p^{ef} + \frac{(e_{31}^{ef})^2}{\varepsilon_{33}^{ef}}\right] - \frac{\varepsilon_{33}^{ef}\beta^2}{h_p} \tag{11.21}$$

Equation 11.22 is obtained by combining Equations 11.18 and 11.20.

$$\kappa(x,t) = \frac{m}{B_B D_G}\left(a_{in}(t) - \ddot{w}_2(t)\right)\left(L - x\right) - \frac{J_0^2}{B_B D_G}\ddot{\theta}(t) + \xi u(t) \tag{11.22}$$

By performing double integration on Equation 11.22 with proper initial conditions, expression 11.23 of deformation $w(x,t)$ of the beam is obtained.

$$w(x,t) = \frac{x^2}{2B_B D_G}\left[\left(L - \frac{x}{3}\right)\left(a_{in}(t) - \ddot{w}_2(t)\right)m - J_0\ddot{\theta}(t)\right] + \xi u(t)\frac{x^2}{2} \tag{11.23}$$

By taking $w_2(t) = w(L,t)$ and approximating $\theta \approx \tan\theta$, Equations 11.24 and 11.25, representing the electromechanical coupling of the structure studied, are obtained.

$$m\ddot{w}_2(t) + \lambda\dot{w}_2(t) + \frac{B_B D_G}{L_{eq}^2}\left[\frac{w_2(t)}{L_B} - \frac{\xi}{2}\left(L_B + L_M\right)u(t)\right] + \frac{J_0}{2L_{eq}^2}\left(L_B + L_M\right)\ddot{\theta}(t) = ma_{in}(t) \tag{11.24}$$

$$L_B\dot{u}(t) + \frac{\beta}{\xi B_B R D_G}u(t) + \beta\ddot{\theta}(t) = 0 \tag{11.25}$$

Auxiliary variables β, ξ, and L_{eq} are developed in the appendix.

MODELING RESULTS

In order to obtain the presented results, the model was evaluated using the Maple™ 10 software. In all the numerical examples, the structural layer and the seismic mass are made of silicon. The piezoelectric layer properties are equal to those of PZT-4 piezoelectric ceramic, pooled in the thickness direction, which is close to the properties of the thin-layered PZT used in MEMS [12]. The top electrode is made of gold. The bottom electrode is neglected in modeling due to its low thickness (100 nm) and its proximity to the neutral axis. The dimensions of the analyzed structure are as follows. The beam width and length are equal to 400 µm, the top silicon layer thickness is 10 µm, and the piezoelectric layer thickness is 1 µm. The mass height and length are 400 µm and the mass width is equal to 410 µm. The top electrode covers the entire surface of the beam and is 400 nm thick. An acceleration of 10 ms^{-2} is applied to evaluate the output power.

Comparison with FEM

Traditional approach to modeling complex electromechanical structures involves the use of models created using the FEM. Despite their accuracy, FEM does not provide a means for efficient structure optimization and the influence of various parameters is often difficult to evaluate. Nevertheless, FEM models are a good source of reference when experimental results are not easily available. In order to verify the validity of the presented model, the analytical results were compared with FEM simulation performed using the ANSYS™ 9 software. The *Solid226* 20-node coupled field elements were used for the piezoelectric layer, the *Solid186* 20-node structural solid elements were used for the silicon layer, and the *Shell181* 4-node finite strain shell elements were used for the top electrode. For simplification, dielectric losses were neglected in FEM simulations and in the analytical analyses that were used for comparison.

In order to compare the analytical model with FEM, a set of simulations was performed. Harmonic FEM analyses were performed for a set of resistive load values connected between the electrodes in order to identify the optimal load value. The maximum value of power was noted along with the corresponding excitation frequency for each load value. The *sparse solver* was chosen for its robustness and efficiency [38]. The resonance frequency of the structure calculated with the analytical model was equal to 1268 and 1208 Hz for the FEM simulation (5% discrepancy). Figures 11.12 and 11.13 compare results of analytical modeling and FEM simulation, respectively, for the power dissipated on an optimal load resistance and the optimal resistance values as functions of the circular frequency ratio for three values of viscous damping.

Both the resistance and power values are in good agreement, while there is a discrepancy in the frequency values; the difference between the resonance and the anti-resonance values (indicating the coupling of the structure) is higher for the analytical modeling. It can be explained by the simplifications made in the beam representation, mainly by the uniform strain assumption. An attempt to precisely model the clamping conditions without the uniform strain assumption would greatly complicate the model [33], and, therefore, is not attempted.

The presented comparison of the analytical model with the FEM shows that, in spite of the slight differences in the numerical values resulting from simplifications made, the influence of various parameters is well represented and the behavior of the energy-harvesting structure can be understood through analysis of the analytical model.

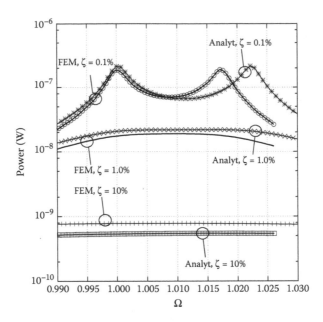

FIGURE 11.12 Power generated on an optimal resistive load versus excitation frequency (expressed as circular frequency ratio) for three levels of viscous damping calculated with the analytical model (analyt) and obtained through finite-element method simulation.

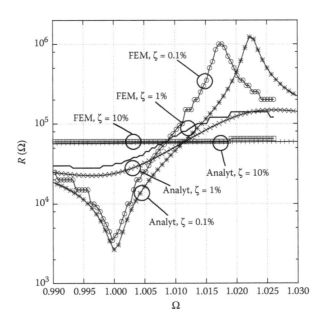

FIGURE 11.13 Optimal resistance values as a function of the excitation frequency (expressed as circular frequency ratio) for three values of viscous damping calculated with the analytical model (analyt) and obtained through FEM simulation.

COMPARISON WITH EXPERIMENTAL DATA

Comparison with experimental data is the ultimate validation of the model accuracy. Nevertheless, given the limited precision of the MEMS technology, not the design parameters of structures but rather the ones extracted from actual manufactured devices should be used (Table 11.5). Variation of the following factors can influence the properties of the manufactured devices:

- Layer thickness and its uniformity
- Material properties, contaminations, and imperfections
- Etching precision, including deep reactive ion etching (DRIE) underetch
- Mask misalignment and precision
- Sacrificial layer residue

As an example, Figure 11.14 presents an SEM image of an MEMS piezoelectric energy harvester with a significant underetch of the seismic mass. This fact, arising from improper tuning of the DRIE process parameters reduces the seismic mass size and thus increases the resonance frequency of the device. Finally, proper measurement techniques should be used taking into account all the connections and parasitic capacitances. For example, in the presented case, the top silicon layer was highly conductive and grounded. Thus, parasitic capacitances created by the contact pads (equal to 20 pF), whose values were extracted from additional test structures on each chip, were taken into account during model verification.

A set of MEMS devices was fabricated to verify the accuracy of the model. The measured resonance frequency of one of the fabricated devices was equal to 1368 Hz (at 0.2 g) compared to 1364 Hz obtained from simulation using the extracted values, proving the model accuracy. Also, both the experimental verification and the modeling results indicate that a resistive load of around 650 kΩ

TABLE 11.5
Extracted Parameters of the Microelectromechanical System Energy-Harvesting Device Used for Model Validation

Parameter	Description	Value
L_p	Beam length	400 μm
H_m	Mass thickness	530 μm
H_s	Top silicon layer thickness	9.94 μm
H_p	Piezoelectric layer thickness	0.91 μm
Q_p	Quality factor of the piezoelectric layer	120
tan δ	Dielectric losses ratio	$1e{-}3$
ζ	Viscous damping ratio	$2.5e{-}3$
h_{te}	Top electrode thickness	0.56 μm
R	Electrical load value	650 kΩ
C	Parasitic capacitance of contacts	20 pF

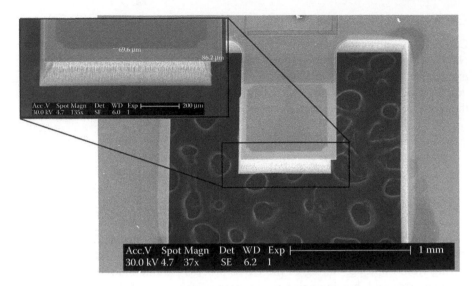

FIGURE 11.14 Underetch of a piezoelectric microelectromechanical system energy generator.

provides the highest power output. Figure 11.15 presents the values of voltage (amplitude) and power generated by the device, obtained with the analytical model and extracted experimentally at positive frequency sweep. A very good match is evident for lower acceleration levels, while for higher accelerations nonlinearities in the mechanical behavior of the device (not taken into account in the analytical model) induce an error. Further discussion on nonlinear behavior of piezoelectric MEMS devices can be found in Reference [39].

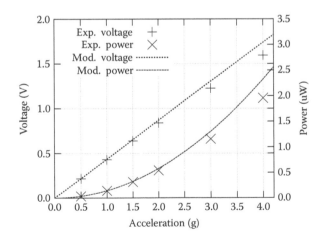

FIGURE11.15 Comparison of the experimental results obtained with a piezoelectric microelectromechanical system energy harvester (Exp) with the modeling results obtained with the analytical model (Mod).

OPTIMIZATION OF THE STRUCTURE

The analytical model can be used to optimize the geometrical structure of the harvester. In order to perform the optimization, the variables have to be identified. If the device is to be fabricated from a standard SOI wafer, the thickness of the seismic mass can be assumed to be equal to 500 μm. In order to keep a good uniformity DRIE all over the wafer, the minimum spacing between vertical walls may not be smaller than 400 μm. This restriction defines the minimal beam length. The maximum length of the structure was constrained to 1.2 mm. The modeling results of the one-dimensional model indicate that the seismic mass should be the biggest possible, and so the beam should be 400 μm long and the mass would therefore be 800 μm long. Furthermore, in order to prevent twisting of the structure the beam should be the widest possible, that is, equal to the mass width. In this case, the structure width becomes irrelevant in optimization and is arbitrarily chosen as equal to 800 μm. All these restrictions leave us with variables linked with the layer thicknesses. The piezoelectric layer thickness is defined by the deposition process. Usually, in order to achieve a crack-free layer with good uniformity, the piezoelectric layer thickness should be between 1 and 4 μm [12]. The silicon layer thickness should not be lower than 5 μm for reliability purposes. The top metalization layer thickness should be as thin as possible (inactive layer with low quality factor), but thick enough to be reliable and permit wire bonding. The minimum thickness of the chromium-gold layer was set at 400 nm.

The analytical model can be used to evaluate the power generated on an optimal load for different configurations of device dimensions, varying the ratio η of the piezoelectric layer thickness to the silicon substrate thickness. The resonance frequency is kept equal to 1 kHz by adjusting the substrate thickness for each dimension set. Figure 11.16 presents the output power generated on a matched resistive load at resonance versus η for different quality factors of the piezoelectric layer at a fixed viscous damping ratio of 0.1% and applied acceleration of 10 ms^{-2}. The piezoelectric layer thickness (h_p) and silicon layer thickness (h_s) are shown as well. The optimal thickness ratio results from interaction of two factors: the device capacity increases with the decreasing thickness of the piezoelectric layer, which limits the output power. On the other hand, the piezoelectric layer has a much lower mechanical quality factor than the monocrystalline silicon substrate. Therefore, by decreasing the piezoelectric layer thickness (the resonance frequency is kept constant by increasing the silicon layer thickness) the overall quality factor of the structure increases and so does the generated power. It can be seen that the quality of the piezoelectric layer strongly influences this relation. While for a low-quality piezoelectric layer (γ = 0.05) the generated power decreases rapidly after reaching an optimum at η = 0.006, for a perfect piezoelectric layer the generated power remains almost constant (limited by viscous damping at 0.1%) after the first maximum and falls because of displacement of the neutral axis into the piezoelectric layer and not because of the global quality of the structure. This statement is in agreement with the general conclusion from modeling that for strongly coupled structures (with two power peaks) the output power can only be increased by decreasing losses.

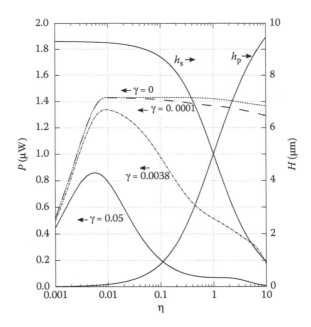

FIGURE 11.16 Analytical modeling results of the output power generated at resonance on a matched resistive load versus the ratio between the piezoelectric layer thickness (PZT-4) and the silicon substrate thickness for different quality factors of the piezoelectric layer γ at input acceleration of 10 m s^{-2}.

The analytical model presented in this section shows very good accuracy in predicting performance of piezoelectric MEMS energy harvesters. Nevertheless, the piezoelectric MEMS structures are strongly coupled, which means that their mechanical behavior is highly influenced by the electrical circuitry connected to it. Therefore, in order to correctly predict the performance of the system, a complete model including the electrical energy processing component should be created. The next section presents such a model created using an analog hardware description language, VHDL-AMS.

COMPLETE SYSTEM MODELING

As discussed in the previous sections, a piezoelectric generator is a complex electromechanical structure. Furthermore, the generated energy has to be rectified and regulated in order to be used by the client electronics. Not only is modeling of the complete system composed of the MEMS device and the AC/DC and DC/DC electronics a very challenging task, but also, additionally, a system designer needs easy-to-use models that capture the essential behavior of the system, allow for efficient optimization, and accurately predict the influence of design modifications [40]. Usually, physical and geometrical parameters of the energy-converting devices are only considered at the very low level of the design flow, typically using simplistic analytical models or FEM tools. Therefore, a designer using this methodology needs to perform separate simulations of mechanical and electrical parts of the system which delivers incorrect results in the case of strongly coupled systems. Recently, advances in analog hardware description languages (A-HDL) enabled the creation of complex mixed signal models that can be used to provide fully simulated electromechanical designs. Furthermore, less advanced users can limit themselves to using a library of models, while others can concentrate on specific components to be simulated on different levels of abstraction. To this end, a significant amount of research has been done covering the subject of multidomain simulation using VHDL-AMS [41,42]. Only recently, complete strongly coupled models with various types of damping have been introduced [24]. Furthermore, MEMS processes are characterized by limited accuracy and material parameter variations, which induce fluctuations in device characteristics and performance. Therefore, efforts are made to balance the effects of those variations already at the design phase to avoid or at least minimize the much more costly improvements of the manufacturing process. To reach this goal, a robust methodology should be capable of producing models that are accurate enough to be predictive, but simple enough not to extend the time to market of the product. In order to accomplish that, a statistical analysis can be performed using A-HDL to assess its sensitivity to process fluctuations.

There are multiple hardware description languages that could be used to perform simulation of an electromechanical system. In our study, we chose to use VHDL-AMS governed by the IEEE-1076.1 standard for its powerful equation notation capabilities provided by simultaneous statements and free quantities. The reader is invited to consult References [24] and [41] for further comparison of A-HDL tools.

This section discusses an example VHDL-AMS implementation of the mixed-signal electromechanical model of the piezoelectric energy generator presented in the previous section connected to an electrical processing circuit. As will be shown, the use of hardware description languages gives further advantages such as the ability to perform transient and statistical analysis.

DESIGN FLOW

In the presented example, the energy-harvesting generator is connected to a power-processing circuit, and so the model must be pin-compatible with the electric network. In order to secure reusability of the created model, a generic interface with all parameters that can be useful for a potential future user should be established. In such implementation, even a basic user can modify system parameters to adapt it to his or her needs, similarly to models of electrical components. Furthermore, the model should correctly handle the widespread values in MEMS; for example, nanometer displacement compared with gigaPascal stresses in the materials. Figure 11.17 presents the major steps that should be taken during model creation. First, a representation of both the electromechanical elements and the purely electrical elements has to be established using the VHDL-AMS notation. Then, the model accuracy has to be adjusted in order to address the trade-off between precision and simulation time to suit the needs of the user. Subsequently, sensitivity of the model on parameter variation can be evaluated. Finally, the results obtained with the model should be compared with actual experimental results using parameters extracted from

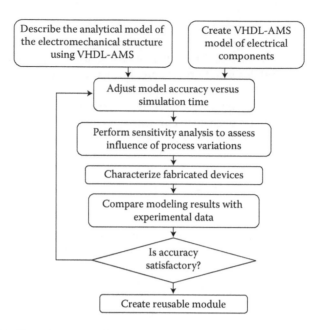

FIGURE 11.17   Flowchart presenting major steps in creation of an analog hardware description language model.

fabricated structures. If the accuracy of performance prediction is sufficient, the model may be transformed into a reusable module and deployed for use. If accuracy is not satisfactory, the model should be adjusted accordingly.

MODEL DEFINITION

1. *Model interface:* The first step in model creation consists of identifying the model parameters. This step is critical, as it sets the degree of reusability of the model and defines the interface used by nonexpert users who want to optimize their structures without understanding how the model works.
2. *Predefined packages:* IEEE provides multiple libraries that can be reused in a custom-created model. Electrical and mechanical systems are covered along with physical constants and mathematical packages that provide common complex functions and constants.
3. *Model parameters:* Parameters of the model are described as constants in the ENTITY section of the model code.
4. *Port interface declaration:* Two choices are possible: conservative flow or signal flow. In cases where an electrical system is to be connected to the model, the conservative flow interface has to be used. Each connector must be associated with a mechanical or electrical discipline. In the presented example, the input to the generator (external acceleration amplitude and frequency) is mechanical and the outputs are purely electrical.

5. *Variable declaration:* Subsequently, through (e.g., current, force) and across (e.g., voltage, displacement) variables used in the model are defined as branch quantities. The declared quantities have to be pin-compatible with the definition of the electrical part of the system in order to allow reusability. The free quantity definition can be used to implement complex variables.

6. *Constitutive equations:* As described in the previous sections, the coupled electromechanical behavior of the structure is described using coupled differential equations. Two domains are modeled: the initial conditions (QUIESCENT_DOMAIN) and the normal operation of the system (time or frequency domain) [43]. The initial conditions imply that the structure is immobile and with no charge on the electrodes.

In order to include losses in the model, and at the same time keep the electromechanical model pin-compatible with the electrical part of the system, free intermediate quantities are used. As presented in the previous sections, the damping is represented as an imaginary part of appropriate values. Therefore, the associated quantity can be defined using a complex number defined in the *math_complex* package as a set of two elements of type *real*. The complex free quantities are used in the equations in the model of the electromechanical system. Then, *branch quantities* are assigned to the real parts of the free quantities and declared as outputs to interface with the rest of the system. It has to be noted though that this method can only be used with sinusoidal signals. The reader is invited to consult works by Boussetta et al. [24] for complete code examples representing a piezoelectric power generator.

EVALUATION

In order to demonstrate the interest of implementing the model using VHDL-AMS, simulation time using various tools was compared. In the first step, the model was implemented with PZT piezoelectric material and viscous damping of 2%. The model was implemented both in Maple and Smash™. As long as the results obtained were identical (the model is the same), the simulation time for VHDL-AMS was equal to 10 ms compared to 1 s for Maple. Furthermore, VHDL-AMS allows an easy extension of the modeled system with electrical components. The same is theoretically possible, but much more complex and thus not practically realizable, with the purely analytical model. Furthermore, time-domain simulation is easy to perform using the VHDL-AMS implementation.

PROCESS VARIATION

A model implemented using VHDL-AMS can be used to assess the sensitivity of the system on parameter variations. If the variations are assumed to be uncorrelated and have Gaussian distribution, the Monte-Carlo analysis can be used. The presented model was implemented using a set of parameters resulting in a structure with resonance frequency equal to 1 kHz, an arbitrary load value of 5 MΩ, and an input acceleration value of 0.2 g. Monte-Carlo analysis was performed assuming parameter

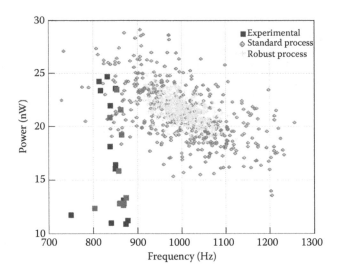

FIGURE 11.18   Results of Monte-Carlo simulation of parameter dispersion of the piezoelectric energy harvester. The output power dissipated on a 5 MΩ at 0.2 g versus the resonance frequency is represented for simulation and experimental testing of manufactured devices.

dispersion of 5% around the nominal values for a standard process, and dispersion of 2% representing a robust process. Figure 11.18 shows simulation results of the dispersion of resonance frequencies and the corresponding generated power levels compared with experimental results. This kind of analysis is especially important in case of resonant structures, where the transducer devices are designed to match their resonance frequency to the dominating frequency in the environment. If process variations are high, the resonance frequency of the resulting device may deviate from the designed value which would significantly reduce its performance.

CONCLUSION

This chapter discussed harvesting of the energy of ambient mechanical vibrations using piezoelectric MEMS devices. In order to take full advantage of the energy present in the environment, the harvesting devices have to be small and inexpensive. MEMS technology can enable this through creation of arrays of microscopic piezoelectric transducers that would be directly integrated with the electronic systems that they power. We presented two models designed to help understand how various factors affect the process of energy harvesting. The first one is a general model of a unidimensional mass–spring system with a piezoelectric element, and the second one is a model of a specific cantilever beam structure, common in MEMS implementations of piezoelectric energy harvesters. Implementation of the cantilever beam model using VHDL-AMS was also discussed to highlight the advantages of global simulation of the entire system, that is, the highly coupled electromechanical structure connected to the purely electrical energy processing circuitry. The modeling results were contrasted with FEM simulation and the experimental results obtained with microfabricated devices. The presented analysis demonstrates that the efficiency of ambient energy harvesting can be greatly improved when the structures and the power processing circuitry are properly optimized.

APPENDIX

List of Symbols and Their Units

Symbol	Meaning	SI Unit
E_i	Electric field component along the i axis	Vm^{-1}
D_i	Electric displacement field component along the i axis	Cm^{-2}
$Y(t)$	Vertical movement of the frame in the one-dimensional model	m
M	Seismic mass value	kg
R	Electrical load resistance	Ω
λ	Viscous damping coefficient	Ns m^{-1}
γ, ζ, $\tan \delta$	Structural, viscous, and dielectric damping ratios, respectively	1
Q	Quality factor	1
K_0, k	Spring constants: ideal and including the structural losses	Nm^{-1}
C_0, C	Capacitance of the piezoelectric element: ideal and including the dielectric losses	F
$u(t)$	Electrical potential difference	V
T_i	Mechanical stress component along the i axis	Nm^{-1}
S_i	Mechanical strain component along the i axis	1
c_{pq}	Elastic stiffness constant	Nm^{-2}
e_{ip}	Piezoelectric constant	Cm^2
ε_{ij}	Permittivity component	Fm^{-1}
ω	Angular frequency	rad s^{-1}
ω_{res}, ω_{ares}	Angular frequency of resonance (zero susceptance) and antiresonance (zero reactance), respectively	rad s^{-1}
a_1	Neutral axis location from the bottom surface of the beam	m
h_{te}, h_p, h_{be}, h_s	Thickness of the top electrode, the piezoelectric layer, the bottom electrode, and the silicon layer, respectively	M
$Y_{te}^{ef}, Y_p^{ef},$ Y_{be}^{ef}, Y_s^{ef}	Effective rigidity of the top electrode, the piezoelectric layer, the bottom electrode, and the silicon layer, respectively	Nm^{-2}
$w(x,t)$	Deformation of the beam at location x and at time t	m
$w(t)$	Displacement of the seismic mass relative to the frame in the one-dimensional model	m
$w_2(t)$	Displacement of the seismic mass center of gravity	m
θ	Angle of rotation of the seismic mass relative to the attachment of the mass to the beam	rad
$a_{in}(t)$, A_{in}	Input acceleration and its amplitude	ms^{-2}
B_B, L_B	Beam width and length, respectively	m
B_M, H_M, L_M	Mass width, height, and length, respectively	m
	Auxiliary variables	
$\xi = \dfrac{\varepsilon_{33}^{ef}\beta}{h_p D_G}$		s$_3$ kg^{-1} m^{-3} A
$\beta = \dfrac{e_{31}^{ef}}{2\varepsilon_{33}^{ef}} h_p \left(2h_s + 2h_{be} + h_p\right)$		m^3 s^{-3} A^{-1} kg
$L_{eq}^2 = \dfrac{L_B^2}{3} + \dfrac{L_M^2}{4} + \dfrac{L_M L_B}{2}$		m^2

REFERENCES

1. B. Calhoun, D. Daly, N. Verma, D. Finchelstein, D. Wentzloff, A. Wang, S.-H. Cho, and A. Chandrakasan, Design considerations for ultra-low energy wireless microsensor nodes, *IEEE Transactions on Computers*, 54(6), 727–740, 2005.
2. M. Hempstead, N. Tripathi, P. Mauro, G.-Y. Wei, and D. Brooks, An ultra low power system architecture for sensor network applications, in *Proceedings of the 32nd International Symposium on Computer Architecture ISCA 2005*, June 4–8, 2005, pp. 208–219.
3. T. Hui Teo, G. K. Lim, D. Sutomo, K. H. Tan, P. K. Gopalakrishnan, and R. Singh, Ultra low-power sensor node for wireless health monitoring system, in *Proceedings of the IEEE International Symposium on Circuits and Systems ISCAS 2007*, May 27–30, 2007, pp. 2363–2366.
4. Toumaz Technology Limited, Sensium Life Platform product brief TZ2050, *Technical Report*, 2011 [Online]. Available at: http://www.toumaz.com/
5. L. Mateu and F. Moll, Review of energy harvesting techniques and applications for microelectronics (keynote address), in *Proceedings of the SPIE Conference Series*, vol. 5837, June 2005, pp. 359–373.
6. S. Roundy, P. K. Wright, and J. Rabaey, A study of low level vibrations as a power source for wireless sensor nodes, *Computer Communications*, 26, 1131–1144, 2003.
7. S. Beeby, M. Tudor, and N. White, Energy harvesting vibration sources for microsystems applications, *Measurement Science and Technology*, 17(12), R175–R195, 2006.
8. T. Sterken, P. Fiorini, K. Baert, R. Puers, and G. Borghs, An electret-based electrostatic u-generator, in *Proceedings of the 12th International Conference on Transducers, Solid-State Sensors, Actuators and Microsystems*, vol. 2, 2003, pp. 1291–1294.
9. P. Glynne-Jones, M. Tudor, S. Beeby, and N. White, An electromagnetic, vibration-powered generator for intelligent sensor systems, *Sensors and Actuators A: Physical*, 110, 344–349, 2004.
10. A. J. du Plessis, M. J. Huigsloot, and F. D. Discenzo, Resonant packaged piezoelectric power harvester for machinery health monitoring, *Proceedings of SPIE*, 5762(1), 224–235, 2005.
11. S. W. Arms, C. P. Townsend, D. L. Churchill, J. H. Galbreath, and S. W. Mundell, Power management for energy harvesting wireless sensors, *Proceedings of SPIE*, 5763(1), 267–275, 2005.
12. N. Ledermann, P. Muralt, J. Baborowski, S. Gentil, K. Mukati, M. Cantoni, A. Seifert, and N. Setter, 1 0 0-textured, piezoelectric Pb(Zr$_x$, Ti$_{1x}$)O$_3$ thin films for MEMS: Integration, deposition and properties, *Sensors and Actuators A: Physical*, 105(2), 162–170, 2003.
13. P. D. Mitcheson, T. C. Green, E. M. Yeatman, and A. S. Holmes, Architectures for vibration-driven micropower generators, *Journal of Microelectromechanical Systems*, 13(3), 429–440, 2004.
14. C. Williams and R. Yates, Analysis of a micro-electric generator for microsystems, in *Proceedings of Transducers'95 and Eurosensors IX*, vol. 1. Stockholm, Sweden, June 25–29, 1995, pp. 369–372.
15. N. duToit, B. Wardle, and S. Kim, Design considerations for MEMS-scale piezoelectric mechanical vibration energy harvesters, *Integrated Ferroelectrics*, 71, 121–160, 2005.
16. E. Lefeuvre, A. Badel, C. Richard, L. Petit, and D. Guyomar, A comparison between several vibration-powered piezoelectric generators for standalone systems, *Sensors and Actuators A: Physical*, 126(2), 405–416, 2006.
17. IEEE, Standard on piezoelectricity, *ANSI/IEEE Std 176-1987*, January 1988.
18. T. Tsurumi, H. Kakemoto, and S. Wada, Dielectric, elastic and piezoelectric losses of PZT ceramics in the resonance state, in *Proceedings of the 13th IEEE International Symposium on Applications of Ferroelectrics ISAF 2002*, 2002, pp. 375–378.

19. J. Peng, H. Luo, T. He, H. Xu, and D. Lin, Elastic, dielectric, and piezoelectric characterization of $0.70Pb(Mg_{1/3}Nb_{2/3})O_3 - 0.30PbTiO_3$ single crystals, *Materials Letters*, 59(6), 640–643, 2005.

20. Piezo Systems Inc, Catalog #8, p.28, 2011, http://www.piezo.com/catalog8.pdf

21. TRS Technologies, Inc., TRS100HD [Online]. Available at: http://www.trstechnologies.com/

22. F. Martin, P. Muralt, M. A. Dubois, and A. Pezous, Thickness dependence of the properties of highly c-axis textured AlN thin films, *Journal of Vacuum and Science & Technology A: Vacuum Surfaces and Films*, 22(2), 361–365, 2004.

23. R. Lanz, P. Carazzetti, and P. Muralt, Surface micromachined BAW resonators based on AlN, in *Proceedings of IEEE Ultrasonics Symposium*, vol. 1, October 8–11, 2002, pp. 981–983.

24. H. Boussetta, M. Marzencki, S. Basrour, and A. Soudani, Efficient physical modeling of MEMS energy harvesting devices with VHDL-AMS, *IEEE Sensors Journal*, 10(9), 1427–1437, 2010.

25. M. Marzencki, Y. Ammar, and S. Basrour, Integrated power harvesting system including a MEMS generator and a power management circuit, *Sensors and Actuators A: Physical*, 145–146, 363–370, 2008.

26. F. Blom, S. Bouwstra, M. Elwenspoek, and J. Fluitman, Dependence of the quality factor of micromachined silicon beam resonators on pressure and geometry, *Journal of Vacuum Science & Technology B: Microelectronics and Nanometer Structures*, 10(1), 19–26, 1992.

27. S. Roundy and P. K. Wright, A piezoelectric vibration based generator for wireless electronics, *Smart Materials and Structures*, 13(5), 1131–1142, 2004.

28. P. Muralt, Recent progress in materials issues for piezoelectric MEMS, *Journal of the American Ceramic Society*, 91(5), 1385–1396, 2008.

29. M. Marzencki, S. Basrour, B. Belgacem, P. Muralt, and M. Colin, Comparison of piezoelectric MEMS mechanical vibration energy scavengers, in *Proceedings of Nanotech 2007*, Santa Clara, California, May 20–24, 2007.

30. J. O. Mur-Miranda, Electrostatic vibration-to-electric energy conversion, MIT PhD Thesis, Massachusetts Institute of Technology, February 2004.

31. P. Muralt, Ferroelectric thin films for micro-sensors and actuators: A review, *Journal of Micromechanics and Microengineering*, 10(2), 136–146, 2000.

32. M. Brissaud, Modelling of non-symmetric piezoelectric bimorphs, *Journal of Micromechanics and Microengineering*, 14(11), 1507–1518, 2004.

33. E. Elka, D. Elata, and H. Abramovich, The electromechanical response of multilayered piezoelectric structures, *Journal of Microelectromechanical Systems*, 13(2), 332–341, 2004.

34. P. Muralt, N. Ledermann, J. Paborowski, A. Barzegar, S. Gentil, B. Belgacem, S. Petitgrand, A. Bosseboeuf, and N. Setter, Piezoelectric micromachined ultrasonic transducers based on PZT thin films, *IEEE Transactions on Ultrasonics, Ferroelectrics, and Frequency Control*, 52(12), 2276–2288, 2005.

35. J. Baborowski, N. Ledermann, and P. Muralt, Piezoelectric micromachined transducers (PMUT's) based on PZT thin films, in *Proceedings of the IEEE Ultrasonics Symposium*, Vol. 2, October 8–11, 2002, pp. 1051–1054.

36. Z. Hao, A. Erbilb, and F. Ayazia, An analytical model for support loss in micromachined beam resonators with in-plane flexural vibrations, *Sensors and Actuators A: Physical*, 109(1–2), 156–164, 2003.

37. V. T. Srikar and S. D. Senturia, Thermoelastic damping in fine-grained polysilicon flexural beam resonators, *Journal of Microelectromechanical Systems*, 11(5), 499–504, 2002.

38. Gene Poole, Ansys equation solvers: Usage and guidelines, Ansys 2002 Conference, http://ansys.net/papers/solver_2002.pdf

39. M. Marzencki, M. Defosseux, and S. Basrour, MEMS vibration energy harvesting devices with passive resonance frequency adaptation capability, *Journal of Microelectromechanical Systems*, 18(6), 1444–1453, 2009.

40. S. D. Senturia, CAD challenges for microsensors, microactuators, and microsystems, *Proceedings of the IEEE*, 86(8), 1611–1626, 1998.

41. F. Pecheux, C. Lallement, and A. Vachoux, VHDL-AMS and Verilog-AMS as alternative hardware description languages for efficient modeling of multidiscipline systems, *IEEE Transactions on* Computer-Aided Design of Integrated Circuits and Systems, 24(2), 204–225, 2005.

42. R. Guelaz, D. Kourtiche, and M. Nadi, Ultrasonic piezoceramic transducer modeling with VHDL-AMS: Application to ultrasound nonlinear parameter simulations, *IEEE Sensors Journal*, 6(6), 1652–1661, 2006.

43. S. G. Sabiro, Mixed-mode system design: VHDL-AMS, *Microelectronic Engineering*, 54(1–2), 171–180, 2000.

12 Self-Powered Wireless Sensing in Ground Transport Applications

Anurag Kasyap and Alexander Edrington

CONTENTS

BACKGROUND

It is commonly accepted that with increasing technology and society's ever-growing consumption of energy, nonrenewable sources of energy could soon be exhausted. In an effort to investigate alternate sources of energy to meet societal demands, research in the past few decades has focused on using alternate forms or renewable resources, such as optical, solar, tidal, and so on. Many companies in the United States, Europe, and Japan are steadily involved in this area with a strong business case for energy scavenging from ambient sources in the future. Many energy-harvesting concepts are already available such as a self-reliant house and a camel fridge, powered by solar energy for operation. Previous studies have successfully shown that energy can be reclaimed from renewable sources such as solar and tidal energy (Saraiva 1989). Solar cells are an existing technology extensively used in self-powered watches, calculators, and rooftop modules for houses. Even though renewable sources can serve as a substitute for the usual power supply resources, energy is still wasted in the form of heat, sound, light, and vibrations that can be further reclaimed, at least partially, for future use. For example, thermal energy was generated from a 0.75×0.9 cm^2 bismuth–telluride thermoelectric junction to produce 23.5 μW for a temperature difference of 20 K (Stark and Stordeur 1999). Qu et al. (2001) designed and fabricated a thermoelectric generator, $16 \times 20 \times 0.05$ mm^3, consisting of multiple micro antimony–bismuth (Sb–Bi) thermocouples embedded in a 50 μm epoxy film capable of producing 0.25 V from a temperature difference of 30 K. Kiely et al. (1991) designed a low-cost miniature thermoelectric generator consisting of a silicon-on-sapphire and silicon-on-quartz substrate. Another thermoelectric power generator based on silicon technology produced 1.5 μW with a temperature difference of 10°C (Glosch et al. 1999). However, in applications where light and thermal energy are not readily available, alternate sources such as mechanical energy offer a good alternative.

Initially, efforts in energy harvesting from vibrations largely focused on human ambulation (Starner 1996) such as breath, heel strike, limb movement, and so on that are useful sources for powering artificial organs (Antaki et al. 1995) and human wearable electronic devices. Subsequently, Kymissis et al. (1998) and Shenck and Paradiso (2001) designed two novel shoe-mounted piezoelectric devices to harness power during heel strike. This could replace or extend the lifetime of conventional portable batteries currently restricted by energy limitations, shelf life, and potential hazards due to chemicals especially for prolonged usage.

SELF-POWERED SENSORS

The development of self-powered sensors has enabled implantation of these devices into various host structures, such as medical implants and embedded sensors in buildings and bridges (Mehregany and Bang 1995). Many of these applications require that the devices be completely isolated from the outside world. These remote devices, along with their accompanying circuitry, have their own power supply that is usually powered by batteries. The strides achieved in battery technology have not sufficiently matched the improvements in integrated circuit technology. Therefore, developing a microscale self-contained power supply offers great potential for applications in remote systems.

The need for self-contained power generators for "self-powered systems" is an important application for energy harvesting, and is rapidly gaining widespread importance (Shenck and Paradiso 2001). Self-powered systems possess an inherent mechanism to extract power from the ambient environment for their operation. The primary features of self-powered systems include power generation, energy extraction, and storage.

Meso-scale energy-harvesting approaches include rotary generators (Lakic 1989), a moving coil electromagnetic generator (Amirtharajah and Chandrakasan 1998), and a dielectric elastomer with compliant electrodes (Pelrine et al. 2001). Single piezoelectric cantilevers (Ottman et al. 2002) and stacks (Goldfarb and Jones 1999) have been investigated for energy-harvesting capabilities but were not operated in a stand-alone, self-powered mode. Another source for power harvesting is mechanical energy from fluid flow. Taylor et al. (2001) designed an energy-harvesting eel that was approximately 1 m long using a piezoelectric polymer to convert fluid flow and vortex-induced strain to generate power. In addition, Allen and Smits (2001) investigated the feasibility of utilizing a piezoelectric membrane in the wake of a bluff body to induce oscillations in the structure generating a capacitance build-up that acts as a voltage source to power a battery in a remote location. Power generation from ocean waves has also been investigated involving very large-scale piezoelectric generators (Smalser 1997).

VIBRATION TO ELECTRICAL ENERGY CONVERSION

Let us now look at a simple schematic of a power generator based on conversion from vibration to electrical energy, as shown in Figure 12.1. The device consists of a spring–mass–damper system acting as a single-degree-of-freedom system with an input vibration that results in an effective displacement $z(t)$.

The following equation is used to represent the behavior of the above system that basically converts the kinetic energy of a vibrating structure to electrical energy by virtue of the relative motion between the base and the inertial mass.

$$M\ddot{z} + R\dot{z} + Kz = -M\ddot{y} \tag{12.1}$$

where z is the relative deflection, y is the input displacement, M is the inertial mass, K is the spring constant, and R is the effective damping in the system that accounts for mechanical and electrical losses. The above model does not include nonlinear effects and is thus valid only under the constraints of the linear system theory. It also does not specify the electromechanical transduction mechanism with which the kinetic energy is converted into electrical power.

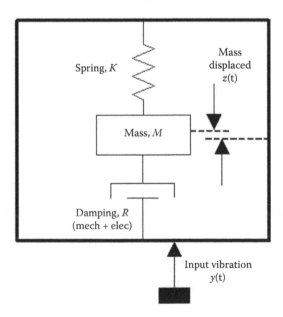

FIGURE 12.1 Schematic of a typical vibration to electrical energy converter.

Vibrational energy reclamation can be achieved conceptually using different transduction mechanisms such as electrostrictive coupling (Uchino et al. 1980), electromagnetic (Hanagan and Murray 1997) and electromechanical coupling (Lee 1990). Among the common linear energy-conserving transducers, the widely used electromechanical transduction mechanisms for energy harvesting involve electromagnetic (specifically electrodynamic), electrostatic, and piezoelectric phenomenon. Electromechanical transducers are classified based on force generation due to the interaction between electric field and charge or magnetic field and current (Hunt 1982) and are commonly represented using equivalent circuits with lumped elements.

Roundy et al. (2003) discuss the theoretical maximum and the practical maximum for the energy densities of various transduction mechanisms, namely piezoelectric, electrostatic, and electromagnetic. The expressions were obtained from basic governing equations of each of the materials using maximum yield stress for the piezoelectric, the electric field for capacitive, and the maximum magnetic field for electromagnetic materials as the respective upper limits. The results indicate that piezoelectric materials possess a practical maximum energy density of 1.77 mJ/cm^3, which is almost four times that of the other transducers. A brief summary on these three energy-harvesting applications is provided next.

In 1996, Williams and Yates designed an electromagnetic generator (5 mm × 5 mm × 1 mm) that had a predicted power output of 1 µW at 70 Hz and 0.1 mW at 330 Hz for an input vibration amplitude of 50 µm. In 1997, Shearwood and Yates designed an electromagnetic generator based on a polyimide membrane 2 µmm in diameter that could generate 3 µW of root mean square power at a resonant frequency of 4.4 kHz. Another electromagnetic generator was developed by El-hami et al. (2001) which consisted of a magnetic core mounted on the tip of a steel beam. For an overall device volume of 0.24 cm^3, an output power of 0.53 mW from a 25 µm input displacement at 322 Hz was produced. In 2000, Li et al. presented a micromachined generator that had a permanent magnet mounted on a spring structure and generated 10 µW at 2 V DC for an input vibration amplitude of 100 µm at 64 Hz from a volume of 1 cm^3.

Electrostatic transduction is the conversion of energy that is produced by varying the mechanical stress to generate a potential difference between two electrodes. An example of this transduction is a simple parallel plate capacitor. In electrostatic transduction, a relative deflection induces charge between the electrodes that can be converted to power. For example, at the microscale, a microelectromechanical system variable capacitor has been designed and fabricated to harvest vibrational energy with a chip area of 1.5 × 1.5 cm^2 and a reported net power output of approximately 8 µW (Meninger et al. 2001).

PIEZOELECTRIC ENERGY HARVESTING

Applying an external electric field across the piezoelectric material induces a mechanical strain in the material, called the "piezoelectric effect." Conversely, when the piezoelectric material is mechanically deformed, the resulting strain produces a voltage. Materials with good piezoelectric properties possess high coupling between the mechanical and electrical domains. This effect can be generated using piezopolymers, such as polyvinyledene fluoride, or piezoceramics, such as lead zirconium titanate (PZT), zinc oxide, aluminum nitride, and barium titanate. For a typical piezoelectric patch, the electric field is often applied vertically across its electrodes, resulting in axial bending stress according to the linear beam theory. Rewriting the constitutive equations that govern the energy-harvesting beam to express the tip deflection in the beam, w, and the charge, q, accumulated in the piezoelectric layer as functions of applied voltage, V, and/or force, F, results in one-dimensional governing equations

$$w = C_{ms} \cdot F + d_m \cdot V \tag{12.2}$$

and

$$q = d_m \cdot F + C_{ef} \cdot V, \tag{12.3}$$

where $C_{ms} = w/F|_{V=0}$ is the short-circuit compliance, $C_{ef} = q/V|_{F=0}$ is the free electrical capacitance, and $d_m = w/V|_{F=0}$ is an effective piezoelectric constant.

Umeda et al. (1996) designed a piezoelectric generator based on impact energy from a freely falling steel ball on a membrane, generating 5 V at 35% efficiency. Ramsay and Clark (2001) proposed a device for bio-microelectromechanical system applications that consisted of a square PZT-5A plate connected to the blood pressure on one side and a chamber with constant pressure on the other. Preliminary results reported an output power of 2.3 μW from a 1 cm × 1 cm × 9 μm plate. It was also reported in their work that the device has a mechanical advantage in converting applied pressure to working stress for piezoelectric conversion, when it functions in the 31-mode than in the 33-mode. White et al. (2001) designed a thick-film piezoelectric composite beam structure that generated 3 μW of power at 90 Hz from ambient vibrations. Their further work made use of a tapered beam design to ensure constant stress distribution to produce 2 μW at 80 Hz for a maximum amplitude of 0.9 mm across an optimal resistive load of 333 kΩ. In 2004, James et al. investigated two applications for two self-powered sensors, namely a liquid crystal display and an infrared link to transmit the data output. The required energy for the prototypes was derived from a 0.17–0.23 g vibrating source at 102 Hz. Hausler and Stein (1984) proposed a device that basically consisted of a roll of polyvinyledene fluoride material that can be attached between body ribs to harness power from regular breathing-induced strain in the material. It was tested on a dog by surgically implanting the device, generating microwatts of power from the breathing.

In 2004, Roundy and Wright designed a piezoelectric vibration generator consisting of a cantilever bimorph bender with a proof mass at its end. Their design was aimed at generating enough energy from 1 cm³ to power a 1.9 GHz radio transmitter from the same vibration source. Their design was predicted to produce 375 μW from a vibration source of 2.5 m/s² at 120 Hz. Sood et al. (2004) developed a piezoelectric micropower generator that is based on a piezoelectric layer deposited and patterned on a membrane consisting of SiO_2 and SiN_x, followed by a ZrO_2 diffusion barrier.

TABLE 12.1

Vibration-Based Energy Harvesters Characterized for Power

Reference	Ambient Source	Size or Mass	Power
Sood et al. (2004)	10 g at 13.9 kHz	170 μm × 260 μm	1.01 μW
Shearwood and Yates (1997)	500 nm at 4.4 kHz	2.5 mm × 2.5 mm × 700 μm	0.3 μW
Meninger et al. (2001)	500 nm at 2.5 kHz	500 mg	8 μW
Li et al. (2000)	100 μm at 64 Hz	1 cm³	10 μW
Roundy et al. (2003)	0.25 g at 120 Hz	28 mm × 3.6 mm × 8.1 mm	375 μW
White et al. (2001)	0.9 mm at 80 Hz		2.2 μW
Marzencki et al. (2005)	0.5 g at 204 Hz	2 mm × 2 mm × 0.5 mm	38 nW
El-hami et al. (2001)	25 μm at 322 Hz	0.24 cm³	0.53 mW
Ching et al. (2002)	200 μm at 60–110 Hz	1 cm³	200–830 μW
Stark and Stordeur (1999)	$\Delta T = 20$ K	67 mm²	20 μW

The two electrodes for the PZT layer are formed using an interdigitated top electrode with Pt/Ti that makes use of the d_{33} mode to extract power. The maximum measured power was 1.01 μW for an optimal load of 5 MΩ and the corresponding voltage was 2.4 V_{DC} (Jeon et al. 2005). Table 12.1 compiles the reported energy harvesters discussed here that generated power from vibration sources using different transduction mechanisms. The columns list the authors and year of each study, the vibration source (mostly resonant in nature), the size of the device, and the overall power harvested.

Although considerable work and research is being carried out in the area of vibration-based piezoelectric energy harvesting, no commercially available products exist that can directly be applied to practical applications. This chapter explains briefly a fundamental understanding of energy-harvesting solutions and addresses some of the basic concerns that when overcome can enable widespread customer adoption by presenting a real-life wireless sensor application.

INTRODUCTION TO ENERGY HARVESTERS

The typical mindset regarding energy harvesting is that its specific function is to provide power, generally in the range of kilowatts to megawatts that can be put back in the electrical grid, for example, to power homes. This can be achieved only using large-scale renewable energy-harvesting technologies such as solar panels, wind turbines, hydrodynamic plants, and so on. Alternatively, when we explore "micro energy harvesting," the main function is not exactly to provide power, but to provide information. These technologies (Beeby et al. 2006) are designed to power a new microelectronics product class, called the autonomous wireless sensor nodes that include low-power microcontrollers, radios, and advanced sensor technologies. The end application dictates the storage option for the solution, which is determined primarily by the duty cycle of the sensor transmissions, peak transmission current, transit time of the vehicle, and so on. It is essential for the product to include both the harvester designed for the environment, and appropriate electronics to meet conditions where sensor transmissions are desired even during inactive periods of the vehicle. Understanding the available energy source and the application requirement is critical in designing the appropriate energy-harvesting solution. In this study, an energy-harvesting solution was designed that is suitable for ground transportation applications capable of generating energy from random vibration.

From a customer's perspective, the goal of an energy harvester is to appear as a battery and its primary value is in its never-dead condition. Even though an energy harvester appears as a battery to the end user, there are some fundamental differences between the two. Batteries are typically characterized by energy density as they are a storage media. However, they have a limited lifetime due to their finite storage capacity. Alternatively, energy harvesters are characterized by power density as they are a conversion medium although they have internal minimal storage required to perform functions. Energy harvesters are truly functional only when they are powering devices and they possess infinite lifetime as long as the ambient energy source is available.

Figure 12.2 shows a comparison of power density (power per unit volume) of a typical AA battery pack consisting of two cells and a candidate energy harvester. As evident in the plot, depending on the deployment time, the advantage over using a battery slowly shifts to the energy harvester. For example, a 2 mW generating energy harvester has a significant advantage over a battery pack if the required product lifetime is one year or longer. In a more realistic situation, a 1 mW harvester, which is achievable through good design tools, can outlast a battery pack in applications beyond the 2–5-year range. These applications are the best-fit scenario for ambient energy-harvesting technologies. If an application requires the device to be powered for a week or two, batteries are still the better option as they can achieve higher power densities for a short period of time. Examples include cell phones and other handheld devices that can be recharged very often and the required lifetime is only a few hours to days. Another factor of comparison is the cost of these devices. Although energy-harvesting technologies offer significant value addition due to

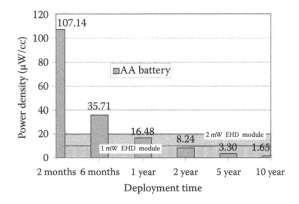

FIGURE 12.2   Understanding the difference between batteries and energy harvesters.

labor, battery maintenance, and replacement costs, their price should be comparable to batteries at higher volumes. Energy harvesters also lag in terms of their size and mass, but with better materials and newer conversion technologies coming through, they can become comparable to batteries in the long run.

ANATOMY OF AN ENERGY-HARVESTING POWER SUPPLY

A typical energy-harvesting power solution primarily consists of five constituent parts. The ambient energy is converted first into its electrical equivalent in the "power conversion" module. Second, the "power processor" module collects the raw electrical signal and converts it into usable DC power. Next, the DC power is appropriately stored in a power storage device. This is followed by the power management part that effectively maintains the thresholds and cutoffs in the electronics to maintain storage without leakage. Finally, the power delivery part is responsible for delivering the right amount of power and current to the actual application as necessary. This part will account for startup currents, required voltage levels, and so on. All these parts are shown in Figure 12.3 as a collective device that acts as a power supply to the wireless platform. Providing only individual parts or just a collection of a few parts is not a desirable solution for the end-user application, particularly if parts such as management and delivery components need to be externally added.

Let us now look at some of the common energy sources that can be harvested to power various wireless systems. The key applications for solar energy in wireless systems are mainly in building automation using indoor lighting and in industrial sensing, for example, in refineries where abundant outdoor sunshine is available. The drawback in using solar energy is that the source is diurnal in nature and is completely unavailable during nighttime or in dark areas. Indoor lighting is not a considerable energy source for harvesting and is only feasible in applications that do not require frequent sensor information. Even thermal energy harvesting has applications in building automation, equipment with heat sources, and temperature differentials. In addition, there are some applications in process management for supply

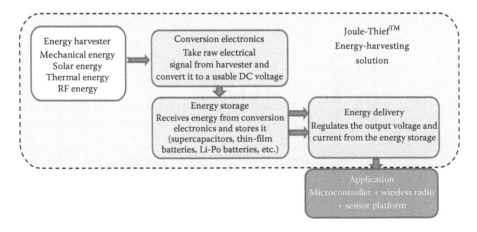

FIGURE 12.3 Simplified block diagram of an energy-harvesting power supply.

chain and transportation markets. Vibration sources occur typically in transportation and machinery condition monitoring. Some mechanical energy-harvesting applications are available in building automation where direct mechanical actuations, for example, switches, can be harvested. Typically, in most applications for any given environment, one energy source is dominant. Sometimes, it is desirable to harvest from a combination of sources to leverage all available energy sources within the application. For energy harvesting to fit in any application, the three keywords are:

1. *Wireless*: It is generally difficult to replace wired power applications where line power supply is available. If the products are wireless, they are typically powered by batteries with finite lifetime. Implementing energy harvesting in these products, independent or in conjunction with batteries, will extend the lifetime of the devices.
2. *Remote*: Energy harvesting is particularly useful in applications where power is required in hard-to-reach areas. Replacing batteries in such applications is labor intensive and generally expensive. Energy harvesting offers potentially infinite lifetime, where these devices can be deployed.
3. *Sensors*: Sensors are a perfect match for energy harvesting as they typically operate in the power ranges achievable by energy harvesters. Furthermore, newer sensor technologies are being developed to be run on battery power and, in many cases, the sensor information is not required continuously. Energy harvesting is very useful in such asynchronous mode operation where power requirements are lower.

EVALUATING CUSTOMER APPLICATION AND FIT FOR ENERGY HARVESTING

Next, the process of selecting the right application and designing an energy-harvesting solution is discussed, using a three-step process by evaluating a customer's needs. This process is a general guideline to establish the sweet spot for energy-harvesting products and can be modified as required to suit specific user needs. A schematic of the procedure is shown in Figure 12.4.

First, details about the end application are gathered which includes the power requirement, energy consumption during transmission, and sleep functions. Other data such as the ON time for the device, the quiescent power, and the frequency of data transmissions are also useful. Once obtained, this information helps determine whether the application is suitable for energy harvesting.

Next, the available energy sources within the environment (Roundy et al. 2003) of the application are explored. Sources such as vibration, solar/light, or thermal need to be evaluated and quantified to estimate the amount of total power that can be harvested. In some cases, the combinations of sources can also be evaluated for energy harvesting. If the ambient source is vibration, factors such as frequency and amplitude content need to be studied. For example, fixed frequency vibration is commonly available for machinery condition monitoring applications. However, most

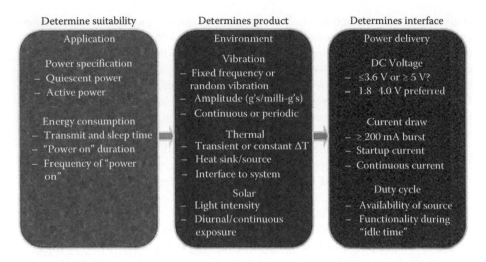

FIGURE 12.4 Evaluation procedure for an application to determine fit for energy harvesting.

vibration is random in applications such as transportation, buildings, and so on. If the source is thermal, information about the temperature difference is required and whether it is a steady state or a transient source. The thermal harvester efficiency is dependent on the heat transfer design through the system, and, hence, it is critical to understand the interface between the source and sink in this phase to effectively maintain sufficiently large temperature differential across the generator. For solar/light sources, information such as outdoor or indoor lighting needs to be gathered. Factors such as location and orientation are absolutely critical in the harvesting potential of the device. This process step enables the right choice for the harvester product.

Finally, the power delivery information for the application needs to be obtained. In this step, information on the required voltage levels for the sensors, current requirements such as startup, quiescent, and so on need to be determined. Another important factor to be considered is the duty cycle, which indicates the power requirements while the energy source is absent. This is particularly true in vehicle applications where they have considerable idle time during which no vibration is available. Another example is solar energy which is available only during the day. This information helps estimate the amount of auxiliary storage required to ensure that the sensors stay powered. This step will determine the type of interface required for the energy-harvesting solution.

In this study, the transportation market for vibration energy harvesting is investigated using the above-discussed process. Applications include container/asset tracking using active radiofrequency identification (RFID), predictive health monitoring, wireless sensors, and so on. The main need in these areas is to obtain real-time automated information about value assets. In addition, critical sensor data about process control and materials during transit are essential to manage the transported goods efficiently. The advantages of energy-harvesting solutions here are that they reduce the overall operating costs and cost of ownership by replacing batteries or elongating the lifetime. They also improve security, protection, and convenience of assets during transit. Furthermore, energy-harvesting devices are a permanent solution for hermetically sealed applications and are highly reliable in extreme environments.

INSIDE THE ENERGY-HARVESTING MODULE

As described before, an energy-harvesting power supply forms the interface between the available energy source and the end application. An energy-harvesting power supply converts ambient mechanical energy sources and provides a two-wire interface to the customer that can be connected to power their application. Appropriate voltage and power levels are designed into the product to suit the application. Inside the power supply, the Ruggedized Laminated Piezo (RLP®) smart energy beam shown in Figure 12.5 forms the actual harvester element (Kasyap et al. 2002). The RLP is prestressed through lamination to place the piezoelectric material in a desirable stress state to allow for greater deflections and strain rates. The shape and loading mechanisms are stress engineered to obtain an optimal strain distribution in the ceramic. It is well known that a cantilever beam is the best configuration for low-frequency vibration and they typically have a linear strain relationship with length in a bending application. However, having a tapered beam shape provides uniform strain in the beam as shown in this picture. The advantage with designing for uniform stress is that it ensures that the whole ceramic is utilized for energy harvesting. The moment-loaded beam configuration is an optimized design for space and therefore increases the power density by three times compared to the tapered beam design.

The next part is the conversion electronics that combine the conversion, management, and delivery functions to power the application. Depending on the source, whether it is solar, thermal, or vibration, the circuit processes the raw electrical signal and converts it into usable power. Typically, for solar and thermal sources, boost converter topologies are widely used and are fairly efficient. However, the circuit design for vibration harvesting is more complicated as maximum power transfer occurs when the impedances match and it is difficult to dynamically track the output voltages for random vibration to harvest at optimal levels. The voltage generated at the RLP due to vibration can be processed using either passive or active electronics collection techniques. The output of the circuit is a standard 1.8–3.6 V industrial

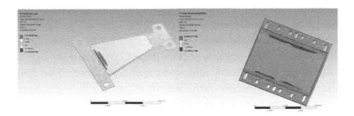

FIGURE 12.5 Uniform stress profiles on RLP smart energy due to shape and load optimization.

voltage with protection from overcharging and excessively discharging the storage batteries/capacitors. Some wireless applications require high current pulses during data transmission, for example, a wireless WiFi sensor node requires a peak current of 200 mA, which is accommodated for in the circuit.

Finally, the energy storage element in the power supply module presents an array of options to select from depending on the applications, ranging from traditional capacitors to super caps and thin-film batteries to high-capacity Saft/Tadiran batteries. The typical operating voltages for these storage devices vary from 1.5 to 4.2 V. Consequently, the conversion circuit acts as a common platform across dissimilar storage options, while maintaining the required voltage and current for the application. Batteries/super caps are still the preferred solution in applications that consume significant energy during startup and network association. Leakage is an unavoidable issue with super capacitors although they have an infinite charge/discharge cycle time. Thin-film batteries are a good option for low-power applications where negligible leakage is desired or in applications where the ambient energy source is absent for an extended period of time.

COMMONLY ADOPTED WIRELESS PLATFORMS

After understanding the energy-harvesting power supply and its constituents, some of the commonly implemented wireless platforms are investigated to establish suitability for transportation logistics applications. Most of the commercially available microcontroller/radio combinations are typically operated using IEEE802.15.4 and 802.11 protocols. They are commonly known as Zigbee and wireless WiFi. Other wireless platforms exist that can be used for these applications, but are not explored in this study. The ultra-high-frequency protocol, also known as the DASH7 (ISO18000-7) standard, is now used for military transportation applications such as active RFID and sensing. Alternatively, high-power wireless applications include GPS, GSM cellular, and so on that require an average power of >5 mW and are used to track position and personal identification.

Some candidate wireless systems selected for this study included products from Texas Instruments, Crossbow, Jennic, and DigiMesh. To evaluate the power requirements for each of these platforms, estimates were first obtained for a typical sensor application that does 20 ms measures and transmits data every 10 s. The sleep current for these devices along with peak transmit current and, therefore, the average harvesting current required are listed in Table 12.2.

It should be noted here that the startup cycle for these modules is not accounted for. It is assumed that the startup energy occurs only once at the beginning and that sufficient reserve energy is available in storage to perform this function. The startup energies are different for these devices as they depend on firmware algorithms being implemented that form the network and maintain association. The Jennic JN5148 module exhibits the lowest power requirement of all platforms tested here. The average power required to sustain this wireless application depends on the operating voltage. They typically vary from 3 to 3.6 V. For example, with DigiMesh, the sleep current is high and, therefore, the harvesting power required is approximately 0.5 mW, which is still attainable through energy harvesting.

TABLE 12.2

Power Requirements for Commercially Available Zigbee Wireless Nodes

System	Vendor	Sleep Current (µA)	Maximum Duty Current (mA)	Harvesting Equivalent (µA)
MSP430 CC2500	TI	8	22	74
Iris Mote (XM2110CA)	Crossbow	8	25	83
DigiMesh 2.4	Digi	10	55	171
JN5148	Jennic	1.25	17.5	52

The advantage of the 802.15.4 platform is that it is a sustainable network where large groups of nodes can exist and communicate with each other to maintain energy efficiency and logistics of goods. Shown in Figure 12.6 is a typical transmission pulse of JN5148 module that has a wake-up, boot, data collection, and transmit functions. This particular sensor node was powered with 3.3 V and had a very low sleep current of 2.5 µA which is negligible. On an average, it required 1.5 mJ per transmission cycle and if the interval is every 5 s, then the power required is 0.3 mW. If the time interval is doubled to 10 s, the power requirement approximately becomes half accounting for the additional sleep current. Hence, in most applications, it is essential to know how often the sensor data are required. It should be noted that any external

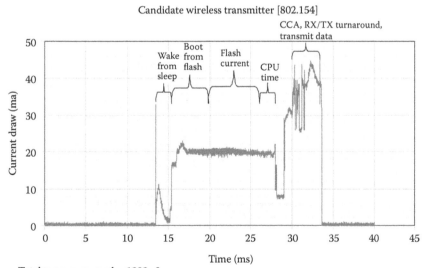

Total tag energy-awake: 1292 µJ
Tag awake once per five seconds
Total tag energy-asleep: 3 VDC *4.5 µA*5 s = 70 µJ
Average power: 1362 µJ every 5 s = 272 µW

FIGURE 12.6   A typical transmission current pulse for the JN5148 module.

sensors connected to the wireless platform may require additional power due to specific operating conditions and have not been accounted for in this estimate.

Another example of a suitable application is a wireless WiFi platform. Here, the primary advantage is that WiFi is available everywhere and the sensor nodes themselves have a unique IP address and can be accessed from anywhere. As expected from the type of application, the power required for WiFi platforms is significantly higher. A typical current pulse during transmission involves wake-up and multiple transmit and receive cycles to ensure that no IP conflicts arise and that configuration settings are identified and verified. Furthermore, the nodes are required to periodically maintain link with the access points and send configuration data to avoid being kicked out from the network. For a 3.3 V supply, the average energy per cycle was measured to be approximately 15 mJ, and to transmit data every 30 s, the required power is 0.5 mW. The actual power is generally higher to accommodate link-up transmissions that occur every 45–60 s. Therefore, it is established through these tests that power requirement for these wireless sensor nodes is approximately 0.3 mW to achieve frequent data transmissions.

GROUND TRANSPORT VIBRATION

It is widely known that the vibration on a transportation vehicle is generally random and, therefore, it is difficult to tune resonant energy-harvesting devices to a certain specific frequency. Furthermore, vibration data depend on the type of vehicle, vehicle speed, road conditions, driving conditions, mounting direction/axis, measurement location, and so on. From the measured vibration data collected over a series of vehicle tests, it was established that the interior vibration in a vehicle is almost insignificant and is difficult to harvest in meaningful amounts without any common frequency content. However, an appreciable amount of vibration is present in the exterior of a vehicle which can be harvested with some common frequency content even though the vibration itself is random. Based on the measurements,

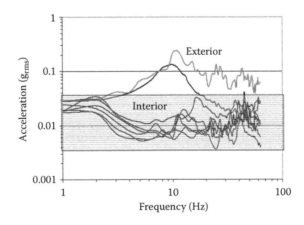

FIGURE 12.7 Frequency spectra of vibration data collected on automobiles and light trucks.

TABLE 12.3
Frequency Ranges of Interest for Vibration
on Automobiles and Their Sources

Suspension-related vibration	2–5 Hz
Tire-related vibration	10–15 Hz
Structure-related vibration	30–40 Hz

it was concluded that some vibration in the 10–15 Hz range can be harvested, which mainly occurs from tire-related vibration, as shown in Figure 12.7. An energy-harvesting module based on the moment-loaded beam discussed earlier was designed to harvest this vibration and other impact events observed in vehicles during transit.

The general vibration profiles experienced in an automobile and their associated sources are summarized as shown in Table 12.3.

CANDIDATE APPLICATIONS AND FIELD TRIALS

A candidate application was selected in a tractor trailer environment for an end user. In the field trials conducted, two energy harvesters were mounted along with a data logger to collect vibration data on the trailer, as shown in Figure 12.8. The energy content generated in the energy harvester module was monitored during the whole trip. The collected data are shown in Figure 12.9 for a period of 1 h which included the time when the truck was moving and the time when it was idle. The energy level in the super capacitor steadily rose from 3.6 to 4.3 J before the truck stopped moving following which there is some leakage in the super capacitor. The average power during transit was calculated to be 390 μW which is sufficient for

FIGURE 12.8 A vibration energy harvester mounted on a tractor trailer.

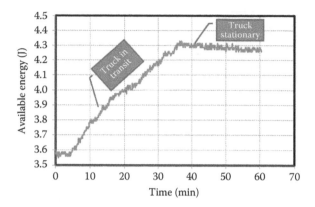

FIGURE 12.9 Energy charging characteristics for a tractor trailer field trial.

active RFID and ground transport applications, specifically the Zigbee sensor node discussed earlier.

Another candidate application was chosen with a need to deploy multiple wireless sensor nodes on high-speed trains to monitor its structural integrity and other parameters such as interior temperature, and so on. Providing wired sensors in this application is difficult as they need to monitor critical parameters in hard-to-reach areas of the bogie. The ground transport energy harvester module discussed earlier was mounted on the train to harvest power necessary to power these applications, as shown in Figure 12.10.

The vibration data were collected using a data logger and reproduced on an LDS vibration shaker to simulate the train vibration. The ground transport module was mounted on the shaker to harvest power from the random vibration. The power thus

FIGURE 12.10 Energy harvester mounted on a candidate train bogey.

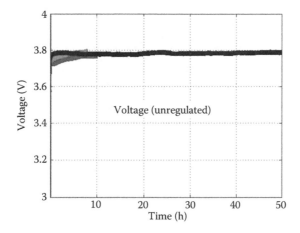

FIGURE 12.11   Voltage balance at the output capacitor to maintain 3.8 V during a train field trial.

produced was dissipated across a 20 KΩ load through an auxiliary electrolytic capacitor while maintaining a steady voltage of 3.8 V, as indicated in Figure 12.11. The vibration in the train was an aggressive profile and after 50 h through the test, the average power generated remained constant at 720 μW, as shown in Figure 12.12. The approximate average power required for the 802.15.4 application discussed earlier is 300 μW for sensor data transmissions every 10 s. As evident from this test, the generated power is sufficient to sustain the wireless network even when external sensors are attached to the node. The aggressive train vibration profile was used to demonstrate the durability of the energy-harvesting device. The excess power generated during transit can be stored either in rechargeable batteries or super capacitors for sensor operation during idle time.

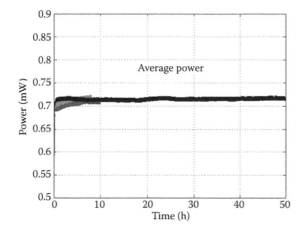

FIGURE 12.12   Power balance by dissipating the harvested energy across a suitable load during test.

CONCLUSION

In conclusion, energy harvesting has already gained acceptance as an alternate power solution for sustainable wireless sensor systems. In particular, energy-harvesting technology offers an extremely viable plug-and-forget attribute for sensors deployed in hard-to-reach areas of remote applications. This attribute further enables mass deployment of sensors to efficiently manage the logistics of goods and critical assets in transportation. An energy-harvesting power supply with a two-wire interface like a battery is desired for many end-user applications. It is critical to evaluate the energy sources within an application environment to understand the fit for energy-harvesting devices. Energy-harvesting solutions are becoming commercially available now and transportation logistics applications present a great opportunity to pursue. A novel stand-alone energy-harvesting power solution was developed and demonstrated functionality with a wireless sensor node operating in two ground transport applications, namely, tractor trailer and train environment.

REFERENCES

Allen, J. J. and Smits, J., Energy harvesting eel, *Journal of Fluids and Structures*, 15, 629–640, 2001.

Amirtharajah, R. and Chandrakasan, A., Self-powered signal processing using vibration based power generation, *IEEE Journal of Solid-State Circuits*, 33, 687–695, 1998.

Antaki, J. F., Bertocci, G. E., Green, E. C., Nadeem, A., Rintoul, T., Kormos, R. L., and Griffith, B. P., Gait-powered autologous battery charging system for artificial organs, *ASAIO Journal*, 41(3), M588–M595, 1995.

Beeby, S. P., Tudor, M. J., and White, N. M., Energy harvesting vibration sources for microsystems applications, *Measurement Science and Technology*, 17, R175–R195, 2006.

Ching, N. N. H., Wong, H. Y., Li, W. J., Leong, P. H. W., and Wen, Z., A laser-micromachined multi-modal resonating power transducer for wireless sensing systems, *Sensors and Actuators A*, 97–98, 685–690, 2002.

El-hami, M., Glynne-Jones, P., White, N. M., Hill, M., Beeby, S., James, E., Brown, A. D., and Ross, J. N., Design and fabrication of a new vibration-based electromechanical power generator, *Sensors and Actuators A*, 92(1–3), 335–342, 2001.

Glosch, H., Ashauer, M., Pfeiffer, U., and Lang, W., A thermoelectric converter for energy supply, *Sensors and Actuators*, 74, 246–250, 1999.

Goldfarb, M. and Jones L. D., On the efficiency of electric power generation with piezoelectric ceramics, *Transactions of ASME, Journal of Dynamic Systems Measurement Control*, 121, 566–571, 1999.

Hanagan, L. M. and Murray, T. M., Active control approach for reducing floor vibrations, *Journal of Structural Engineering*, 123(11), 1497–1505, 1997.

Hausler, E. and Stein, E., Implantable physiological power supply with PVDF film, *Ferroelectronics*, 60, 277–282, 1984.

Hunt, F. V., *Electroacoustics: The Analysis of Transduction, and its Historical Background*, 2nd edn. Cambridge: Harvard University Press, 1982.

James, E. P., Tudor, M. J., Beeby, S. P., Harris, N. R., Glynne-Jones, P., Ross, J. N., and White, N. M., An investigation of self-powered systems for condition monitoring applications, *Sensors and Actuators A*, 110, 171–176, 2004.

Jeon, Y. B., Sood, R., Jeong, J.-H., and Kim, S.-G., MEMS power generator with transverse mode thin film PZT, *Sensors and Actuators A*, 122(1), 16–22, 2005.

Kasyap, A., Lim, J.-S., Johnson, D., Horowitz, S., Nishida, T., Ngo, K., Sheplak, M., and Cattafesta, L., Energy reclamation from a vibrating piezoceramic composite beam, *9th International Congress on Sound and Vibration (ICSV9)*, Orlando, Florida, July 2002.

Kiely, J. J., Morgan, D. V., Rowe, D. M., and Humphrey, J. M., Low cost miniature thermo-electric generator, *Electronics Letters*, 27(25), 2332–2334, 1991.

Kymissis, J., Kendall, C., Paradiso, J., and Gershenfeld, N., Parasitic power harvesting in shoes, *Second IEEE International Conference on Wearable Computing*, IEEE CS Press, Los Alamitos, California, 1998, pp. 132–139.

Lakic, N., Inflatable boot liner with electrical generator and heater, U.S. Patent No. 4845338, 1989.

Lee, C. K., Theory of laminated piezoelectric plates for the design of distributed sensors/actuators. Part I: Governing equations and reciprocal relationships. *Journal of the Acoustical Society of America*, 87, 1144–1158, 1990.

Li, W. J., Chan, G. M. H., Ching, N. N. H., Leong, P. H. W., and Wong, H. Y., Dynamical modeling and simulation of a laser-micromachined vibration-based micro power generator, *International Journal of Nonlinear Sciences and Numerical Simulation*, 1(5), 345–353, 2000.

Marzencki, M., Basrour, S., Charlot, B., Spirkovich, S., and Colin, M., A MEMS piezoelectric vibration energy harvesting device, *PowerMEMS Conference*, December 2005, pp. 45–48.

Mehregany, M. and Bang, C., MEMS for smart structures, Smart structures and materials: Smart electronics, *Proceedings of SPIE*, March 2–3, 1995, San Diego, California; *SPIE*, 2448, 105–114, 1995.

Meninger, S., Mur-Miranda, J. O., Amirtharajah, R., Chandrakasan, A. P., and Lang, J. H., Vibration-to-electric energy conversion, *IEEE Transactions on Very Large Scale Integration (VLSI) Systems*, 9(1), 64–76, 2001.

Ottman, G. K., Hofmann, H. F., Bhatt, A. C., and Lesieutre, G. A., Adaptive piezoelectric energy harvesting circuit for wireless remote power supply, *IEEE Transactions on Power Electronics*, 17(5), 669–676, 2002.

Pelrine, R., Kornbluh, R., Eckerle, J., Jeuck, P., Oh, S., Pei, Q., and Stanford, S., Dielectric elastomers: Generator mode fundamentals and applications, Smart structures and materials: Electroactive polymer actuators and devices, *Proceedings of SPIE*, 4329, 148–156, 2001.

Qu, W., Plotner, M., and Fischer, W.-J., Microfabrication of thermoelectric generators on flexible foil substrates as a power source for autonomous microsystems, *Journal of Micromechanics and Microengineering*, 11(2), 146–152, 2001.

Ramsay, M. J. and Clark, W. W., Piezoelectric energy harvesting for bio MEMS applications, *Smart Structures and Materials: Industrial and Commercial Applications of Smart Structures*, 4332, 429–438, 2001.

Roundy, S. and Wright, P. K., A piezoelectric vibration based generator for wireless electronics, *Smart Materials and Structures*, 13, 1131–1142, 2004.

Roundy, S., Wright, P. K., and Rabaey, J. M., A study of low level vibrations as a power source for wireless sensor nodes, *Computer Communications*, 26, 2003, 1131–1144.

Saraiva, J. A. G., Alternative energy systems, *Mediterranean Electrotechnical Conference Proceedings*, Piscataway, New Jersey, 1989, pp. 77–78; available from IEEE Service Center (Cat no. 89CH2679-9), Piscataway, New Jersey.

Shearwood, C. and Yates, R. B., Development of an electromagnetic micro-generator, *Electronics Letters*, 33(22), 1883–1884, 1997.

Shenck, N. S. and Paradiso, J. A., Energy scavenging with shoe-mounted piezoelectrics, *IEEE Micro*, 21(3), 30–42, 2001.

Smalser, P., Power transfer of piezoelectric generated energy, U.S. Patent No. 5703474, Patent and Trademark Office, Washington, D.C., 1997.

Sood, R., Jeon, Y. B., Jeong, J.-h., and Kim, S.-G., Pizoelectric micro power generator for energy harvesting, *Technical Digest of the 2004 Solid-State Sensor and Actuator Workshop*, Hilton Head, South Carolina, 2004.

Stark, I. and Stordeur, M., New micro thermoelectric devices based on bismuth telluride-type thin solid films, *Proceedings of the 18th International Conference on Thermoelectrics*, IEEE, Piscataway, New Jersey, 1999, pp. 465–472.

Starner, T., Human-powered wearable computing. *IBM Systems Journal*, 35(3–4), 618–629, 1996.

Taylor, G. W., Burns, J. R., Kammann, S. M., Powers, W. B., and Welsh, T. R., The energy harvesting eel: A small subsurface ocean/river power generator, *IEEE Journal of Oceanic Engineering*, 26(4), 539–547, 2001.

Uchino, K., Nomura, S., Cross, L. E., Jang, S. J., and Newnham, R. E., Electrostrictive effect in lead magnesium niobate single crystals, *Journal of Applied Physics*, 51, 1142–1145, 1980.

Umeda, M., Nakamura, K., and Ueha, S., Analysis of the transformation of mechanical impact energy to electric energy using piezoelectric vibrator, *Japanese Journal of Applied Physics*, 35(5B), 3267–3273, 1996.

White, N. M., Glynne-Jones, P., and Beeby, S. P., A novel thick-film piezoelectric micro-generator, *Smart Materials and Structures*, 10, 850–852, 2001.

Williams, C. B. and Yates, R. B., Analysis of a micro-electric generator for microsystems, *Sensors and Actuators A*, 52, 8–11, 1996.

Index